Y0-AAF-442

Irrigation at High Altitudes: The Social Organization of Water Control Systems in the Andes

Printed in the United States of America
ISBN 0-913167-66-5

Library of Congress Cataloging-in-Publication Data
Irrigation at high altitudes : the social organization of water control systems in the Andes / edited by William P. Mitchell and David Guillet.
p. cm. - (Society for Latin American Anthropology publication series ; v. 12)
Includes bibliographical references.
ISBN 0-913167-66-5
1. Irrigation—Social aspects—Andes Region. 2. Indians of South America—Andes Region—Agriculture. 3. Indians of South America—Andes Region—Social conditions. 4. Irrigation farming—Andes Region. 5. Freshwater ecology—Andes Region. 6. Water rights—Government policy—Andes Region. 7. Andes Region—Social conditions. 8. Andes Region—Politics and government. I. Mitchell, William P., 1937- II. Guillet, David, 1943- . III. Series.

F2230.1.I77I77 1994 94-20688
333.91'315'.98—dc20 CIP

Irrigation at High Altitudes: The Social Organization of Water Control Systems in the Andes

Edited by **William P. Mitchell and David Guillet**

Inge Bolin
Stephen G. Bunker
Paul H. Gelles
David Guillet
William P. Mitchell
Karsten Paerregaard
Linda J. Seligmann
Jeanette E. Sherbondy
John M. Treacy
Bruce Winterhalder
Karl S. Zimmerer

Foreword by Jeffrey David Ehrenreich

Volume 12
Society for Latin American Anthropology Publication Series
Jeffrey David Ehrenreich, General Editor

A publication of
The Society for Latin American Anthropology
and The American Anthropological Association

1993

In Memory of
John M. Treacy
Friend and Colleague

CONTENTS

Cover Photo: Paul Gelles

FOREWORD

The Society for Latin American Anthropology Series provides scholars with the opportunity to reach a large audience of anthropologists with the results of their collaborative research efforts. This volume, the twelfth in the series, includes the work of eleven scholars who together have vast ethnographic and archaeological experience in the Andes region of South America. It is my pleasure to present the members of the Society for Latin American Anthropology with *Irrigation at High Altitudes: The Social Organization of Water Control Systems in the Andes,* a volume edited by William P. Mitchell and David Guillet.

The work for this book began in a symposium that was organized for the International Congress of Americanists held in Amsterdam, 1988. The intention of the symposium was to bring together scholars of the Andes region who had focused in their research on questions and issues surrounding systems of irrigation. Mitchell reports that invitations went out to specialists in highland regions and the lack of funds unfortunately prevented some Latin American scholars from participating. It was thought that highland irrigation systems were distinct from those of the lowlands and thus in need of separate treatment. Also, it was significant that despite the obviously central role such irrigation systems play in the lives and cultures of the region, little systematic comparative analysis had so far appeared in the literature. It was decided that participants should focus on contemporary systems, and, with the exception of the material on Inca Cuzco, this restriction was carried through into the volume. After the congress, Mitchell and Guillet undertook the organizing and editing of the volume.

One of the principal efforts of the editors was directed at integrating the papers of the volume's diverse authors into a coherent whole. To this end, papers of the volume were shared between all the authors as the book was evolving. Thus, unlike so many edited volumes in which essays that have been produced more or less independently from each other wind up together, the papers contained in this collection have been read along the way by the book's other authors. The result is an edited volume with a greater than usual sense of unity. There are common themes, questions, and issues–if not common theoretical agreements–that cut across the essays. The essays speak both to each other and with consideration of each other. The volume represents a genuine effort at scholarly collaboration which illuminates a central theme of research across a wide ethnographic region. I believe that Latin American specialists will find this book to be an important contribution to their scholarly libraries.

As is the case with all such enterprises as this volume, many people apart from the authors have contributed their time, energy, and talents in producing the final product you have before you. At Cornell College, where the manuscript was prepared for publication, two of my assistants deserve special recognition. Cheryl Dake worked diligently in typing the revisions made by authors. Annelise Earley prepared all of the tables and figures in the book, never losing her sense of humor as the process unfolded. Speaking for the Society for Latin American Anthropology, I am extremely grateful for their invaluable contributions to this project.

Jeffrey David Ehrenreich
Mount Vernon, Iowa, March, 1994

CHAPTER ONE

Introduction: High Altitude Irrigation

DAVID GUILLET
Catholic University

&

WILLIAM P. MITCHELL
Monmouth College

Irrigation is crucial to life in the Andes. This need for water has always been obvious on the coastal plain, one of the driest deserts in the world. In this region, any human settlement beyond the barest minimum requires irrigation, essential to both household and farm, throughout the year. Pre-Columbian populations built some of the largest and most complex irrigation systems of the premodern era in this zone. They constructed canal systems that united entire valleys and sometimes connected the watersheds of more than one valley (see Kosok 1965). These irrigation systems were so prominent and important that they provided Wittfogel (1957) and Steward (1949, 1955) evidence to support their models of hydraulic society and the evolution of ancient states.

Highland irrigation is no less vital. Its importance to agriculture, however, has been masked by the semimoist environment and the modest size of the canal systems. Because it rains torrentially in most areas during the major growing season of crops, farmers irrigate only briefly, a brevity that can leave the impression that irrigation is unimportant. This limited water, however, is necessary for the cultivation of most crops, especially maize, the single most important food in both ancient and modern diets (Mitchell 1976, 1991a; Murra 1960). These systems, moreover, provide potable water for households and livestock during much of the year.

Irrigation in the mountains is found in two distinct ecological zones: valley bottoms and highland slopes. Water for slope irrigation usually comes from small streams, springs, and the runoff from high altitude seeps, glaciers, and permanent ice caps. Because this water is scarce, its use is regulated, and farmers are unable to irrigate year-round. Valley bottom systems normally rely on small rivers, which have enough water to allow farmers to irrigate whenever they want. Taking advantage of favorable thermal conditions, most farmers in these valleys are thereby able to grow two crops a year, one in the rainy season and another in the dry season. Valley bottom systems have been studied less extensively than those on the slopes, an emphasis that is reflected in the chapters of this book.

Unlike those on the coast, highland irrigation systems (both on the slopes and in the valleys) are normally small and decentralized. Water is transmitted to fields by means of open canals either directly from the water source or indirectly via small reservoirs. Canal and reservoir construction and maintenance, water allocation and distribution, and the resolution of conflicts are often managed at the village or interhamlet level. The small size of the systems, however, does not mean that they lack complexity. Many highland irrigation structures are engineering marvels, as Seligmann and Bunker demonstrate in this volume.

Local management of irrigation does not always ensure the egalitarian distribution of water. The control of water is a key element in local stratification, and the battle over water rights is a major theme in many communities (see Gelles and Mitchell *intra*). Hacendados and powerful townspeople have often dominated the distribution of water, frequently giving water to peasants only in return for labor.

Even in those systems controlled by peasants, farmers often fight over water, and sometimes resort to force, especially when water is distributed on a first come/ first served basis. Since only powerful peasants are able to mobilize significant support, they are able to obtain more water for themselves and their allies. At times of heavy demand, however, peasants use formalized distribution mechanisms that assign water to plots of land (not persons) in an organized sequence, thereby alleviating most disputes (see Gelles and Mitchell *intra*). This distribution is still not egalitarian, however. Since farmers control different amounts of land, they also control different amounts of water.

The wide geographic range of highland irrigation systems is impressive. In a 1977 survey of 99 percent of the officially recognized peasant communities in Peru, 59 percent (1,589) reported that they employed irrigation, and of these, 61 percent (970) distributed the water communally.[1] The greatest number of communities with irrigation (DCCN 1980: table 36) were in the departments of Cuzco (321), Lima (226), and Ayacucho (214). To these systems of canal irrigation should be added other forms of water control, including raised fields, sunken fields, and drained fields that had been and still are used to manage moisture and raise productivity.

Along valley bottoms many communities construct flood-control devices to prevent rainy season inundation of their fields (Mitchell 1976). Recent evidence also indicates at least one post-Columbian use of a filtration gallery, or *qanat*, in the highlands (Paucartambo), as well as a number of others on the coast (Barnes and Fleming 1991).

Variation in Andean drainage and irrigation systems is as great as that in any area of the world. Water flow patterns, physical irrigation infrastructure, and irrigation management differ greatly from village to village, even within the same drainage basin.[2] Such system diversity suggests complex interactions of environment, technology, economy, and sociopolitical organization. Nonetheless, the Andes are united by a remarkable homogeneity of culture and society, especially when compared to the diversity found in the other great mountain habitat of the world, the Himalayas (Orlove and Guillet 1985).

Since the prefarming period, Andean communities have drawn on comparable cultural and evolutionary roots. They have also had parallel conquest histories, first by local states, then by the Huari, Inca, and Spanish empires (see Sherbondy *intra*). In consequence, even though the central Andean cultural region extends through three nation-states (Bolivia, Ecuador, and Peru) and is partly found in two others (Argentina and Chile), only two indigenous languages (Quechua and Aymara) and Spanish are commonly spoken.[3] In addition, despite social and cultural differences, most highland communities share similar concepts of kinship, community organization, and religion. These resemblances extend to many of their beliefs and practices relating to irrigation. The ideology and performance of the irrigation cleaning and blessing (known as *yarqa aspiy* or *yarqa ruway*), for example, are remarkably alike in those communities where they have been described.

Variation among irrigation systems in a setting of relative cultural homogeneity allows unique opportunities for carefully controlled ecological and social comparisons (Eggan 1954; Ragin 1987). Although the authors in this volume have not deliberately set out to make such comparisons, they all consider the relationship between environment and social organization. They maintain this focus even though they also reflect their differing theoretical perspectives.

Four of the papers (those of Gelles, Guillet, Paerregaard, and Treacy) concern the Colca Valley in the department of Arequipa. Another four (those of Bolin, Seligmann and Bunker, Sherbondy, and Zimmerer) pertain to the Department of Cuzco. The paper by Mitchell examines a valley in the department of Ayacucho, an area to the north and west of these two zones, and Winterhalder's paper covers a region encompassed by a northeast-southwest transect through the Andean cordillera of southern Peru (through the departments of Cuzco, Puno, and Arequipa). This geographic distribution reflects fortuitous patterns of research during the last twenty years, rather than any methodological design. More scholars have worked in Cuzco than other areas of highland Peru. In recent years, attention has shifted to the Colca Valley. It would be useful to have additional material from the north, the eastern escarpment, the arid highland valleys of the western escarpment outside the Colca Valley, and the valley bottom areas of the inter-Andean valleys. We hope this volume will stimulate such comparative research.[4]

We have organized this volume not on the basis of geography but on thematic coherence. Winterhalder's discussion of Andean environmental variation (Chapter 2) and Sherbondy's analysis of the hydraulic organization of the Inca (Chapter 3) set the climatic and historical stage for the succeeding articles. Treacy's and Zimmerer's studies (Chapters 4-5) add to Winterhalder's discussion by providing material on the hydraulic technology used on the western and eastern escarpments of the Andes. The majority of the remaining papers deal with the political economy of irrigation. Bolin, Guillet, and Gelles (Chapters 6, 7, and 10) emphasize the relationship between the state and the local polity. Paerregaard (Chapter 8) focuses on the importance of ritual in unifying an extremely decentralized irrigation system. Seligmann and Bunker, Gelles, and Mitchell (Chapters 9, 10, and 11) consider the forces (ecological, economic, political, and social) favoring and militating against the maintenance and expansion of contemporary Andean irrigation.

VARIATION IN HIGHLAND ANDEAN WATER CONTROL SYSTEMS

Winterhalder presents data on precipitation, temperature, and the onset of thermal and rainy seasons from 69 weather stations on a cross-Andes transect, oriented northeast to southwest, perpendicular to the axis of the cordilleras. He provides an indispensable data set: precise information on the environmental variation across the Andes, quantifying the more qualitative classifications previously available to us. He demonstrates that the predictability of rainfall and minimum temperatures decreases with increased altitude; temperatures and rainfall are less predictable at higher altitudes, than they are at lower altitudes. His data illuminate the reasons for the geographic distribution of Andean irrigation systems, and he provides us with two case studies, using his environmental data to show why one community irrigates and the other does not. The importance of Winterhalder's material extends beyond the study of irrigation systems, as it illuminates the reasons for the distribution of Andean agropastoral adaptations in general, as well as the consequent patterning of population and settlements.

He groups his data into three zones: the eastern escarpment, the altiplano (we prefer the term "inter-Andean valleys"), and the western escarpment. Annual precipitation, greatest in the eastern escarpment, decreases to between 400 and 800 millimeters in the inter-Andean valleys and drops to a range of 200 to 400 millimeters in the valleys that descend the western slopes of the cordillera. The rainy season declines from 5-6 to 3-4 months as one moves from the high slopes of the eastern escarpment across the altiplano to the high slopes of the western escarpment.

Winterhalder derives several propositions from his analysis. In the inter-Andean valleys and the slopes of the western escarpment, irrigation extends the growing season and helps stabilize the irregular and often late-arriving rainfall. In the western escarpment, irrigation also augments the moisture available to crops during the growing season. In both areas, soils dampened by irrigation lower the frost risk caused by low minimum temperatures. In the eastern escarpment, where rainfall is higher because of rising air masses from the adjacent humid Amazon basin, the role of irrigation in modifying the timing and quantity of water for agriculture is of less importance. Technology to provide drainage, however, is needed in the moist areas of this zone.[5]

These broad generalizations do not preclude substantial variation caused by local geomorphology and topography. The different exposures of the eastern escarpment, for example, create a multiplicity of microareas that vary with respect to sunlight and rain. One region may be in a rain shadow and require irrigation, whereas another, only a short distance away, has too much moisture and demands drainage. Similar variation, although perhaps less extreme, is found in the inter-Andean area and in the western escarpment. Nonetheless, the broad generalizations apply: communities located in the eastern escarpment are more likely to need drainage than those located in any of the other zones (Bruce Winterhalder, personal communication, February 20, 1992).

The papers in this volume portray areas in all three of Winterhalder's zones. Bolin and Zimmerer deal with communities in the eastern escarpment; Gelles, Guillet, Paerregaard, and Treacy consider towns in the western escarpment (all from the Colca Valley). Cuzco, described by Sherbondy, is located in the inter-Andean area. Huanoquite, the village discussed by Seligmann and Bunker, is found on the border between the inter-Andean valleys and the eastern escarpment. Mitchell's contribution from Ayacucho deals with an inter-Andean valley outside of Winterhalder's research zone, one that is farther west than all the other areas and probably much drier for that reason.

ETHNOHISTORIC MODELS OF IRRIGATION ORGANIZATION

Sherbondy's discussion of the valley of Cuzco provides striking insight into the importance of water and irrigation not only in the social organization of the Incas but also as found in social organization throughout the Andes today. The Inca drew on preexisting social forms to create their institutions of empire (Aveni and Silverman 1991; Murra 1986; Zuidema 1967). They, in turn, used these reworked institutions to reorganize the local societies that they conquered, employing the organization of the Inca capital of Cuzco as a model (Zuidema 1964). Consequently, even though the Inca empire was of short duration, lasting less than one hundred years before its conquest by the Spanish in 1532, the patterns described by Sherbondy for the Inca had (and still have) widespread applicability.

Sherbondy demonstrates that irrigation constituted an important aspect of local Inca administrative organization (see also Zuidema 1986). In the Valley of Cuzco, irrigation was fundamental to the division of the valley into two moieties, or *saya,*[6] as well as to the system of *ceque* lines (Sherbondy 1982) and the organization of the local groups known as *ayllus* and *panacas*. Except for the *ceque* system and the *panacas,* these social units have persisted (albeit with many changes).[7]

Of special importance is Sherbondy's demonstration of the hydrographic foundation of the dual divisions, or *saya,* so basic to Andean social organization.

These divisions, which are prominent in ceremonial and political organization, have close ties to irrigation. In the valley of Cuzco, the Huatanay River divided the land into two social units (*saya*): Hanan Cuzco, the right bank, faced upriver, higher in sociopolitical rank than Hurin Cuzco, the left bank. Each unit had a separate irrigation system and its own water official assigned to monitor boundaries and oversee the system. Many contemporary communities maintain a similar pattern. In the village of Quinua in the department of Ayacucho, for example, the dual division into Hanan Sayoc and Lurin Sayoc is based on the hydrography of the mountain slope, and each division uses a different drainage and irrigation system (Mitchell 1976).

The importance of irrigation to Andean social organization is demonstrated by Sherbondy's analysis of the social changes introduced by the Spanish. Although the Spanish dismembered the highest levels of Inca political organization, they retained much of the local structure associated with irrigation, native production, and access to native labor (see also Rowe 1980:94). As expediency governed the transition to colonial administration, the Spanish continued to use traditional water officials, water distribution procedures, and the irrigation calendar. The Inca *ayllus* and royal *panacas* were incorporated into the colonial parishes as social groups with corporate rights to land and water, even as they were stripped of their power as independent state institutions.

VARIATION IN HIGHLAND WATER CONTROL SYSTEMS

The remaining papers in the volume concern contemporary communities. In most of the Andes irrigation is crucial to maize cultivation, although this may not be the case in the wetter areas of the eastern escarpment.[8] In many areas, especially the Colca and Cuzco Valleys, this intensive system of cultivation is associated with terracing, a technology that we refer to as the irrigation-terracing-maize complex.

Not only is maize the premier ceremonial crop in the Andes (Murra 1960), but it also provides more protein and nearly twice the calories per unit of land than do potatoes, the second most important crop (Mitchell 1991a:80). Peasants eat maize at most meals, consuming it in fresh, parched, and milled forms. They also prepare maize beer to serve at fiestas and ceremonial work groups (Mitchell 1991b). Intrinsic to the system of social stratification, this beer is a critical prestation; the wealthy use it to get labor, and the poor demand it as an entitlement. Because maize is also easily stored and transported, its cultivation is of critical importance to the welfare of the Andean people.

Maize, however, grows only in very restricted altitude zones, ones that are not too high and cold or too low and dry. In most zones, moreover, maize requires irrigation. The same variety of maize can take six to nine months to mature, depending on altitude, but the rainy season at most altitudes of the interandean valleys and western escarpment is considerably shorter. Farmers also consider the thermal season in calculating when to plant maize. If they plant at the beginning of the rainy season (November, December, or January depending on area), maize would still be growing in June, when cold nights can cause frost damage. This risk is lessened, and maize can mature before the frosts begin, by advancing the sowing to the relatively dry month of September by means of irrigation (Mitchell 1976, and *intra*).

The importance of seeding maize at the right time–not so early that it will dry out, and not so late that it risks late season frost–encouraged the native peoples of the Andes to create an elaborate agrarian and ritual calendar so important that it has survived the conquest largely intact (Villanueva and Sherbondy 1979: ix; Sherbondy *intra*).

Irrigation is also used to plant maize and other foods (especially cash crop potatoes) a month or so before the main planting. This dry season planting, often called the *michka,* has a limited distribution (Mitchell *intra*) because it must be irrigated every week or so, a much greater need for water than in the rainy season crop cycle (see Mitchell, Winterhalder, and Zimmerer *intra*). Few slopes in the inter-Andean Valleys or on the western escarpment have enough water to plant this crop, so that dry season foods are usually grown only in valley bottoms with abundant irrigation water (Mitchell *intra,* and n.d.).

Terracing is not always found with slope irrigation, but where it is found, it facilitates the irrigation of maize by providing a field platform onto which water can be led (see Donkin 1979; Guillet 1987; Treacy *intra*; Winterhalder *intra*). In addition to inhibiting erosion, terraces also promote the downward flow of cold air, thereby reducing the risk of frost and permitting the cultivation of maize on otherwise frost-susceptible slopes (Guillet 1987).[9]

Although maize cultivation had been widespread in the Andes before the Inca, the Inca intensified maize production by expanding irrigation and the associated system of terraces (see Murra 1960; Rowe 1946; and Wachtel 1982). In conjunction with the cultivation of potatoes and edible chenopods (Troll 1968; Guillet 1983:565), this irrigation-terracing-maize regime achieved remarkable levels of productivity, doing so in an unpredictable and otherwise low-yielding mountainous environment. The Inca introduced their system of intensive maize farming into the areas they conquered. In the Mantaro Valley, for example, local peoples moved to lower altitudes after the Inca conquest, where they began both to produce and consume more maize (Hastorf 1990; Hastorf and Earle 1985). Preliminary results of research from northwest Argentina similarly indicate that the Inca conquest initiated increased irrigation and maize production (Terence N. D'Altroy, personal communication, October 31, 1991).

The papers on the Colca Valley, found in the western escarpment in the department of Arequipa, are particularly useful in providing information on the irrigation-terracing-maize complex. Treacy, in his contribution to the volume, analyzes the interaction of irrigation and terracing in the Colca Valley village of Coporaque, focusing on the landscape engineering employed to extract and conserve available moisture. His description of this irrigation technology is one of the few available and is a notable contribution on that basis alone.[10] He also addresses several theoretical issues, making his paper an even more substantial contribution.

In Coporaque, farmers employ a metaphor of "teaching water" to describe the process of hydraulic management. To them, water is untutored and elusive and in need of training to make it behave correctly. They do so by employing an impressive technology that ranges from the hydraulic features (both temporary and permanent) found in individual fields to permanent terraces, feeder canals, control valves, secondary distribution canals, and reservoirs that extend over a wide territory. Of these, agricultural terracing is by far the most important because of its role in the irrigation-terracing-maize complex.

Treacy emphasizes the role of terraces in controlling water flow, a consideration that leads him to broaden the definition of "terrace" to include any field that has been artificially flattened to retard the flow of water. Most scholars and development agencies have focused on the spectacular bench terraces of worked stone found in the Cuzco Valley, the Colca Valley, and elsewhere. Treacy argues, however, that these structures are not the only type of terrace, a point that Mitchell (1985) has also made for the Ayacucho valley, where even informants had overlooked the plain and rough terraces found in the village. As we have argued above, and as Winterhalder (*intra*) shows, terraces do more than slow the flow of water (see Guillet 1987), but Treacy's point is well taken. There is more to a terrace than spectacular walls.

Treacy's contribution also sheds light on the issue of the skills needed for the construction of Andean canal systems. Were the builders of the complex irrigation system in Coporaque the equivalent of hydraulic engineers with specialized knowledge? Treacy concludes they were not. The basic skills for such construction are shared by all farmers, although some are better at it than others. Canal building is the responsibility of local people who use patient "teaching" methods, perhaps organized by someone who is most skilled at managing flows, but not necessarily a specialist. He concludes that ordinary farmers were probably the builders of irrigation systems in pre-Colombian times as well.

As we will see below, Seligmann and Bunker take a different approach in their paper. They conclude that the people of Huanoquite (department of Cuzco) would have required specialists to construct their irrigation system. Perhaps the two views are not as divergent as they seem. The irrigation system in the Colca Valley is not as complex as that described by Seligmann and Bunker in Huanoquite, which entailed moving water long distances over different drainage systems and around mountains. Even bench terraces, although spectacular, do not seem to require specialized skills available only to specialists, a point that can be made not only for Peru but also for the dramatic terraces of the mountain provinces of the Philippines (Bacdayan 1974).[11] The largely tacit skills of an entire social group may transcend the skills of a single person (Bruce Winterhalder, personal communication, March 3, 1992). Consequently, we conclude that specialized engineering skills were sometimes required in the Andes, but probably infrequently.

Irrigation in Tapay, at the other end of the Colca Valley, is organized very differently from that found in surrounding communities. Instead of elaborate bench terraces and common canals, Tapay is characterized by a decentralized collection of fourteen very small irrigation systems. Each system services a small cluster of fields, each with its own separate source of water and other infrastructure. The local irrigation group also distributes water and resolves conflicts independently without any recourse to a hierarchical or centralized organization of power.

Paerregaard argues that this acephalous collection of irrigation structures is unified by means of a complex schedule of ritual offerings (*pagos*). These offerings are proffered to water spirits at the levels of the hamlet, moiety, and village, in a repetitive cycle. The ceremonies progress along the system, advancing toward the same geographical location on Seprigina Mountain, where the critical rituals of the separate groups are celebrated at the same time. Different irrigation groups thereby come together, unifying an otherwise decentralized system. The attempt of a small group of evangelical Protestants to challenge these celebrations is used by Paerregaard to demonstrate the importance of water ceremonial in creating Tapay unity.

In this analysis, Paerregaard concurs with Treacy's conclusions about the irrigation cleaning and celebration in Coporaque. In Coporaque, during the annual cleaning of the irrigation system, the entire population ascends to the headwaters of the canal system. Splitting into moiety halves, the farmers of Urinsaya clean one branch, and those of Anansaya, the other. They descend, working as they proceed, and meet to rest and celebrate at the point where the two canals join. Then, united, they clean the main canal. For Treacy, this event symbolically represents the principle of *chu'llay,* the melding and subsuming of two sources of water into a unitary flow. The canal cleaning and ritual in Coporaque thus acts in much the same way as in Tapay: it helps create a sense of communal unity.

Environmentally very different from the Colca Valley, the eastern escarpment is characterized (according to Winterhalder) by a thermal season in the higher elevations of 8 to 12 months and a rainy season of 4 to 6 months. Absolute amounts of precipitation are significantly greater than in the inter-Andean valleys or the western escarpment. These patterns suggest that irrigation would not always be useful for extending the growing season or for augmenting available moisture. Rather, in the wet areas of this zone, one would expect to find landscape modifications and water control systems designed to drain water, especially for the cultivation of tuber crops. Recent discoveries of a host of indigenous drainage technologies to cultivate fertile but waterlogged soils support this expectation. Unfortunately, most of these drainage systems are represented only by abandoned archeological remains (Denevan et al. 1987; Erickson 1983; but see Flores Ochoa and Paz Flores 1984).

In his contribution, Zimmerer provides one of the few accounts of a functioning drained field system. Zimmerer analyzes a drained field complex in Colquepata, a community located in the intermediate altitudes of the Paucartambo highlands of southern Peru. His contribution is a significant description of wetland farming and of the role of social organization in the adoption of new technology. His paper also includes a meticulous–and thereby important–description of agricultural labor.

Colquepata cultivators drain swales (low marshy areas) to grow potatoes and other cultivars. They reduce excess moisture to agriculturally appropriate levels by draining water from the fields into gullies and rivulets through a network of canals. Household and reciprocal labor is sufficient for construction and maintenance of the fields; conflicts are few, and the system requires no centralized scheduling or coordination. The small-scale labor organization of these drained fields approximates that found in simple canal systems, such as that of Tapay (Paerregaard *intra*).

Zimmerer's paper specifically describes the social context of technological innovation. In Colquepata, farmers use drained field cultivation to grow potatoes as a cash crop. However, the economic gain of high potato prices was not in itself sufficient to develop this new technology. Colquepata farmers were able to use preexisting agricultural skills and patterns of labor organization. They obtained the workers needed for this labor-intensive innovation through already-existing reciprocal-labor (*ayni*) and wage-labor (*jornal*) networks. These labor systems not only provided workers but also created the social conditions necessary for common patterns of construction and the rapid diffusion of the new technology throughout the community.

It is the social and physical environment, however, that explains why the people of Colquepata (but not other communities) were able to adopt the new farming methods. Unlike farmers in surrounding communities, those in Colquepata controlled their own land and labor. They were thereby free to innovate, and they were probably more willing to undertake the capital investments such innovation requires. Colquepata is also closer than surrounding communities to the Cuzco market, which significantly reduces their transport costs, further encouraging them to direct their efforts to creating new fields for cash cropping.

The lands of Colquepata also benefitted from a suitably wet habitat that allowed potato production during the dry season with a relatively simple technology. This contrasts with Mitchell's data from Ayacucho, where peasants have been unwilling to build new dams and expanded canal systems. The differences in the two areas may partly be the result of the differential labor costs (in construction, maintenance, and use) of drained field cultivation versus irrigated cultivation. Irrigated cultivation in the very dry Ayacucho Valley is more labor intensive than drainage systems, or even irrigation, in moister areas of the Andes.

IRRIGATION SYSTEMS AND THE STATE

One of the most notable processes of this century has been the general expansion of state power. Faced with the need to feed growing populations, states have increasingly attempted to improve the use of water by codifying and updating water management law and by creating improved extension services to acquaint farmers with new laws and procedures. As part of this effort, they have expanded agricultural ministries to implement the new laws and have encouraged agricultural universities and research stations to develop better methods of water transport, water distribution, and conflict resolution. The efficacy of these managerial innovations is a much discussed aspect of contemporary irrigation (Hunt 1989; Sampath and Nobe 1986; Utton and Teclaff 1978).

Peru is an excellent place to analyze the impact of state intervention on locally organized irrigation. At the beginning of this century, irrigation management had been left largely in local hands. The peasant political leadership known as the *varayoc* were responsible for organizing peasant corvée labor and often supervised local irrigation systems (Mitchell 1976). Although part of the local prestige hierarchy, these leaders also served the interests of the municipality and powerful townspeople. Used as a police force in a form of indirect rule, the *varayoc* obtained labor and other resources for elites (Mitchell 1991a:149-155). In the latter half of this century, the *varayoc* have been disappearing at a rapid rate, many of their functions (including those associated with irrigation) being directly assumed by the state (Mitchell 1991a:149-155, 163-177).

Efforts in modern Peru to improve water management date from the enactment of a national water code in 1902. Designed in response to upstream-downstream problems endemic to the large-scale river-based irrigation systems of the coast, the law had a limited impact in the highlands. This isolation of the highlands from national water control was to change. In 1969, the reformist government of President Velasco enacted by decree a Water Law creating special-purpose local irrigation associations linked to a larger water management hierarchy. This law took the control of irrigation systems from local municipal bodies and transferred that control to the water users themselves, reversing a trend of town domination over water that had begun with the Spanish conquest (Guillet 1989).

The papers of Bolin, Gelles, and Guillet consider the implementation of the 1969 Water Law in highland communities. Bolin deals with the impact of the water law and of development schemes on irrigation along three canal systems in the Vilcanota Valley. Her data demonstrate the disparate responses of highland peasants to the irrigation law. Only the people along one of the canal systems that she studied have implemented the required "Regulations for Water Users" to any great extent, and even that was done incompletely. The communities along another of her canal systems have complied with the rules in a minor way, and the people along the third canal system have ignored the rules altogether.

Bolin finds that, contrary to common assumptions, governmental agencies in the Vilcanota Valley have had some positive influences on local irrigation systems. The communities along the three canal systems vary in the extent of their control–that is, autonomy–over irrigation and in the success of their irrigation management. Using Weber's model of autonomy, she concludes that, "although autonomy is essential to the organization of irrigation at the local level, irrigation management does not necessarily improve in proportion to increasing autonomy." The adequacy of the water supply and the nature of outside control are additional variables that determine such success. If water is scarce, state intervention may be crucial to provide irrigators with resources necessary to act autonomously. In Bolin's study, the community with the most local control over irrigation was also the community with the greatest amount of conflict over water; here, the important variable was adequacy of the water supply, not local control of the irrigation system.

Bolin finds that irrigators must be involved in the planning of any irrigation scheme if it is to be successful. Intervention works best when local people request development and are incorporated into the decision-making process early on, a conclusion that Gelles also reaches in his paper.

Guillet devotes his entire paper to the 1969 law and the variation in its implementation in five communities of the Colca Valley. Although the state has successfully reorganized and reinvigorated irrigation associations in many of the agricultural villages of the valley, traditional systems of water management persist, from heavily "Incaic" Yanque (where the local irrigation systems are integrated with moiety and supramoiety organization) to the acephalous Tapay, which lacks management levels above the local irrigation groups.

Guillet suggests that the persistence of traditional irrigation organization in the Colca Valley is best explained as a rational response to farming in a stressful environment. These traditional strategies have allowed the people of the area to maintain a productive agricultural system for more than a millennium. Consequently, these arrangements should not be seen as mechanisms to combat capitalist penetration of the area (or only as this), but as systems that are valid in their own right.

By focusing on the adaptive aspect of irrigation, Guillet finds that variation in irrigation structure is in part explained by the characteristics of local water sources. Communities with moiety irrigation systems are adapting to local hydrography, a pattern like that described by Mitchell (1976) for the Ayacucho Valley. Similarly, the many springs that provide water in Tapay help explain the acephalous organization of irrigation in that community.

Guillet's and Bolin's analyses concur in several respects. In finding that the state has been unable to implement fully the management forms dictated by the Water Law, they suggest that the success of state intervention depends on its compatibility with local ecological and social needs. In effect, the state negotiates with, rather than dictates to, highland communities in the process of "rationalizing" their irrigation systems. Even following state intervention, most water management decisions remain local. Upper levels of the state-imposed hierarchy are weak and ephemeral in comparison with the viability and endurance of the community and local irrigation group.

They both conclude that the 1969 Water Law has had a beneficial effect on irrigation management by strengthening special-purpose irrigation associations. These associations now have the autonomy to elect their own authorities, officials who control water allocation, resolve disputes, and organize infrastructural improvements and maintenance. Mitchell's paper supports their contention that under certain circumstances state intervention empowers local communities. In the department of Ayacucho he found that peasants had used earlier law to wrest control of water from haciendas and to place it in the hands of an elected water board. The peasants were effective in their struggle because of the political alliances they had forged with the elected government of President Belaunde.

THE DEGRADATION AND REJUVENATION OF HIGHLAND IRRIGATION SYSTEMS

It has become clear that peasants respond not only to local pressures but also to those emanating from beyond the local social system. In many areas of the Andes and elsewhere, local food production has declined. Understanding the reasons for this decline is of paramount importance in this period of food shortages and growing population. The contributions of Seligmann and Bunker, Gelles, and Mitchell illuminate the reasons why various communities in the Andes have had difficulty in creating and repairing their irrigation infrastructure.

In their contribution, Seligmann and Bunker analyze the degradation of the irrigation system of Huanoquite in the department of Cuzco. They demonstrate that the original construction of this canal system required much greater knowledge and labor than is available today. This technologically sophisticated irrigation system is one of the most complex ones described in the Andean literature, and its construction was an impressive achievement. Today the people of Huanoquite can no longer maintain their elaborate canal system, much less reconstruct it. Not only are they unable to field an effective labor force for canal work, but they have also lost much of the environmental knowledge and many of the skills necessary for practical repairs. Continued losses of such abilities can only lead to further degradation of the system.

Seligmann and Bunker attribute these shortcomings to political fragmentation caused by the Spanish conquest and subsequent domination of the region by haciendas. The Agrarian Reform further fragmented Huanoquite by creating competing communities and by ignoring the need to create mechanisms to recruit communal labor. Unlike irrigation systems in the Colca Valley that get water from high altitude springs and snowmelts, the Huanoquite system draws its water directly from a river. Because communities tap into the flow along the way, upstream and downstream users cannot have access to water at the same time, a situation that creates potential conflict.

Like many other communities throughout Peru and the Third World, Huanoquite has shifted from communal to individual productive strategies. Peasants expend their time and energy in market production, migration, and the education needed for cash production. They have little time to work on the irrigation system or even to learn how to make the repairs. Mitchell reaches comparable conclusions for peasants in Ayacucho, and we suspect that analogous processes are undermining rural production generally.

Seligmann and Bunker also explore the impact of oral traditions in creating and sustaining the sense of hopelessness that inhibits the ability of Huanoquiteños to maintain their irrigation system. According to these traditions, the ancient people constructed the irrigation system with the aid of supernatural power. Since the contemporary residents of Huanoquite do not have this supernatural power, they are–realistically, from the point of view of the myths–unable to rebuild it. This is an interesting account of the way in which mythology helps sustain the effects of ecological, economic, and political forces. Myth provides positive reinforcement so long as the myths reflect real processes. When they do not, when myth loses its link to social reality, the reinforcement is negative.

Gelles tells us a story that is both similar and different, for the people of the Colca Valley community of Cabanaconde have experienced both failure and success in expanding their irrigation system. They have failed in those efforts in which local and regional elites have dominated the process, with resulting inter-class conflict. They have succeeded in those cases where the peasantry has been able to mobilize to advance its own interests without effective opposition from other communities or from elites.

The people of Cabanaconde have tried to reclaim lands since at least 1916. The local rich had made many of the earliest attempts to reclaim fallow land, but they did so through seizure rather than through purchase, a familiar story in the Andes. The peasantry successfully opposed these usurpations by refusing to irrigate the reclaimed lands. Other attempts, in this case by the peasantry, to obtain more water to reclaim abandoned land were impeded by regional politics. Competing claims over water–a rather common Andean problem–prevented the people of Cabanaconde from reconstructing an abandoned Inca canal. A proposed source of water used by a neighboring village and regional elites was effectively denied to Cabanaconde through intimidation and court action.

The picture changed in the 1980s, when Cabanaconde farmers, threatened by severe drought, illegally and forcefully linked into the Majes canal. This canal was an international development project that was designed not to provide water for the highlands, but to bring highland water to the politically powerful coast. This allocation of development funds to the export sector on the coast instead of to the peasant sector in the highlands is routine in Peru and must be seen as one of the problems obstructing highland agriculture (Mitchell 1991a:97-101).

After the people of Cabanaconde seized the initiative, the Majes personnel acquiesced and assigned water to Cabanaconde to avoid any further confrontation. Cabanaconde has used this water to reclaim abandoned terraces and other lands. Although most of this land reclamation has succeeded, some has failed. The continuing conflict between rich and poor over the land seizures earlier in the century has prevented the people of Cabanaconde from bringing all of their potentially reclaimable lands into production.

Gelles also portrays the role of the peasant method of water distribution (what he calls "the local model") in creating peasant identity and power. This system not only has important ritual meaning but is also controlled by the peasants, a control that facilitated peasant resistance to the land seizures of the local rich earlier in the century. The peasantry has opposed the full implementation of the 1969 Water Law, fearing they would lose their control of water allocation. It is one of the ironies of invented tradition (Hobsbawm and Ranger 1983), however, that this "local model" has historical roots in the hegemonic institutions imposed on Cabanaconde by the Inca and Spanish empires.

Gelles affirms–as Bolin, Guillet, Mitchell, and Seligmann and Bunker also do–that the expansion of the water supply must be seen not only as a technical problem but also as a political and social one. He concludes his paper by suggesting that the success of a land and water reclamation project depends on its appeal to the interests of the local majority instead of the local rich. Development efforts, therefore, must use the institutions and beliefs of that majority, especially the recognized peasant community and the local model of water distribution.

Although we find the notion that the majority should determine its own destiny congenial, it would be interesting to determine the limits of Gelles's proposition. It is conceivable that under certain circumstances the majority may be unaware of its interests. The majority can also be manipulated, divided into groups acting at the behest of elite factions.

In his paper, Mitchell examines the ecology and political economy of irrigation. Like Seligmann and Bunker, he is perplexed by the seeming irrationality of underused farming resources where there is great poverty and hunger. His data show that peasants in the department of Ayacucho could produce more food if they expanded their irrigation systems, but they have not done so. The relatively arid Ayacucho Valley is one of the poorest departments in Peru (Banco Central de Reserva 1981; McClintock 1984, 1988; Mitchell 1991a:127). That the Sendero Luminoso guerilla movement had its start in this department in the early 1980s is only one of the many manifestations of the area's poverty.

Why have Ayacuchanos not expanded irrigation under conditions of economic stress? Mitchell proposes that to understand this apparent anomaly one must examine the ecology and political economy of production, which has encouraged nonfarm rather than farm work. He uses data from the District of Quinua, located about an hour by truck from the city of Ayacucho, to develop his thesis.

According to Mitchell, Quinuenos have been unable to expand irrigation because the energy investment to do so would be too costly in the dry conditions prevailing in the Ayacucho Valley, and because the relative returns on the investment would be lower than the returns on nonfarm work. Ayacucho peasants have also been constrained by the pressures of local population growth and by national and international economic conditions. Spurred by this population growth and the relative decline in the value of farm production since the 1940s, Quinuenos have migrated and entered nonfarm occupations (craft manufacture, petty trade, highway repair, and so forth) at an ever increasing rate. Since even expanded farm production would not provide them with enough food, farming has become increasingly supplemental to total household income.

Migration and nonfarm work have produced serious labor shortages, and Quinuenos are forced to choose among competing labor demands. Such labor stress, common in the Andes (Brush 1977; Collins 1988; Stadel 1989), further constrains local production. Quinuenos have responded to this strain on labor by constructing schools–the infrastructure for their new commercial roles–rather than by expanding irrigation–the infrastructure of farming. Like the rest of us, peasants must make choices apportioning scarce labor among competing demands.

Mitchell's analysis is supported by Dobyns' 1962 study of recognized communities (Dobyns 1964:57, 62, 92 and passim). In this study, only 21.2 percent of the communities had built new irrigation canals, whereas 83.6 percent of them had built new schools, and 44.7 percent of them had constructed new roads, in spite of the fact that the most frequent problem reported in the survey was aridity. The patterns Mitchell finds in Ayacucho appear to be widespread.

Mitchell's paper also highlights the impact of larger spheres on the local community. Many of the processes militating against irrigation in Ayacucho stem from decisions of national governments to favor export over local food production.

The International Monetary Fund, the World Bank, and other international bodies also tend to act in a context favoring exports over domestic food production (George 1990:96; Mitchell 1991a:98-101). Major agricultural exporters have also played a role, providing subsidies to Peru and other Third World countries to purchase their grains, thereby reducing demand for local production.

It is clear that local communities are woven into a global system that often constrains and buffets them. This world system has often fostered rural poverty, and in Peru it has encouraged the decline of rural farm production and the massive migration of peasants to Lima. The development of the Shining Path guerilla movement, the spread of cholera, and the strength of the drug trade are additional problems that can be attributed to the same processes. The losses are not Peru's alone, however. Violence and the diseases of poverty pay no heed to political boundaries. No nation is an island, and John Donne's bell tolls for us all. We write this in March 11, 1992, the month that passengers on a flight from South America to the United States were infected with cholera and on the day that the former editor of "El Diario" (New York) was gunned down in a New York restaurant, presumably by Colombian narco-gangs, in retaliation for his work in opposition to the drug trade.

Efforts to rejuvenate degraded systems of irrigation and water control must proceed on the basis of a firm understanding of the causes of degradation and a realistic appraisal of the factors that inhibit technical solutions to pressing problems of land and water. It is clear from the papers throughout the volume that technical issues are only part of the solution. Such social considerations as class relationships and the international economic system are equally important. Peasants make economic decisions that are more or less rational. They put effort into food production only when it pays them to do so.

The future success of Andean irrigation requires understanding of both peasant technologies and the ecological and economic forces constraining production. We hope this volume contributes to such knowledge, as well as to increased understanding of Andean society and irrigation in general.

NOTES

Acknowledgments. This volume began as a symposium organized for the International Congress of Americanists in 1988. The discussants at the symposium, Barbara Price and Robert Hunt, helped frame the papers with lively and illuminating comments. We also wish to thank the contributors to the volume for their efforts in the preparation of the papers and in their responses to the demands of editing.

We are very grateful to Jane Freed for her editorial assistance, to Cheryl Dake and David Syring for their help in assembling the manuscript, to Annelise Earley for preparing the figures, and to Jeffrey David Ehrenreich, General Editor of the Society for Latin American Anthropology Publication Series, for his patience and help.

Both Guillet and Mitchell have worked equally in the preparation of the conference on which the volume is based and, until the final editing, in the development of the book. In the final months of production, however, Guillet was in Spain, so that Mitchell took over major responsibility for the editing and consequently is listed first as volume editor, while Guillet is designated first author of the Introduction. Nonetheless, it is difficult to assign priority–the two editors are responsible for the overall shape of both the volume and the introduction.

We have dedicated this volume to the memory of John Treacy. His tragic death not only cut short the life of a productive and respected scholar but also left many who knew him with a great sense of loss.

We appreciate the thoughtful comments of Monica Barnes, Inge Bolin, Jane Freed, Paul Gelles, Barbara Jaye, and Bruce Winterhalder on earlier versions of this paper. The Monmouth College Grants and Sabbaticals Committee has helped in supporting some of the costs of preparing this manuscript.

1. Originally known as "indigenous communities" (*comunidades indígenas*) and later as "peasant communities" (*comunidades campesinas*), these communities are corporate entities established by law (Dobyns 1964). In 1962 there were 1,600 such communities, and in 1977 there were 2,728. The 1977 survey reached 2,716, or 99 percent of these registered communities (DCCN 1980).

2. For information on highland irrigation and drainage systems see Denevan et al. 1987; Donkin 1979; Guillet 1987; Lechtman and Soldi 1981; Mitchell 1976; Ravines 1978. In 1986 the Peruvian journal *Allpanchis* devoted two issues to highland irrigation (Nos. 27 and 28).

3. Some small groups also speak other aboriginal languages, such as Uru, Kauke, Tupe, and Chipaya. See volume 2 of *The Handbook of South American Indians* (Steward 1946) for descriptions of the various linguistic and cultural groups in the central Andes.

4. The Shining Path guerrilla movement, the military response, and the cocaine trade have made research increasingly difficult in Peru. We can only hope that the situation improves, not only for future research but for the sake of Peruvians themselves.

5. See Ravines (1978:107-188) for other descriptions of Andean irrigation technologies.

6. Although the term "moiety" is sometimes restricted to a division into two kin groupings, in the Andes scholars use the term to designate the frequent division of highland communities into two territorial groups, or *saya*. These units, pronounced differently throughout the Andes, are variously known as "Anan Cuzco" and "Urin Cuzco" in Cuzco, "Hanan Sayoc" and "Lurin Sayoc" in Ayacucho, and "Anansaya" and "Urinsaya" in the Colca Valley.

7. In their contribution to this volume, Seligmann and Bunker suggest that the *ceque* system served as a mnemonic device "for storing local environmental knowledge." In their analysis, the *ceque* system organized environmental features in a manner that "continues to provide spatial, historical, religious, and ecological orientation" to local populations, even though the Inca empire and the *ceque* system itself have long since disappeared.

8. Unfortunately, because there are very few ethnographies describing the communities of the eastern escarpment, we do not have enough information to make a clear statement about agriculture in this zone. It is likely, however, that the rainy areas of this zone may differ significantly from areas elsewhere in the Andes.

9. In the principality of Andorra and other nearby areas of the Pyrenees, Mitchell has observed nonirrigated terraces with stone retaining walls, some used for agriculture and others used for animal grazing. Both types function to prevent erosion, the animal terraces mitigating what would otherwise be substantial hoof erosion on the steep slopes. Terraces may also produce turbulence in the downward flow of air, mixing cold and warmer air and thereby reducing the risk of frost (Bruce Winterhalder, personal communication, March 3, 1992).

10. For comparable studies of western escarpment irrigation communities see Echeandía 1981; Gelles 1986; Guillet 1992; and Montoya et al. 1979.

11. Speaking of a western Bontoc group in the Philippines, Bacdayan describes the construction of a new dam and 25 kilometer canal "through the mountains with no sophisticated surveying equipment." The task was made possible only through the effective recruitment of labor and "the people's detailed familiarity with the territory from years of hunting and foraging" (1974:252).

REFERENCES CITED

Aveni, Anthony F., and Helaine Siverman
1991 "Between the Lines; Reading the Nazca Markings as Rituals Writ Large." *The Sciences* (July/August):36-42.

Bacdayan, Albert S.
1974 "Securing Water for Drying Rice Terraces: Irrigation, Community Organization, and Expanding Social Relationships in a Western Bontoc Group, Philippines." *Ethnology* 13:247-260.
Banco Central de Reserva
1981 *Mapa de pobreza del Perú 1981*. Lima: Banco Central de Reserva.
Barnes, Monica, and David Fleming
1991 "Filtration-Gallery Irrigation in the Spanish New World." *Latin American Antiquity* 2:48-68.
Brush, Stephen B.
1977 "The Myth of the Idle Peasant: Employment in a Subsistence Economy." In *Studies in Peasant Livelihood*. Rhoda Halperin and James Dow, editors, pp. 60-78. New York: St. Martin's Press.
Collins, Jane
1988 *Unseasonal Migrations: The Effects of Rural Labor Scarcity in Peru*. Princeton: Princeton University Press.
DCCN (Dirección de Comunidades Campesinas y Nativas)
1980 *Comunidades campesinas del Perú Información básica*. Lima: Dirección de Comunidades Campesinas y Nativas, Ministerio de Agricultura y Alimentación.
Denevan, William M., Kent Mathewson, and Gregory Knapp, editors.
1987 *Pre-Hispanic Agricultural Fields in the Andean Region*. International Series 359. Oxford: British Archaeological Reports (BAR).
Dobyns, Henry F.
1964 *The Social Matrix of Peruvian Indigenous Communities.* Cornell Peru Project Monograph. Ithaca: Department of Anthropology, Cornell University.
Donkin, Robin A.
1979 *Agricultural Terracing in the Aboriginal New World*. Tucson: University of Arizona Press.
Echeandía, Juan
1981 *Tecnología y cambios en la comunidad de San Pedro de Casta*. Lima: Universidad Nacional Mayor de San Marcos.
Eggan, Fred
1954 "Social Anthropology and the Method of Controlled Comparison." *American Anthropologist* 56:743-763.
Erickson, Clark L.
1983 *An Archaeological Investigation of Raised Field Agriculture in the Lake Titicaca Basin of Peru.* Ph.D dissertation, University of Illinois, Urbana.
Flores Ochoa, Jorge, and Percy Paz Flores
1984 "El cultivo en qocha en la puna sur andina." In *Contribuciones a los estudios de los andes centrales*. Shozo Masuda, editor, pp. 59-100. Tokyo: University of Tokyo Press.
Gelles, Paul
1986 "Sociedades hidraulicas en los Andes; algunas perspectivas desde Huarochirí." *Allpanchis* 27:99-147.
George, Susan
1990 *A Fate Worse than Debt* (revised edition). New York: Grove Weidenfeld.
Guillet, David
1983 "Toward a Cultural Ecology of Mountains: The Andes and the Himalayas Compared." *Current Anthropology* 24:561-574.
1987 "Terracing and Irrigation in the Peruvian Highlands." *Current Anthropology* 28(4):409-430.
1989 "The Struggle for Autonomy: Irrigation and Power in Highland Peru." In *Human Systems Ecology*. Sheldon Smith and Ed Reeves, editors, pp. 41-57. Boulder, CO: Westview Press.
1992 *Covering Ground: Communal Water Management and the State in Highland Peru.* Ann Arbor: University of Michigan Press.

Hastorf, Christine
1990 "The Effect of the Inka State on Sausa Agricultural Production and Crop Consumption." *American Antiquity* 55:262-290.

Hastorf, Christine, and Timothy Earle
1985 "Intensive Agriculture and the Geography of Political Change in the Upper Mantaro Region of Central Peru." In *Prehistoric Intensive Agriculture in the Tropics*. International Series 232. Ian Farrington, editor, pp. 569-595. Oxford: British Archaeological Reports (BAR).

Hobsbawm, Eric, and Terence Ranger, editors
1983 *The Invention of Tradition*. Cambridge: Cambridge University Press.

Hunt, Robert C.
1989 "Appropriate Social Organization? Water Users Associations in Bureaucratic Canal Irrigation Systems." *Human Organization* 48:79-90.

Kosok, Paul
1965 *Life, Land and Water in Ancient Peru*. New York: Long Island University Press.

Lechtman, Heather, and Ana María Soldi
1981 *Runakunap kawsayninkupaq rurasqankunaqa. La tecnología en el mundo andino*. Mexico, DF: Universidad Nacional Autonoma de Mexico.

McClintock, Cynthia
1984 "Why Peasants Rebel: The Case of Peru's Sendero Luminoso." *World Politics* 37:48-84.
1988 "Peru's Sendero Luminoso Rebellion: Origins and Trajectory." In *Power and Popular Protest: Latin American Social Movements*. Susan Eckstein, editor, pp. 61-101. Berkeley: University of California Press.

Mitchell, William P.
1976 "Irrigation and Community in the Central Peruvian Highlands." *American Anthropologist* 78:25-44.
1985 "On Terracing in the Andes." *Current Anthropology* 26:288-289.
1991a *Peasants on the Edge: Crop, Cult, and Crisis in the Andes*. Austin: University of Texas Press.
1991b "Some Are More Equal Than Others: Labor Supply, Reciprocity, and Redistribution in the Andes." *Research in Economic Anthropology* 13:191-219.
n.d. "Multizone Agriculture in an Andean Village." In *Ayacucho Archeological-Botanical Project*, Volume 1, Richard S. MacNeish, editor. Ann Arbor: University of Michigan Press. In press.

Montoya, Rodrigo, María J. Silveira, and Felipe J. Lindoso
1979 *Producción parcelaría y universo ideológico: el caso de Puquío*. Lima: Mosca Azul.

Murra, John V.
1960 "Rite and Crop in the Inca State." In *Culture and History: Essays in Honor of Paul Radin*. Stanley Diamond, editor, pp. 393-407. New York: Columbia University Press.
1986 "The Expansion of the Inka State: Armies, War, and Rebellions." In *Anthropological History of Andean Polities*. John V. Murra, Nathan Wachtel, and Jacques Revel, editors, pp. 49-58. Cambridge: Cambridge University Press.

Orlove, Benjamin S., and David W. Guillet
1985 "Theoretical and Methodological Considerations on the Study of Mountain Peoples: Reflections on the Idea of Subsistence Type and the Role of History in Human Ecology." *Mountain Research and Development* 5(1):3-18.

Ragin, Charles C.
1987 *The Comparative Method*. Berkeley: University of California Press.

Ravines, Rogger, editor
1978 *Tecnología andina*. Lima: Instituto de Estudios Peruanos.

Rowe, John Howland
1946 "Inca Culture at the Time of the Spanish Conquest." In *Handbook of South American Indians,* Volume 2. Julian Steward, editor, pp. 183-330. Washington, DC: Bureau of American Ethnology. Bulletin 143.

1980 "Inca Policies and Institutions Relating to the Cultural Unification of the Empire." In *The Inca and Aztec States, 1400-1800: Anthropology and History*. George A. Collier, Renato Rosaldo, and John D. Wirth editors, pp. 93-118. New York: Academic Press.

Sampath, R. K., and Kenneth C. Nobe, editors
1986 *Irrigation Management in Developing Countries: Current Issues and Approaches*. Boulder, CO: Westview Press.

Sherbondy, Jeanette Evelyn
1982 *The Canal Systems of Hanan Cuzco*. Ph.D dissertation, University of Illinois, Urbana.

Stadel, Christoph
1989 "The Perception of Stress by *Campesinos*: A Profile from the Ecuadorian Andes." *Mountain Research and Development* 9: 35-49.

Steward, Julian H.
1949 "Cultural Causality and Law: A Trial Formulation of the Development of Early Civilizations." *American Anthropologist* 51:1-27.

Steward, Julian H., editor
1946 *Handbook of South American Indians, Volume 2, The Andean Civilizations*. Bureau of American Ethnology, Bulletin 143. Washington, DC: Smithsonian Institution.
1955 *Irrigation Civilizations: A Comparative Study*. Washington, DC: Pan-American Union.

Troll, Carl
1968 "The Cordilleras of the Tropical Americas; Aspects of Climatic, Phytogeographical and Agrarian Ecology." In *Geo-Ecology of the Mountainous Regions of the Tropical Americas*. Carl Troll, editor, pp. 15-56. Bonn: Ferd Dummlers Verlag.

Utton, Albert E., and Ludwik Teclaff, editors
1978 *Water in a Developing World*. Boulder, CO: Westview Press.

Villanueva, Horacio, and Jeanette Sherbondy
1979 "Cuzco: Aguas y poder." *Archivos de Historia Rural Andina*, 1. Cuzco: Centro de Estudios Rurales Andinos Bartolomé de Las Casas.

Wachtel, Nathan
1982 "The Mitimas of the Cochabamba Valley: The Colonization Policy of Huayna Capac." In *The Inca and Aztec States: 1400-1800, Anthropology and History*. George A. Collier, Renato I. Rosaldo, and John D. Wirth, editors, pp. 199-235. New York: Academic Press.

Wittfogel, Karl A.
1957 *Oriental Despotism: A Comparative Study of Total Power*. New Haven: Yale University Press.

Zuidema, R.T.
1964 *The Ceque System of Cuzco; The Social Organization of the Capital of the Inca*. Leiden: E.J. Brill.
1967 "El Origen del Imperio Inca." *Universidad; Organo de Extension Cultural de la Universidad de Huamanga (Ayacucho)* 3(9).
1986 "Inka Dynasty and Irrigation: Another Look at Andean Concepts of History." In *Anthropological History of Andean Polities*. John V. Murra, Nathan Wachtel, and Jacques Revel, editors, pp. 177-200. Cambridge: Cambridge University Press.

CHAPTER TWO

The Ecological Basis of Water Management in the Central Andes: Rainfall and Temperature in Southern Peru

BRUCE WINTERHALDER
University of North Carolina, Chapel Hill

The noble proportions of the Peruvian Andes and their position in tropical latitudes have given them climatic conditions of great diversity. . . . The greatest variety of climate is enjoyed by the mountain zone. Its deeper valleys and basins descend to tropical levels; its higher ranges and peaks are snow-covered. Between are the climates of half the world compressed . . . with extremes only a day's journey apart.
–Isaiah Bowman, *The Andes of Southern Peru*

INTRODUCTION

The peasant farmer in the Andes studies the clouds with an interest that is palpable. Sufficient and timely precipitation determines the abundance of the harvest and therefore the degree to which the family's diet and income will be adequate for the coming year. Weather is as important as soil, seed, and labor in agricultural production, but it is less amenable to control. In the marginal agricultural lands of the high Andes its effects can be sharp and its pattern capricious.

In this paper I describe seasonal patterns of precipitation and temperature in the southern Peruvian portion of the central Andes. The analysis is based on data collected from various weather services in the mountains of southern Peru. It relies on concepts (predictability) that portray better than the standard ranges and averages how weather and climate actually affect agriculture and human adaptation.

As observed by Bowman (1968[1916]) and virtually every other geographer or anthropologist who has spent time there, the Andes confront the human ecologist with an astounding diversity of ecozones and environmental conditions. The agro-ecological regions encompassed within a transect across these mountains offers rich materials for comparative analysis. Although the data presented here indirectly may aid in the analysis of prehistoric and historic developments in the Andes, I focus on comparative features of climate as evident in the weather recorded over approximately the last twenty-five years.

In the context of the present volume, I envision irrigation as an attempt to mediate between the sometimes fickle patterns of weather and the always exacting schedule of agricultural crops. To understand irrigation design, management, and effectiveness, we must know how these are influenced by the quantity, patterning, and predictability of the precipitation that initiates the terrestrial portion of the hydrological cycle. Thus, our study of the clouds takes up some of the same issues that inspire the attention of the Andean peasant.

THE ANDEAN LANDSCAPE

From Juliaca (near the Peruvian end of Lake Titicaca), the road to Sandia heads northeast across the Altiplano. Past the town of Putina, it gently ascends an upvalley grade following a small river. Leaving the stream, the road banks abruptly upward through a series of switchbacks to a 4,650 meter crest. Here it offers a spectacular view northeast across the Carabaya basin. The vista is breathtaking, aesthetically and physiologically. This part of the Carabaya is known as the Ananea plateau. It is a high plain elevated even above the Altiplano. Far across the plateau to the northeast lies a single row of glaciated summits (the Cordillera Oriental, Nevado Nacarro, and Nevado Ananea), and beyond them, the great cauldron of the Amazon Basin. The broken line of peaks marks the sharp break between the flat, high puna and the steep eastern escarpment, the plunging descent of the mountains into the lowland tropical rain forests.

If one looks backward toward the southwest, toward the landscape just traveled, the road descends to the Altiplano through foothills of exposed sedimentary formations. The contorted stratigraphy is twisted into whorls of brown, red, and mauve that rise here and there above the peneplain, testimony to the immense pressures that folded and buckled the earth to create these mountains. Not visible, but in this same direction 250 kilometers distant, is the western edge of the Altiplano, the Cordillera Occidental. Beyond this chain of high snow covered ridges the western escarpment descends to the cool, coastal deserts that border the Pacific.

This spot above the Carabaya plain is a good geographical vantage from which to consider the climate of these mountains.

The central or tropical Andes form a massive barrier perpendicular to the prevailing easterly winds of the southern hemisphere. This positioning determines the large-scale features of their climate. Air that is moisture laden from the Atlantic and its trip across the Amazon Basin forms into fog and rain clouds as it ascends the slopes of the eastern escarpment. From April through August this upslope flow affects only the lowermost elevations, but as the humidity increases and winds intensify in September and October–coincident with the southward displacement of the seasonal trade winds (Sarmiento 1986:12)–the precipitation belts extend upslope and eventually lap onto and across the Altiplano. They reach as far to the southwest as the upper portions of the western escarpment. Beyond this point, nearly depleted of moisture and more retentive of the residual water vapor because of downslope movement, expansion, and warming, the winds quickly lose their capacity to release moisture. Precipitation attenuates in quantity and duration. At the lowest elevations on the western escarpment, coastal meteorological stations may go years without recording even a trace of precipitation.

By March the winds weaken. The clouds and precipitation retreat northeast. The Altiplano rainy season ends. The scattered bunch grasses of the high plains turn dry and brown. The night air becomes colder. In this tropical zone, seasonal changes of temperature depend more on cloud cover than on changes in solar insolation. Hence they coordinate with, in a sense are secondary to, the clouds and rain.

Within the compass of this broad pattern there is considerable and only partially predictable variation. The rains may come early or late, in regular or irregular monthly distribution, and in quantities varying from drought to flood. The deeply incised, topographically spectacular relief of the Andes complicates the pattern with localized rain shadow and orographic effects, and with exposure to frost from the downslope pooling of cold air. Incessant wind, intense radiation, thin soils and steep slopes further differentiate the landscape by reducing the physiological availability of water to plants. Desiccation threatens when the rains are too little, erosion when the rains are too much.

At the spring periods of transition (roughly, October and April), the seasonal pulse of rainfall onto the upper escarpments and central basins of the Andes is visible from the Carabaya overlook. One can look through the gaps between the peaks onto the top of the clouds boiling upward from the humid forests below. Occasionally, great tongues of fog will push through the passes and roll like a flow of white vaporous lava onto the plains. Depending on the season, these upwellings of cloud signal the advance or the hesitant retreat of the great pulse of moisture that gives birth to the high Andean rainy season.

GEOECOLOGY, HISTORY AND ADAPTATION

The natural elements of climate are prominent in Andean literature, perhaps partly because they are overwhelming to scientific visitors. The humans and the domesticated animals and plants inhabiting the zone push hard against real physiological limits set by climate and altitude. Compared to temperate zones, it is not a salubrious environment for the production of food, although some of the same qualities that make it difficult have been fashioned into the raw materials of human adaptation (Murra 1984).

Carl Troll (1960, 1968) introduced the term *geoecology* to describe the constellation of environmental features that dominate life in the southern hemisphere's tropical high mountains. He has comprehensively described their broad geographical outlines (see also Bowman 1968[1916]; Winterhalder and Thomas 1982; Molina and Little 1981). A key climatic observation is that of the *diurnal temperature climate* (Bowman 1968:19). By virtue of tropical latitude, the central Andes[1] experience minimum seasonal variation in temperature and day length. For instance, at 10° from the equator the summer-to-winter solar insolation difference is only 20 percent. At 50° from the equator it nearly is 400 percent (Sarmiento 1986:11). However, because of high altitude the mountain valleys and plateaus experience pronounced day-to-night thermal differences. The same clear, thin atmosphere that makes the overhead solar insolation intense during the day does little to impede the rapid radiative loss of ground level heat at night. Diurnal variation can be as great as 25°C during the dry season. During the wet season this effect is somewhat reduced. Moist air and cloud cover act to retain warmth, especially at night. Despite this, nighttime frosts can occur in any month of the year in high-elevation agricultural zones.

Prehistoric Patterns

Secular changes as well as short-term fluctuations in the mountain climate play a major role in archaeological explanations of Andean prehistory. Considering the long term, Kent (1987) notes that three types of prehistoric evidence–shrinkage of the Quelccaya glacial ice cap near Cuzco, retreat of Amazonian forests, and a shift of the diet of populations located along the southern shore of Lake Titicaca (Chiripa, Bolivia) toward greater dependence on lacustrine resources–converge to suggest a period of aridity on the Altiplano from 750 B.C. to A.D. 350. These environmental indicators parallel sociopolitical shifts in the highland polities of Tiwanaku and Pukara. These societies apparently contracted and consolidated their Altiplano realms while developing relationships of trade and exchange with settlements on the lower altitude ecozones of the two escarpments. Kent argues that the centralization of political control and its reorientation along a geographical axis of ecological diversity were both stimulated by prolonged drought.

Similarly, Cardich marshalls evidence in favor of the thesis that climate variations "rather than changes in settlement pattern, land tenure and other human and cultural factors were primarily responsible for major fluctuations in population density during pre-Columbian times" (1985:296-297). He correlates major events in the horizon/period framework of Peruvian prehistory with shifts between deterioration and amelioration of highland climate. By comparing archaeological evidence and historical observations with current distribution of crop zones, he shows that agriculture in the highlands previously has extended as much as 250 meters higher than the current limit at about 4,050 meters.[2] He notes the possibility that prehistoric fluctuations that depressed temperature caused a similar 250 meters magnitude shift *below* current levels. If so, the (500 meters) range of these changes would be enough to cause major dislocations of agriculture and population over much of the Altiplano. Like Paulsen (1976), Cardich attributes Inca expansion to worsening climatic conditions in the 15th century. According to the independent estimate of Schoenwetter (1973; see also Browman 1987a:7), conditions favorable to agriculture extended in the past to altitudes 350 meters above their current limits.

Paulsen (1976) and Isbell (1978; cf. Conrad 1981) have argued that Andean polities expanded to buffer short-term environmental perturbations. In this view, political evolution in the uncertain highland environment was driven by the need to encompass dispersed and therefore independently varying sources of production. The ability to control environmental zones on both of the escarpments was especially attractive. One can find elements of this argument underlying the "archipelago" model of Andean ecopolitical organization (Murra 1972; 1984). In addition, fluctuations of rainfall are among several hypotheses advanced to explain the extensive postcontact abandonment of terraced land in the Andes (Donkin 1979; Denevan et al. 1987; Guillet 1987a: 417; Malpass n.d.).

Whatever the eventual appraisal of these hypotheses linking climate change and political development, pre-Columbian Andean societies prospered because they developed unique political and technological means to wrest a living from their high altitude territories. According to Murra, this involved

> three distinct steps . . . two are essentially climatological and agronomic feats; the third, more complex, was expressed in the social structural and economic arrangements which handled the other two (1984:120).

The first of these steps was the achievement of high agricultural productivity and reduced risk in what appears otherwise to be a marginal and uncertain environment. The second was the use of strong solar insolation and the cold to freeze-dry meat and tubers for accumulation and prolonged storage. And the third was spatially integrative political and economic structures that coordinated the production and distribution of resources from dispersed, altitudinally differentiated ecozones–the "verticality" or "archipelago" model (Murra 1972; see also Orlove 1977).

Whether viewed as a prime mover of sociocultural development or as a resource available to peoples whose history ultimately was moved by other factors, hypotheses involving climate figure prominently in Andean history and prehistory.

Peasant Production

Nowhere is the impact of climate more evident than in the lives of contemporary Andean peasants. Although they make the observation in different terms, ethnographers working in the Andes routinely note the great immediacy of climate for agricultural production. Their texts often present vivid descriptions of the vagaries of severe drought, floods, and frost in the highlands:

> In 1956, 1957, and 1964 there were serious droughts; in 1962 there was a serious flood; and in 1960 and 1963 there were February floods followed by severe frosts and a premature end to the rainy season. In 1965 part of the pampa of Taraco (province of Huancané) was covered by several feet of water from the Ramis River, stranding hundreds of the Indians' conical huts, while flooding others; in the pampa of Ilave a bridge was closed because of the high water. In both cases, however, it was explained that these conditions were by no means as serious as those during the 1962 and 1963 floods (Dew 1969:39-40).

Major droughts are known from 1982-1983, 1964-1966, 1956-1957, and 1942-1944 (Browman 1987a; Brown 1987; Guillet 1987a). Their consequences can be severe. For instance, in the 1982-1983 drought, highland crop production fell by 60 percent to 70 percent (Browman 1987a: 7-8). Similar decrements in production are cited for Ilave by Brown (1987), caused not by a severe and seasonally prolonged drought but by irregular monthly distribution of rainfall (1976-1977) abetted by a hard February frost (see also Kent 1987:298). Monthly variation like this can be devastating. It has its greatest impact during the germination and seedlings period (Mitchell *intra*).

More general estimates of the risk and impact of drought and frost vary in their severity and manner of deposition. In the higher agricultural lands some part of the "crops are lost to frosts or droughts every third year on the average" (Browman 1987a:13). Bernabe Cobo (reported in Flores Ochoa 1987:278) reported in the seventeenth century that harsh climatic environment on the Altiplano resulted in a bad crop two years out of three. An economic review using data on sale and barter around Lake Titicaca notes that "fluctuations in harvest are influenced more by climate than by economic conditions" (Orlove 1986:95). Commenting on the same article, Guillet (1986:101) cites the importance of environmental factors such as droughts in an understanding of the agrarian crisis in the Andes.

The environmental dimensions of this agricultural risk are the subject of a statistical analysis by Lhomme and Rojas (1986). For three stations on the Bolivian Altiplano (Charaña, 4,057 meters; Copacabana, 4,018 meters; and La Paz, 3,632 meters) they have calculated the annual pattern of incidence of three types of agricultural risk: desiccation, frost, and hail.[3] The joint probability of at least one of these climatic insults falls below 50 percent for only three to four months (December through March) at Copacabana and La Paz. It remains above 90 percent year-round for Charaña. Agriculture at these altitudes is risky, and based on these stations, the difference between 4,000 meters and 4,050 meters is large. It is not surprising, then, that "the principle economic organizing strategy of Andean arid land producers is risk management" (Browman 1987a:2).

CURRENT KNOWLEDGE: A CLIMATE OVERVIEW

The central Andes derive key features from their tropical latitude, but it is altitude, exposure, and relief that dominate local climate patterns. Because extensive collection of weather data began only in the mid-1960s (see Tables 2.1, 2.2 and 2.3), detailed analyses of extant weather records are rare, and there are few analytical summaries of regional or local climates. Existing generalizations are broad-brush. Although lacking an extensive database, Johnson (1976) is the most complete review. It is the source of most of the following generalizations.

November through April is recognized as the rainy reason in the Southern Hemisphere mountains of Peru. Precipitation is increasingly concentrated in a seasonal burst in the central months of this period as one moves southeast down the axis and, coincidentally but to a lesser extent, west and across the central Andes. Depending on the location, the buildup of precipitation occurs gradually in October or November in a series of pulses, whereas the end of the season is more abrupt and typically occurs in April. Compared to Ecuador and Columbia, the Andes of southern Peru (at 12 to 18 degrees south latitude) experience to a greater degree the effects of "increased continentality and the transition to the extratropical zone" (Sarmiento 1986:31). For instance, they occasionally are subjected to prolonged cold spells because of northward outbreaks of polar air masses.

Eastern Escarpment

The Amazon Basin provides near-ideal conditions for saturating the easterly winds with water vapor. It offers unimpeded movement to the Atlantic trade winds, moisture soaked landforms, luxuriant vegetation, and warmth for transpiration and evaporation. Rainfall is substantial over the tropical basin but higher yet on the eastern flanks of the Andes, the first orographic barrier to the moisture laden winds. Thus, Pucallpa at 151 meters averages 146 centimeters of rainfall, whereas a short distance away, Yurac, which lies at 295 meters and adjacent to the Andean foothills, gets 490 centimeters (Johnson 1976:155). Rainfall diminishes abruptly beyond the initial mountain flanks and then more gradually as one moves upslope (a general pattern of tropical mountains; Lauer, cited in Sarmiento 1986:13).

There are two complicating factors that can reverse the trend: aspect and pronounced vertical relief. Lee slopes, and especially those in the deep valleys that lie perpendicular to the cross-Andean airflows, get their highest precipitation at the high-elevation ridges. These valleys may be quite xeric at lower elevations where rain shadow effects are most intense. In some higher altitude areas, massive outcrops of cordillera can intensify the local orographic blockage of airflows and lead to heavy localized precipitation immediately downslope from the topographic obstruction.

Temperature lapse rates (Johnson 1976:173) are given as 6.8°C/1,000 meters for the first 1,000 meters, 3.2°C/1,000 meters for the next 1,500 meters, and 6.5°C for

TABLE 2.1 WESTERN ESCARPMENT WEATHER STATIONS

Station	Abbr	Latitude & Longitude		Altitude (m)	Year Start	Year Stop
Orcopampa	PRC	15° 16"	72° 21"	3779	1951	1984
Imata	IMA	15° 50"	71° 05"	4436	1935	1986
Yanque	YAN	15° 39"	71° 39"	3417	1951	1984
Cotahuasi	COT	15° 12"	72° 54"	2685	1964	1985
Cabanaconde	CAB	15° 37"	71° 59"	3287	1951	1984
Sibayo	SIB	15° 29"	71° 27"	3847	1947	1986
Angostura	ANG	15° 11"	71° 38"	4155	1962	1984
Crucero Alto	CRU	15° 46"	70° 55"	4400	1964	1985
Pulhuay	PUL	15° 05"	72° 26"	4600	1964	1984
Chinchayllapa	CHI	14° 56"	72° 43"	3950	1964	1986
Aplao	APL	16° 05"	72° 30"	510	1964	1986
Chuquibamba	CHU	15° 50"	72° 40"	2880	1954	1986
Pampa de Majes	PdM	16° 21"	72° 10"	1440	1950	1986
Punta Islay	ISL	17° 01"	72° 07"	43	1954	1976
Punta Atico	PAT	16° 14"	73° 42"	20	1954	1985
El Frayle	FRA	16° 09"	71° 11"	4015	1964	1986
Andagua	AND	15° 30"	72° 21"	3589	1951	1983
Huanca	HUA	16° 02"	71° 33"	3080	1964	1984
Arequipa	ARE	16° 22"	71° 33"	2520	1896	1984
Ayo	AYO	15° 41"	72° 16"	1956	1951	1984

Notes, Tables 1 through 3: Latitude is south; Longitude is west; Year/Start = year of the initiation of weather records; Year/Stop = last year in the recorded sample of weather records.

the elevations between 2,500 and 4,500 meters, with the lowered rate at intermediate altitudes apparently a result of inversions and the effects of pronounced topographic relief associated with deep valleys.

Altiplano

In the central portions of the Andes, the onset of the rainy season can vary by several months from year to year. Localized showers are common because much of the rainfall is convective in origin (often in the form of intense thunderstorms [Schwerdtfeger 1976]), and site-to-site variation in the pattern of precipitation is high. Compared to stations located to the northwest along the Andean axis, in southern Peru annual precipitation in the highlands is increasingly confined to the November through March period, totals are lower, and they are more variable (as

TABLE 2.2 EASTERN ESCARPMENT WEATHER STATIONS

Station	Abbr	Latitude & Longitude		Altitude (m)	Year Start	Stop
Calca	CAL	13° 20"	71° 57"	2926	1965	1981
Cirialo	CIR	12° 43"	73° 11"	900	1968	1978
Quillabamba	QUI	12° 53"	72° 44"	950	1964	1981
Maranura	MAR	12° 57"	72° 40"	1500	1970	1978
Pisac	PIS	13° 26"	71° 51"	2971	1963	1983
Ccatcca	CCA	13° 37"	71° 34"	3700	1965	1983
Urubamba	URU	13° 18"	72° 07"	2863	1964	1986
Yucay	YUC	13° 19"	72° 04"	2830	1968	1983
Machu Picchu	MAC	13° 09"	72° 31"	2080	1964	1977
Huyro	HUY	12° 57"	72° 35"	1700	1964	1981
Occobamba	OCC	12° 48"	72° 20"	1700	1964	1982
Ollachea	OLL	13° 46"	70° 29"	2850	1964	1986
Quincemil	QUI	13° 12"	70° 46"	619	1959	1984
Tambopata	TAM	14° 13"	69° 12"	1280	1964	1987
San Gabon	SGB	13° 27"	70° 28"	820	1964	1987
Paucartambo	PAU	13° 16"	71° 37"	2830	1964	1982
Sina	SIN	14° 30"	69° 17"	3000	1963	1978
Cuyo Cuyo	CYO	14° 28"	69° 32"	3414	1963	1987
Limbani	LIM	14° 10"	69° 43"	3200	1964	1987

Notes, see Table 1.

measured by the coefficient of dispersion).

In general, mean maximum temperatures in the Andes have a modest and shallow peak during the months of September, October, and November. This period gets the enhanced warming which results from the southward track of the sun before the full onset of the rainy season introduces more frequent and continuous overcast. Between December and March the cloud cover reduces insolation and thus daytime high temperatures. To an even greater extent it mitigates nighttime lows by impeding the loss into space of long-wave radiation. Thus, daily minimum temperatures are lowest (and diurnal variation is at its greatest) during the dry season. Skies are clear and dry and nights are long, conditions that facilitate escape of long-wave radiation. Diurnal variation is least during the wet season. Pronounced, latitudinally induced seasonality begins to affect the region only in its southernmost extension.

TABLE 2.3 ALTIPLANO WEATHER STATIONS

Station	**Abbr**	**Latitude**	**& Longitude**	**Altitude (m)**	**Year Start**	**Stop**
Chuquibambilla	CHU	14° 47"	70° 45"	3910	1931	1987
Arapa	ARA	15° 08"	70° 07"	3880	1964	1987
Ayaviri	AYA	14° 53"	70° 36"	3900	1965	1987
Azangaro	AZA	14° 55"	70° 11"	3863	1964	1986
Huancane	HUA	15° 12"	69° 45"	3890	1964	1987
Huaraya-Moho	HUM	15° 23"	69° 29"	3881	1957	1987
Puno	PUN	15° 50"	70° 01"	3825	1964	1987
Cabanillas	CAB	15° 39"	70° 22"	3850	1964	1987
Ilave	ILA	16° 06"	69° 38"	3880	1964	1987
Salcedo	SAL	15° 53"	70° 00"	3852	1950	1971
Juliaca	JUL	15° 29"	70° 09"	3825	1966	1987
Santo Tomas	SNT	14° 27"	72° 05"	3660	1965	1972
Llally	LLA	14° 56"	70° 53"	3890	1964	1981
Lampa	LAM	15° 22"	70° 22"	3892	1964	1987
Pucara	PUC	15° 08"	70° 50"	3910	1970	1987
Nuñoa	NUN	14° 29"	70° 38"	4135	1963	1987
Acomayo	ACO	13° 56"	71° 42"	3250	1965	1983
Munami	MUN	14° 46"	69° 57"	3949	1965	1987
Putina	PUT	14° 55"	69° 53"	3920	1966	1987
Crucero	CRU	14° 20"	70° 02"	4460	1966	1987
Sicuani	SIC	14° 17"	71° 13"	3550	1957	1984
Progreso	PRO	14° 41"	70° 22"	3950	1964	1987
Anta	ANT	13° 28"	72° 09"	3435	1965	1982
Chinchero	CHI	13° 24"	72° 03"	3762	1954	1976
Orurillo	ORU	14° 44"	70° 31"	3920	1966	1987
Santa Rosa	SRS	14° 37"	70° 47"	4000	1966	1987
Ananea	ANA	14° 40"	69° 32"	4600	1963	1987
Antauta	ANT	14° 20"	70° 25"	4150	1963	1975
Granja Kayra	KAY	13° 34"	71° 54"	3219	1931	1986
Cuzco	CUZ	13° 32"	71° 58"	3399	1945	1984

Notes, see Table 1.

Western Escarpment

In central and southern Peru, the steep western escarpment is a transition region between the wetter highlands and the arid coastal zone. Already depleted of moisture during the ascent of the eastern escarpment and traverse of the Altiplano, the precipitation potential of the air is reduced by downslope warming and increased water retention capacity. On the ground, the availability of water is further diminished by evapotranspiration stemming from warmth and lack of cloud cover. In the lowest elevations, the high humidity of the cool coastal air produces dense low-level fogs (*garuas*) that blanket the zone from May through October. Between the elevations of 100 meters and 800 meters these fogs are dense enough to sustain unique moisture-capturing vegetation mats (*lomas*). These capture the airborne moisture to create a localized band of vegetation receiving the equivalent of 10 to 20 centimeters of precipitation.

Although El Niño episodes can result in sporadic, heavy precipitation on the north coast of Peru, they do not appear to have this effect south of Chimbote (about 9° south latitude). Temperatures along the littoral zone are cool because of the modulating effect of the cold oceanic Humboldt current. Peak annual temperatures on this escarpment are reached at about 1,500 meters. Above this, the lapse rate again becomes negative with increasing altitude.

AN APPLICATION

Regional patterns of irrigation are an excellent vantage from which to examine the climate patterns of the southern Peruvian Andes in greater detail. Other questions (for example, comparisons of sectorial fallowing systems, crop complexes, methods of subsistence risk reduction, the interaction of agriculture with pastoralism, and so on) could motivate a similar analysis, but irrigation is especially appropriate given the abundant comparative materials of the present volume. The following discussion presumes that a narrow analytical question can facilitate description of a complex data set that is relevant to a much broader range of problems. Patterning of meteorological conditions combined with information on agriculture should help one understand the distribution, types, and functioning of irrigation systems in this region.

I begin with a very simple observation. The agricultural terraces in the Colca Valley, located on the high western slopes of the southern Peruvian Andes, were built with an imbedded irrigation network (Denevan et al. 1987; Guillet 1987a). About 250 kilometers to the northeast, directly across the axis of the cordillera, lie the agricultural terraces of the Sandia Valley. These terraces, located high on the eastern Andean escarpment, were built without provision for irrigation. Today, irrigation remains common in the Colca Valley, and rain-fed agriculture remains standard in the Sandia Valley.

How might we explain this difference? Both escarpment regions have long histories of occupation and often were linked by their mutual incorporation into polities centrally located on the Altiplano (Goland 1988; Julien 1983). Both have an agricultural substratum of magnificent stone-faced terraces, built by peoples obviously sophisticated in agricultural engineering. Both share similar kinds of potential water sources: direct precipitation as well as melt from the snows and glaciers of adjacent summits. Until the recent impact of market production in the Colca, both have grown the same crops: Andean tubers in the highest elevations and maize at lower altitudes. The superficial facts and constraints of cultural development, landscape, and physical geography, as well as cropping regimes, do not appear to offer a ready answer to this inquiry.

Climatic differences appear more promising. Annual rainfall in the two regions differs substantially. It is 290 to 500 millimeters in the Colca Valley and about 600 to 900 millimeters at comparable elevations (3,000 to 4,000 meters) in the Sandia Valley. Here is an answer: the Colca is deficient in precipitation, and the Sandia region is not. But irrigation can serve functions besides that of simply augmenting precipitation. It would be premature to end the investigation with the most obvious possibility–that is, irrigation occurs where absolute amounts of rainfall otherwise are insufficient for crop production. Other meteorological or ecological conditions, such as the timing of precipitation or its predictability, or the moisture retention capacity of soils, also are of potential importance to our inquiry.

This question is like others that could be asked about the regional patterning of agricultural systems in the Andes: its resolution will require an ecologically sophisticated understanding of the highland environment (Winterhalder 1980). With that in mind, the following analysis approaches Andean climate from the perspective of agricultural systems, production risk, and, more generally, human adaptation.[4]

The Problem

I define irrigation as the collection, movement, and dispersal of water to enhance agricultural production. It can be distinguished from drainage, which entails the removal of excess water from an agricultural site. Irrigation can function to modify the timing, quantity, predictability, or quality of the water available for agriculture, relative to incident precipitation or subsurface sources. Understanding the rationale for irrigation and tactics of water management requires that we be able to distinguish among these functions and relate them to temporal and spatial qualities of the water source. The functions may not be entirely separable. For instance, if the total quantity of precipitation during the cropping season is marginal for agriculture, it may also be the case that crops will be unusually susceptible to irregularities in the week-to-week distribution of rains, to delays in their onset or to their premature conclusion.

Hypothetically, then, irrigation might have one or more of these beneficial effects:

(1) *Timing:* it can extend the period of water availability, at the initiation or conclusion of the usual period of precipitation (for example, to facilitate the growing

of more slowly maturing crops or perhaps to obtain multiple crops);

(2) *Quantity:* it can augment incidental precipitation to thresholds required by certain crops (just as drainage might be used to diminish retention of precipitation or subsurface moisture sources);

(3) *Predictability:* it might be used to more evenly distribute or otherwise manage irregular availability of water, by the day or longer time periods, through the provision of storage reservoirs or by drawing from unsynchronized sources; and, finally,

(4) *Quality:* it might be used to change the chemical composition of local water supplies or soils, by introducing, diluting, or flushing certain chemicals, or to change the local microclimate, by elevating humidity or reducing the possibility of damage from frost.

The function of timing implies that water must be available when other ecological and agro-economic conditions are salubrious. Temperature is especially important, but so also are market factors, household schedules, and competing activities that may reduce labor availability or raise labor costs. Water supplies must coincide with the thermally favorable period for crop production even if rainfall does not. Likewise, irrigation might be used to begin a crop early, before the normal seasonal onset of precipitation, to benefit from market conditions or so that the producer can take advantage of temporary labor opportunities.

Each of these four hypotheses about the function of irrigation implies that water is moved from a source to a favorable crop zone. In the process, irrigation might be used to disperse a concentrated source or concentrate a dispersed one. It can be adapted to the needs of food crops or used to enhance forage for livestock production (see Browman 1987b). The spatial extent of an irrigation network is not a defining feature: self-contained irrigation systems can cover many kilometers (Farrington 1980) or be so small as to be encompassed within the 100 or so meter radius of the ingenious Qocha systems of the Altiplano (Flores Ochoa 1987). Of course, irrigation and other forms of water management (for example, drainage, impounding) can entail other, nonecological factors, such as economic or political control, or even military tactics (see the papers throughout this volume). However, I will restrict my focus to its agro-ecological aspects. Using meteorological data, I will attempt to predict which of the above functions underlie the regional distribution and operation of some Andean irrigation systems.

THE SAMPLE

The data reported here were collected predominantly by the Servicio Nacional de Meteorología e Hidrología (SENAMHI) in Peru. They were photocopied or transcribed at SENAMHI offices in Lima, Cuzco, Puno, and Arequipa, during many visits extending over six years. The stations are located in the departments of Cuzco [n = 23, 52 percent of 44 stations listed], Puno [n = 29, 41 percent of 70 stations listed], and Arequipa [n = 20, 23 percent of 87 stations listed]. The station samples in each department were selected based on the quality and time depth of the records available and on their geographic and altitudinal representativeness. All stations recorded daily precipitation. Most also collected maximum and minimum temperature, and some recorded other measures, such as number of days with precipitation, relative humidity, vapor pressure, hours of sunlight, evaporation, wind direction, wind speed, and cloud cover.

The data were available and were coded in the form of monthly sums (precipitation) or averages (maximum and minimum temperature). Of greatest interest are total monthly precipitation (millimeters) and mean monthly maximum and minimum temperature (°C). The average duration of the records available from these stations is 19 years. In total, the sample represents 1,309 station-years of meteorological information. Most of the stations were established in the late 1960s. In the results that follow, three stations have been eliminated from the original sample because of spotty or suspect data. Tables 2.1 through 2.3 provide a partial description of the stations, organized by the three major biogeographical zones in southern Peru: the western escarpment, the eastern escarpment, and Altiplano. The data have been coded and analyzed on a microcomputer using Lotus 1-2-3 or Quattro "spreadsheet" software. All statistics were calculated with Systat.

ANALYSES AND METHODS

Some of the analyses will rely on familiar descriptive measures (for example, annual precipitation and mean monthly or annual maximum and minimum temperatures). These measures are of interest because of the paucity of published meteorological information for the southern Peruvian Andes[5] and because of the geographical comparisons possible with a regional data set. Two measures are novel and require some explanation.

Predictability

Predictability is a measure of the regularity of a periodic phenomenon, such as seasonal rainfall or temperatures (Colwell 1974; Stearns 1981). It refers to the statistical predictability of a phenomenon, as distinct from the cognitive ability to forecast its occurrence. Technically, predictability is the statistical certainty of the state of a phenomenon (such as quantity of rainfall) given information about time (for example, the month of the year). If, for the whole year, knowing the month or season allowed one to make a certain statement about the quantity of rainfall, then precipitation would be completely predictable.

Predictability can arise in two different ways, termed constancy and contingency. *Constancy* measures the evenness of the phenomenon at each time interval in the period of interest. Thus, if the same amount of rain falls in each month of the year for all years recorded, precipitation would be maximally predictable because of complete constancy. *Contingency* is best thought of as repetition of seasonal pattern. If the quantity of rainfall were different in each month of the year but in a pattern that was precisely the same every year, predictability would be maximal due to contingency. *Predictability* is an aggregate measure of regularity arising from the combined effects of constancy and contingency. It is measured on a scale of 0 to 1, as a sum of varying amounts of constancy and contingency (evenness and seasonality, respectively).[6]

To give an Andean example, monthly precipitation at the city of Cuzco has a predictability of 0.45. This implies that the distribution of monthly rainfall is highly irregular from year to year. Constancy at Cuzco is 0.12; contingency equals 0.33. Thus, the limited regularity of monthly precipitation at Cuzco is due predominantly to the repetition of a highly seasonal pattern. In general, predictability is most useful when applied comparatively to assess the relative differences in the regularity of meteorological or other conditions at differing sites.

Low predictability means that there is little regularity to a phenomenon. Surfeit or deficit or some state between occurs with little pattern. Environmental irregularities (especially drought, flooding, and frost) and the production risks they entail are paramount concerns of Andean peasants. Predictability is of obvious and immediate importance to the analysis of agricultural systems and production tactics, although it rarely has been assessed in human ecology studies (Winterhalder 1980). I have applied this measure in assessing two factors: monthly rainfall and monthly minimum temperatures.

Onset and Conclusion

The second novel measure refers to the onset and conclusion of the rainy and thermal seasons. These have been defined as follows:

Onset: the first month of the rainy season in which a majority of years recorded have rainfall greater than 64 millimeters; the first month of the thermal season in which a majority of years have mean minimum temperatures greater than 0°C.

Conclusion: the last month of the rainy season in which a majority of years recorded have rainfall greater than 64 millimeters; the last month of the thermal season in which a majority of years have mean minimum temperatures greater than 0°C.

By these criteria Cuzco has a rainy season of 5 months, extending from November through March, and a thermal season of 12 months. The criteria used here–64 millimeters of precipitation and 0°C–are somewhat arbitrary from an agronomic perspective, but they provide a convenient index for regional, intersite comparisons.

RESULTS

I have organized the data in two formats. The first (Figure 2.1a) positions all 69 stations on a cross-Andes transect, oriented northeast to southwest perpendicular to the axis of the cordilleras. This places all the stations as if they occupied a single cross section of the mountains, stretching from the Amazon Basin across the Altiplano to the coast. It probably is the most general geographical comparison that

FIGURE 2.1a LOCATION OF SAMPLE WEATHER STATIONS.

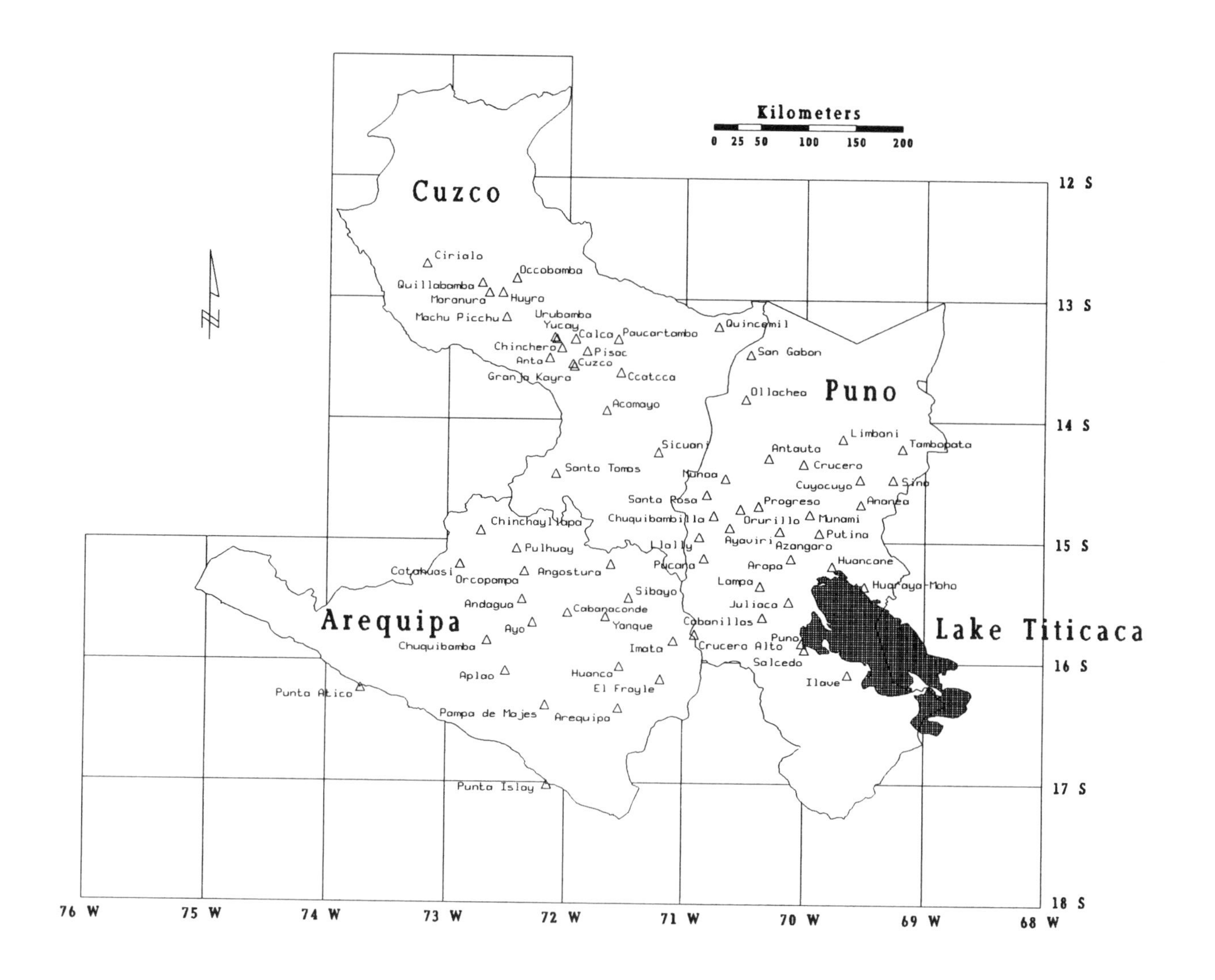

FIGURE 2.1b TOPOGRAPHICAL POSITION OF SAMPLE WEATHER STATIONS.

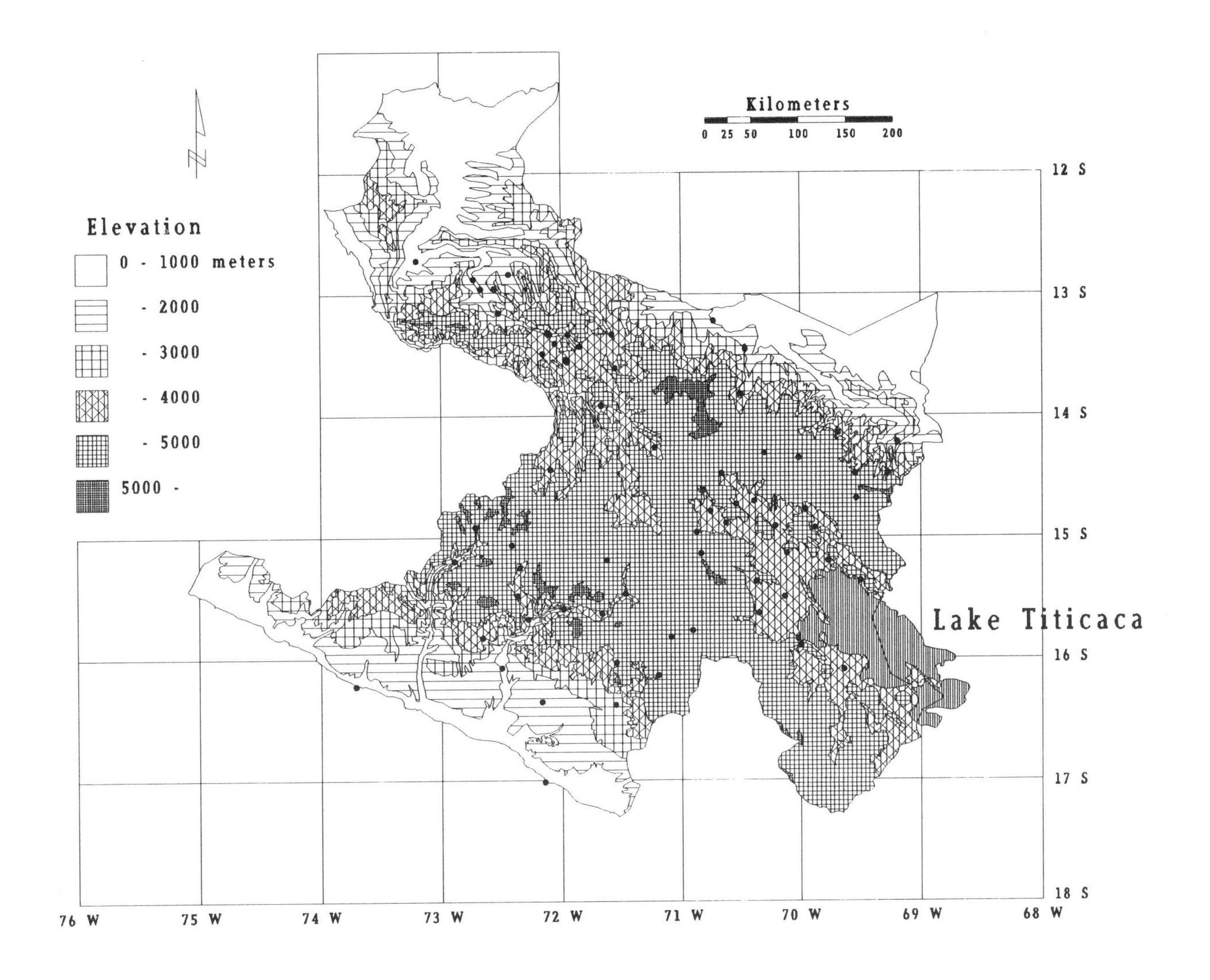

can be made with this data. The second format (Figure 2.1b) allows a more focused analysis. In it, I group data falling into the three geo-ecological zones of the Andes: the eastern escarpment, Altiplano, and western escarpment, using definitions as follows:

Eastern escarpment stations lie to the northeast of the divide of the Cordillera Oriental. They stretch from the headwaters into the lower reaches of the valleys draining into the Amazon Basin. The highest of these stations is Ccatcca at 3,700 meters, and the lowest is Quince Mil at 619 meters.

Altiplano stations occupy the interior basins and valleys between the eastern and western cordilleras. Most lie in the Titicaca Basin or the region surrounding Cuzco. The lowest is Granja Kayra, near Cuzco (3,232 meters) and the highest is Ananea (4,600 meters).

Western escarpment stations lie to the southwest of the divide of the Cordillera Occidental. These stations reach from sea level (Punta Atico at 20 meters) to 4,600 meters (Pulhuay).

All western escarpment stations lie within the department of Arequipa, whereas the departments of Cuzco and Puno both contain Altiplano and eastern escarpment stations. The data for stations in each of these zones are summarized in Tables 2.4 through 2.6.

Precipitation

Because of the visual density of the information in the tables, I will the discuss the results mainly by reference to a series of figures and regression analyses.

Figure 2.2 is a plot of the altitude of the transect stations. If the vertical scale were reduced, this figure would almost perfectly replicate the topographical cross section of the southern Peruvian Andes at the latitude of Lake Titicaca (see also Troll 1968:45; reprinted in Winterhalder and Thomas 1982). The saw-toothed character of the northeastern end of the transect reflects the tendency of eastern escarpment drainages to run for some distance parallel to the axis of the Andean chains. Thus, some stations set well into the province of the mountains nonetheless are situated at low elevations in valley bottoms. As noted earlier, the rain-bearing winds ascending this escarpment often must cross a series of valleys and ridges. In contrast, valleys on the western escarpment typically drain perpendicular to the mountains, directly and more or less continuously downslope. On this side of the cordilleras, elevation is better correlated with geographic position. The climatic impact of these topographic differences between the escarpments will become apparent below.

Figure 2.3 shows average annual precipitation along the transect. Some features of this data will be familiar from the earlier summary of central Andean meteorology: the very high precipitation of the eastern escarpment, montane zone; the high degree of variability on the middle and upper portions of the eastern escarpment; and the gradual and relatively continuous decrease of precipitation as one descends the western escarpment. An unexpected result is the very even distribution of precipitation across the Altiplano and into the Cordillera Occidental.

TABLE 2.4 WESTERN ESCARPMENT WEATHER DATA

Station	Avg P (mm)	Max P (mm)	Min P (mm)	Yrs (#)	Max T (°C)	Min T (°C)	P/C/M Precip	Onset Concl Rains	P/C/M Min Mon T	Onset Concl Thermal
Orcopampa	505.7	871.0	220.5	29			0.37/0.11/0.26	12 3		
Imata	569.3	893.3	182.8	34	12.8	-6.8	0.34/0.08/0.27	12 3	0.48/0.12/0.35	
Yanque	422.7	715.7	217.4	31			0.35/0.10/0.25	1 3		
Cotahuasi	297.9	577.6	110.1	13	22.4	8.3	0.44/0.12/0.32	1 3	0.36/0.25/0.11	7 6
Cabanaconde	399.0	807.7	105.2	33			0.49/0.25/0.24	1 3		
Sibayo	556.1	798.3	254.5	29	18.0	-1.6	0.35/0.06/0.29	12 3	0.56/0.05/0.51	11 4
Angostura	757.2	969.8	437.3	22	17.3	-10.1	0.41/0.07/0.34	12 3	0.59/0.08/0.51	1 3
Crucero Alto	579.2	914.9	230.7	17			0.40/0.10/0.30	12 3		
Pulhuay	631.6	1251.1	178.3	13			0.39/0.13/0.26	1 3		
Chinchayllapa	702.9	1021.5	189.3	20			0.36/0.13/0.23	12 3		
Aplao	6.8	13.6	2.0	12	26.9	11.8	0.68/0.55/0.13	0	0.64/0.23/0.42	7 6
Chuquibamba	173.0	389.0	36.5	9	16.0	6.1	0.61/0.38/0.23	0	0.68/0.48/0.20	7 6
Pampa de Majes	4.5	18.5	0.0	21	26.0	12.1	0.59/0.51/0.08	0	0.72/0.47/0.25	7 6
Punta Islay	6.7	21.2	0.0	6	20.7	16.4	0.73/0.59/0.14	0	0.66/0.40/0.26	7 6
Punta Atico	4.4	33.8	0.0	14	21.6	16.2	0.84/0.77/0.07	0	0.76/0.49/0.27	7 6
El Frayle	299.4	441.1	150.9	14	13.3	-4.2	0.43/0.11/0.31	1 2	0.50/0.18/0.32	1 3
Andagua	451.5	747.8	157.0	32			0.50/0.26/0.24	1 3		
Huanca	168.4	1039.4	40.7	17			0.55/0.34/0.21	0		
Arequipa	97.7	465.7	2.1	36	22.2	6.6	0.60/0.40/0.20	0	0.50/0.32/0.17	7 6
Ayo	87.1	203.1	26.4	31			0.48/0.29/0.19	0		

Notes, Tables 4 through 6: **Avg P** = Average annual precipitation; **Max P** = Maximum recorded annual precipitation; **Min P** = Minimum recorded annual precipitation; **Yrs** = number of complete years in sample; **Max T** = average monthly maximum temperature; **Min T** = average monthly minimum temperature; **P, C, M /Precip** = Predictability, constancy and contingency of monthly precipitation; **Onset Concl/Rains** = Onset and conclusion of the rainy season, 1 = January through 12 = December; **P, C, M/Min Mon T** = Predictability, constancy and contingency of monthly minimum temperature; **Onset Concl/Thermal** = Onset and conclusion of thermally favorable season for agriculture, 1= January through 12 = December.

TABLE 2.5 EASTERN ESCARPMENT WEATHER DATA

Station	Avg P (mm)	Max P (mm)	Min P (mm)	Yrs (#)	Max T (°C)	MinT (°C)	P/C/M Precip	Onset Rains	Concl Rains	P/C/M Min Mon T	Onset Thermal	Concl Thermal
Calca	514.3	668.5	357.3	14	22.1	5.6	0.44/0.12/0.32	12	3	0.46/0.13/0.32	7	6
Cirialo	1248.1	1631.7	887.5	10	31.1	19.6	0.48/0.15/0.33	11	4	0.91/0.81/0.10	7	6
Quillabamba	981.6	1312.0	667.3	16	30.6	16.7	0.50/0.23/0.27	10	4	0.57/0.46/0.10	7	6
Maranura	945.7	1246.9	792.6	7	29.4	17.5	0.52/0.22/0.30	11	3	0.77/0.54/0.23	7	6
Pisac	669.5	1414.3	219.8	18			0.37/0.08/0.29	12	3			
Ccatcca	594.7	718.3	499.3	15	15.3	1.1	0.49/0.11/0.38	12	3	0.56/0.19/0.37	9	4
Urubamba	391.4	592.4	70.6	20	22.2	6.2	0.38/0.12/0.26	1	1	0.59/0.15/0.44	7	6
Yucay	512.7	601.9	397.5	12	23.1	6.8	0.47/0.08/0.39	12	3	0.64/0.18/0.45	7	6
Machu Picchu	1996.3	2381.3	1571.3	11	22.3	9.9	0.58/0.28/0.30	8	4	0.63/0.44/0.19	7	6
Huyro	1766.5	2170.4	673.5	16	25.1	13.6	0.51/0.23/0.29	10	4	0.76/0.56/0.19	7	6
Occobamba	1863.5	3008.6	1301.3	15	25.9	12.9	0.52/0.25/0.27	9	4	0.57/0.47/0.10	7	6
Ollachea	1282.4	2574.2	829.4	19	17.9	6.9	0.37/0.12/0.25	10	3	0.54/0.39/0.15	7	6
Quincemil	7140.6	10024.3	5270.5	18	27.7	18.2	0.66/0.51/0.15	7	6	0.72/0.59/0.13	7	6
Tambopata	1526.3	1926.8	1280.0	23	25.0	14.7	0.58/0.35/0.22	9	4	0.77/0.55/0.23	6	6
San Gabon	5600.0	9115.2	3919.6	23			0.58/0.44/0.14	6	6			
Paucartambo	522.4	787.4	264.4	9	18.6	6.9	0.41/0.10/0.31	12	3	0.53/0.17/0.36	7	6
Sina	1720.3	2316.9	982.4	13			0.49/0.22/0.27	9	4			
Cuyo Cuyo	823.8	1696.5	536.0	16			0.50/0.21/0.28	12	3			
Limbani	951.0	1199.4	536.7	15			0.50/0.19/0.30	10	3			

Notes, see Table 4.

TABLE 2.6 ALTIPLANO WEATHER DATA

Station	Avg P	Max P (mm)	Min P	Yrs (#)	Max T (°C)	MinT	P/C/M Precip	Onset Concl Rains	P/C/M Min Mon T	Onset Concl Thermal
Chuquibambilla	691.7	1016.8	281.1	56	16.9	-2.7	0.46/0.14/0.32	12 3	0.55/0.18/0.37	12 3
Arapa	734.5	1302.1	435.0	21	15.9	2.4	0.38/0.11/0.27	12 3	0.72/0.26/0.46	9 5
Ayaviri	687.5	2164.2	333.9	17	16.0	-0.6	0.43/0.16/0.27	12 3	0.58/0.18/0.39	10 4
Azangaro	550.6	702.4	382.6	14	15.8	1.0	0.45/0.13/0.32	12 3	0.73/0.20/0.53	9 4
Huancane	681.5	1026.0	424.6	23	14.5	0.5	0.38/0.11/0.27	12 3	0.53/0.15/0.38	10 4
Huaraya-Moho	945.1	1383.3	615.4	20	14.5	2.9	0.43/0.11/0.32	11 4	0.74/0.26/0.49	8 5
Puno	724.3	1072.5	403.9	23	14.3	2.6	0.37/0.08/0.29	12 3	0.77/0.24/0.53	8 5
Cabinillas	652.8	852.7	218.9	21	16.7	1.6	0.43/0.10/0.33	12 3	0.66/0.11/0.55	9 4
Ilave	800.2	1192.4	335.3	14	14.6	1.1	0.36/0.10/0.26	12 3	0.62/0.14/0.47	9 4
Salcedo	658.3	1085.2	355.7	21	15.8	1.3	0.41/0.10/0.31	12 3	0.56/0.24/0.32	9 5
Juliaca	610.6	845.2	396.8	17	17.0	-0.7	0.42/0.13/0.29	12 3	0.60/0.09/0.51	11 4
Santo Tomas	742.4	1160.3	513.5	5			0.55/0.18/0.37	12 3		
Llally	751.0	1104.6	407.4	9	15.2	-1.5	0.44/0.09/0.36	12 3	0.66/0.24/0.42	12 4
Lampa	783.6	1620.7	522.7	20	16.3	-0.8	0.42/0.10/0.31	12 3	0.55/0.09/0.46	11 4
Pucara	794.1	1275.6	457.5	18			0.45/0.13/0.33	12 3		
Nuñoa	688.0	932.8	185.5	24			0.43/0.14/0.29	12 3		
Acomayo	880.0	1111.0	355.1	18	20.6	4.9	0.51/0.14/0.37	11 3	0.73/0.25/0.48	7 6
Munami	589.1	1044.2	320.6	18	16.0	0.6	0.43/0.15/0.27	12 3	0.51/0.31/0.20	9 4
Putina	683.3	880.6	445.7	21			0.44/0.13/0.31	11 3		
Crucero	906.1	1422.6	557.0	21			0.45/0.15/0.30	11 4		
Sicuani	597.1	960.5	122.5	20	19.4	2.6	0.38/0.12/0.26	12 3	0.50/0.24/0.26	8 5
Progreso	595.1	823.8	355.3	20	15.4	2.0	0.47/0.11/0.36	12 3	0.72/0.26/0.46	9 5
Anta	742.0	887.7	591.3	16	18.0	3.4	0.49/0.11/0.38	11 3	0.61/0.11/0.50	9 5
Chinchero	799.9	1113.8	420.5	12	20.4	6.8	0.42/0.11/0.31	11 3	0.28/0.09/0.20	7 6
Orurillo	750.8	1179.1	521.8	21			0.42/0.11/0.31	12 3		
Santa Rosa	1094.6	1383.8	757.8	9			0.52/0.13/0.39	11 4		
Ananea	658.6	874.0	486.9	21			0.47/0.17/0.30	12 3		
Antauta	709.1	1365.3	492.8	10			0.41/0.18/0.23	12 3		
Granja Kayra	676.7	923.2	477.7	21	24.8	3.8	0.47/0.09/0.38	12 3	0.68/0.25/0.43	8 5
Cuzco	745.0	982.1	389.5	35	19.7	4.3	0.45/0.12/0.33	11 3	0.67/0.21/0.46	7 6

Notes, see Table 4.

FIGURE 2.2 THE ALTITUDE OF THE SAMPLE STATIONS AS ARRAYED ON THE NORTHEAST (NE) TO SOUTHWEST (SW) TRANSECT ACROSS THE SOUTHERN PERUVIAN ANDES.

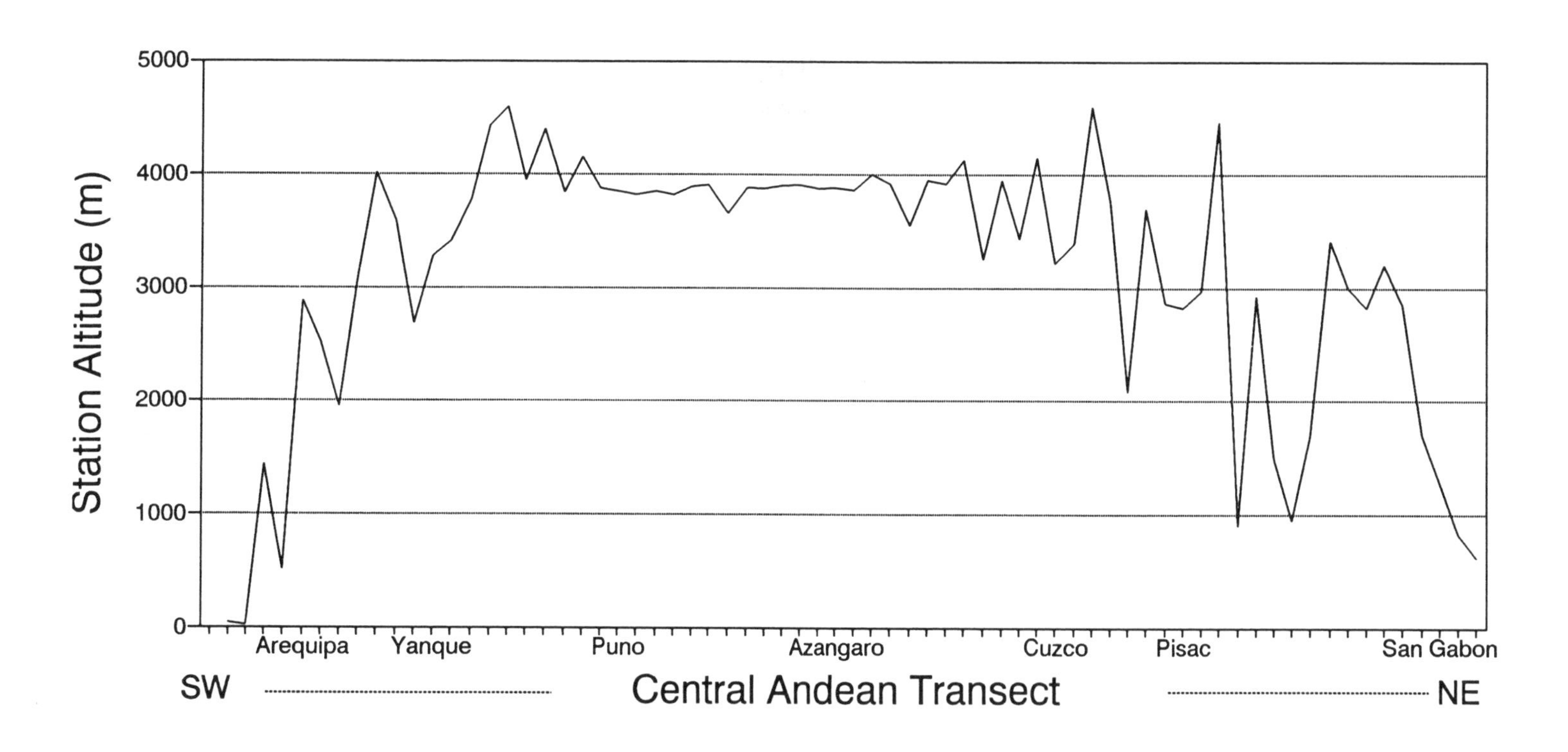

FIGURE 2.3 MEAN ANNUAL PRECIPITATION OF SAMPLE STATIONS AS ARRAYED ON THE NORTHEAST (NE) TO SOUTHWEST (SW) TRANSECT ACROSS THE SOUTHERN PERUVIAN ANDES.

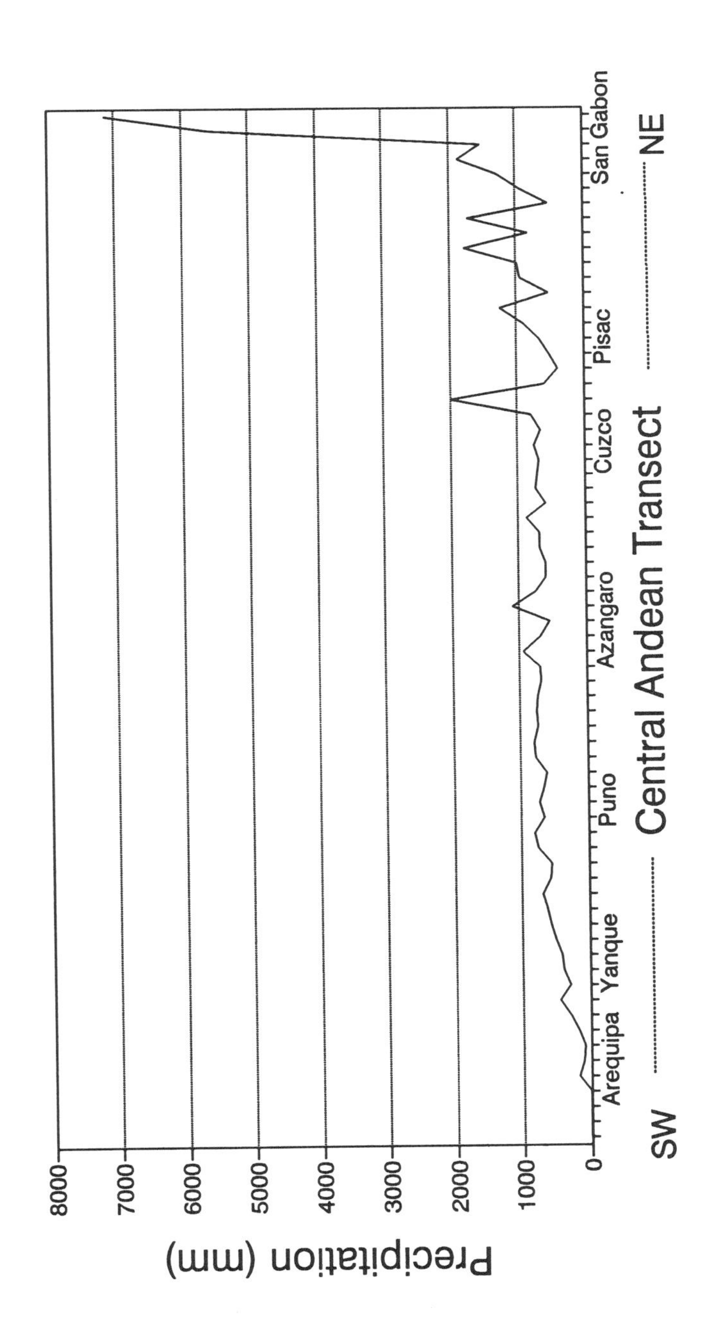

The classic illustrations of Troll (1968; reprinted in Molina and Little 1981; Winterhalder and Thomas 1982) describe climatic zonation in the interior Andean basin going from northeast to southwest as "moist" and then "dry" puna belts.[9] Although these descriptive terms imply differences in the absolute amount of precipitation, the present data suggest that geo-ecologists need to look for other factors to explain these ecological zones–perhaps differing soils, temperature (acting to limit the physiological availability of water to plants), or the predictability of precipitation. With respect to the focal question of this inquiry, note that irrigation in the middle-to-upper altitude portions of the western escarpment can draw for a water source on the relatively high precipitation of adjacent Cordillera Occidental highlands.

Predictability of monthly precipitation (Figure 2.4) is high for the lowest stations of the eastern escarpment mainly because of constancy. At San Gabon and Quince Mil, for instance, it rains with fairly great regularity nearly year-round. Predictability diminishes gradually but with significant variation from the middle-upper altitudes of the eastern escarpment through the upper altitudes of the western escarpment. It increases dramatically as one moves downslope on the western escarpment, reaching peak values at the coast where with high predictability it seldom rains in any month of any year. From the perspective of our inquiry, irrigation to reduce irregularities in monthly precipitation would have its maximum benefit in the upper reaches of the western escarpment and on the adjacent Altiplano.

Greater resolution is possible if we examine this evidence by region. Precipitation as a function of altitude on the western escarpment is shown in Figure 2.5, along with two regression lines (see Table 2.7). A linear regression on this data has a Pearson correlation coefficient of $r = 0.87$, with an explained variance of $r^2 = 0.76$. However, visual inspection suggests that a polynomial regression [average annual precipitation = constant x altitude^2] might provide a better fit and it does ($r = 0.97$; $r^2 = 0.94$). The meteorological basis of this polynomial relationship between altitude and precipitation on the western escarpment is unclear, but the statistical concordance is striking. With the polynomial model, altitude alone accounts for 94 percent of the station-to-station variation in annual precipitation. The polynomial relationship depicted in Figure 2.5 again highlights the potential benefit of being able to capture by means of irrigation the moisture found in the highest zones of this escarpment. Relative to lower zones, precipitation occurs there in abundance.

The predictability, constancy, and contingency of monthly precipitation on the western escarpment is shown in Figure 2.6. Predictability decreases from about 0.8 to less than 0.4 as altitude increases. Constancy or evenness of the annual pattern declines even more sharply. By contrast, contingency increases upslope. At the uppermost stations, nearly all the limited regularity of monthly precipitation is caused by a pronounced seasonal pattern. As with amount of rainfall, there is a fairly regular relationship between predictability measures and elevation. Altitude alone explains 80 percent of the interstation variance in predictability ($r = 0.89$; Table 2.7).

Rainfall on the eastern escarpment has a less regular relationship to altitude (Figure 2.7), because of the effects of valley topography. A linear regression model

FIGURE 2.4 THE PREDICTABILITY OF MONTHLY PRECIPITATION AT THE SAMPLE STATIONS AS ARRAYED ON THE NORTHEAST (NE) TO SOUTHWEST (SW) TRANSECT ACROSS THE SOUTHERN PERUVIAN ANDES.

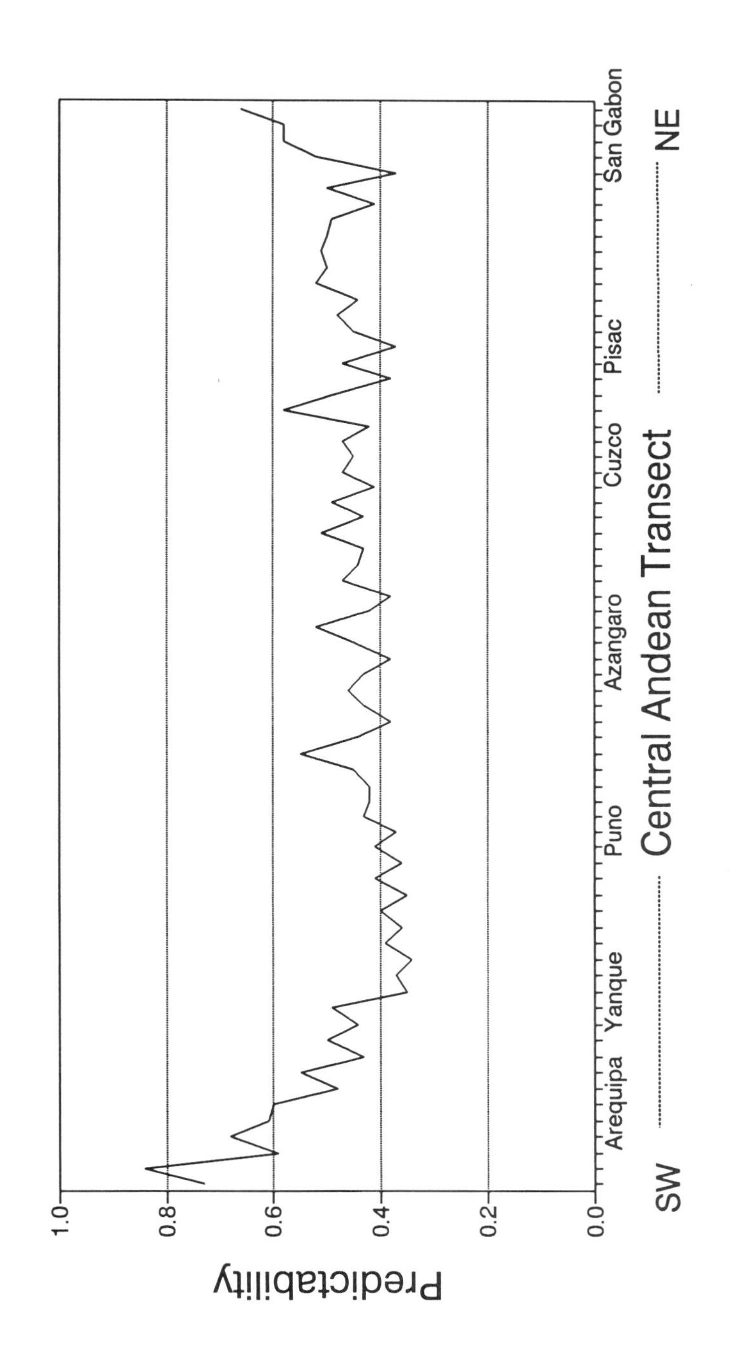

TABLE 2.7 REGRESSION ANALYSES OF CLIMATIC FACTORS ON ALTITUDE (MEASURED IN 1000s OF METERS), BY GEOECOLOGICAL ZONE

	Western Escarpment	Eastern Escarpment	Altiplano
AVERAGE ANNUAL PRECIPITATION			
n	20	19	30
r (Pearson)	0.874	0.627	0.036
r^2	0.764	0.393	0.001
Constant	-116.3	4107.6***	676.6*
Factor	154.0***	-1115.0**	14.0
MEAN MAXIMUM TEMPERATURE (°C)			
n	11	14	21
r (Pearson)	0.742	0.898	0.847
r^2	0.551	0.806	0.718
Constant	24.8***	33.3***	53.0***
Factor	-2.08**	-4.50***	-9.57***
MEAN MINIMUM TEMPERATURE (°C)			
n	11	14	21
r (Pearson)	0.943	0.978	0.616
r^2	0.890	0.957	0.379
Constant	17.6***	23.3***	24.5**
Factor	-5.24***	-5.90***	-6.07**
PREDICTABILITY OF MONTHLY PRECIPITATION			
n	20	19	30
r (Pearson)	0.893	0.624	0.168
r^2	0.798	0.390	0.028
Constant	0.752***	0.599***	0.536***
Factor	-0.088***	-0.048**	-0.0025
PREDICTABILITY OF MONTHLY MEAN MINIMUM TEMPERATURE (°C)			
n	11	14	21
r (Pearson)	0.617	0.674	0.130
r^2	0.380	0.454	0.017
Constant	0.693***	0.825***	0.856
Factor	-0.044*	-0.088**	-0.063

Note: *** Significant at 0.001; ** Significant at 0.01; * Significant at 0.05. All statistics used the MGLH module of Systat.

FIGURE 2.5 WESTERN ESCARPMENT: MEAN ANNUAL PRECIPITATION AS A FUNCTION OF ALTITUDE, WITH SUPERIMPOSED POLYNOMIAL REGRESSION (SEE TABLE 2.7).

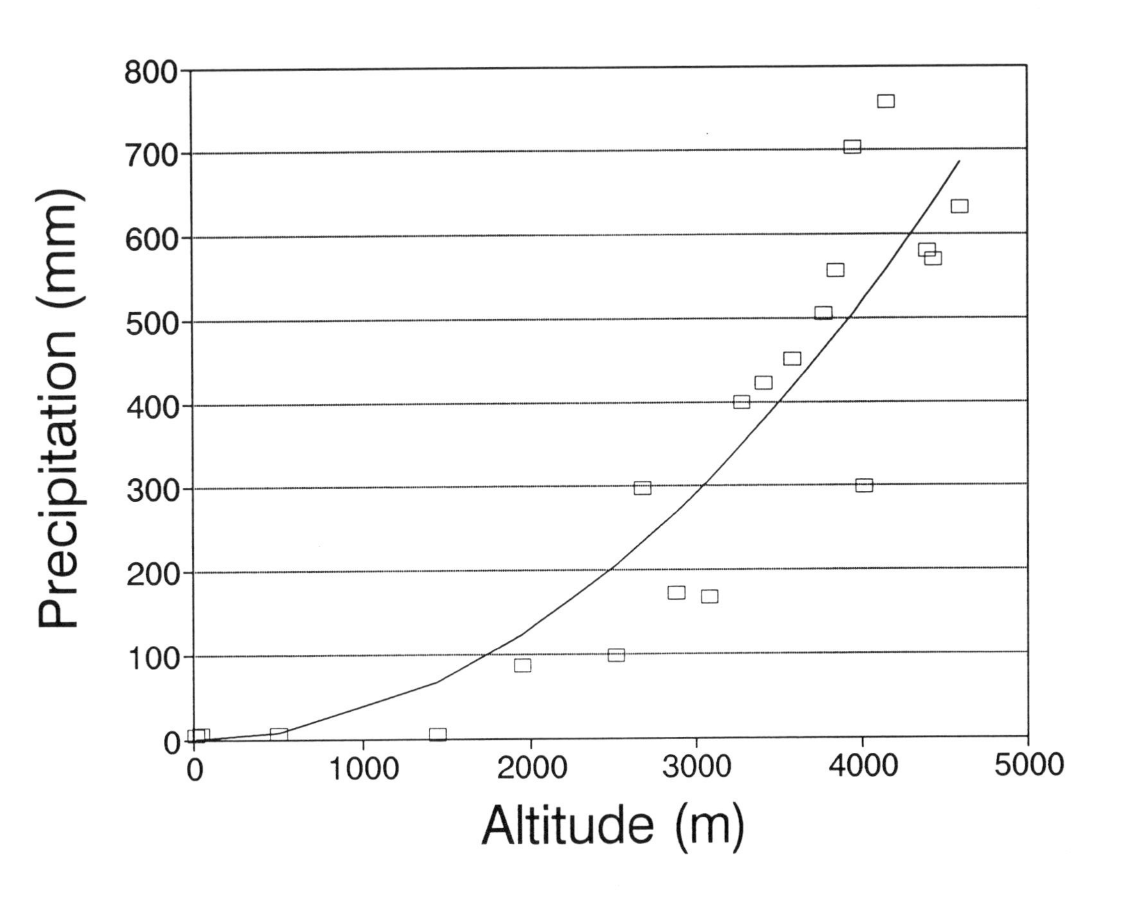

FIGURE 2.6 WESTERN ESCARPMENT: PREDICTABILITY (P), CONSTANCY (C) AND CONTINGENCY (M) OF MONTHLY PRECIPITATION AS A FUNCTION OF ALTITUDE, WITH SUPERIMPOSED LINEAR REGRESSIONS (SEE TABLE 2.7).

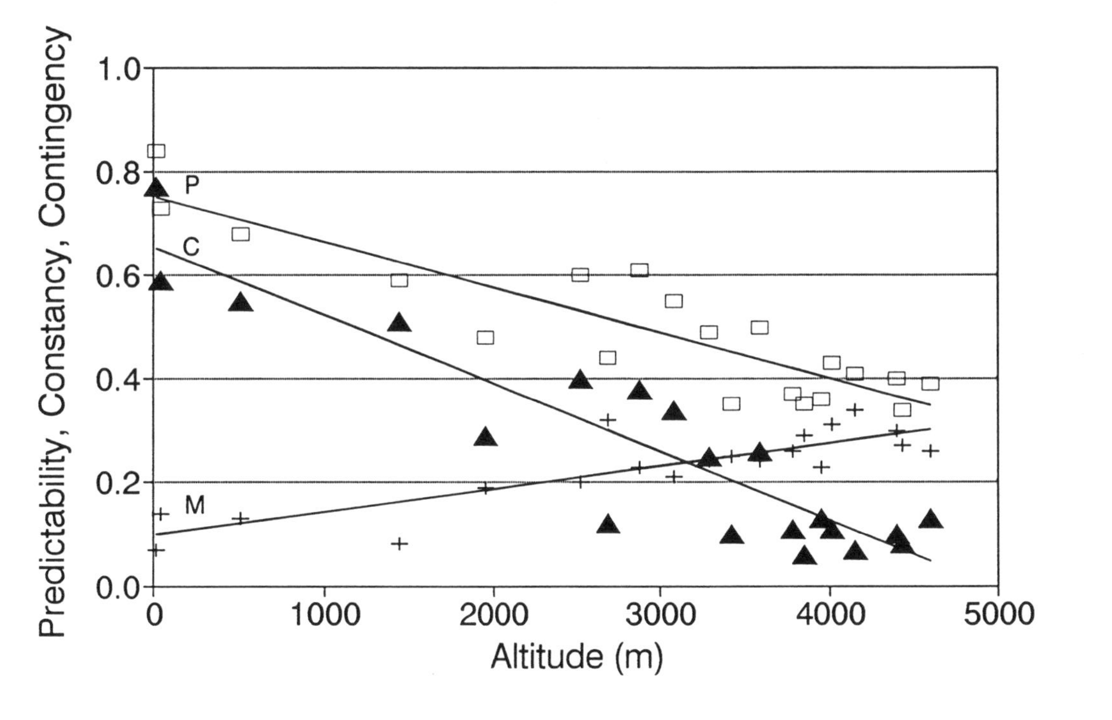

FIGURE 2.7 EASTERN ESCARPMENT: MEAN ANNUAL PRECIPITATION AS A FUNCTION OF ALTITUDE, WITH SUPERIMPOSED LINEAR REGRESSION (SEE TABLE 2. 7).

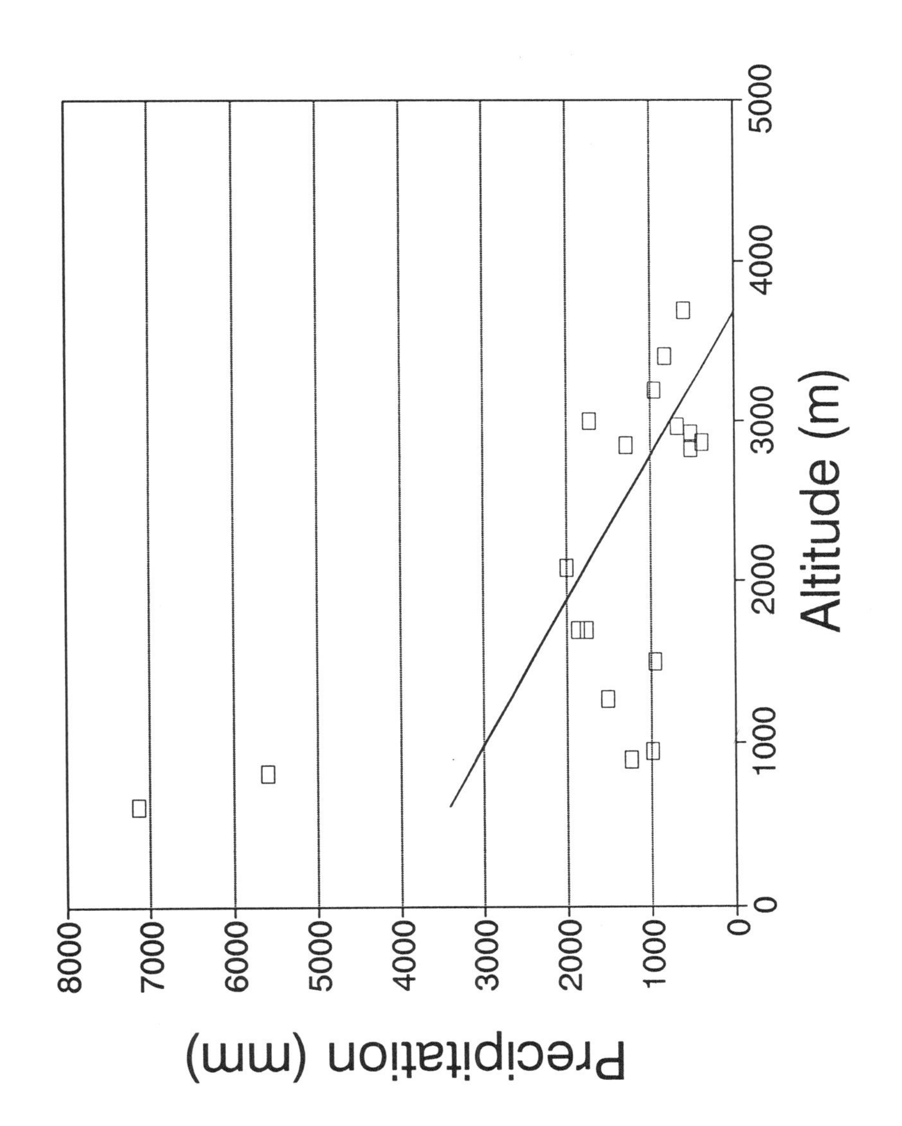

indicates that precipitation decreases upslope ($r = 0.63$; $r^2 = 0.39$), but visual inspection suggests that the very lowest stations on the mountain flanks (Quince Mil and San Gabon) are meteorologically distinct from the stations occupying the higher elevations.[10] Altitude accounts for less than 40 percent of the variance in precipitation among these eastern escarpment stations. Hence, although there is a weak overall trend toward lower rainfall as one moves upslope, rain shadow effects may cause localized reversal of this relationship in deep valleys. The high ridges, shoulders, and flanks of the valley may get more rainfall than the lower slopes and valley bottom, which are warmer because of low elevation and drier because of their isolation from the moisture bearing winds aloft.

Predictability of monthly precipitation on the eastern escarpment decreases upslope ($r = 0.62$; $r^2= 0.39$) as does the constancy component (Figure 2.8). As on the western escarpment, at the highest altitudes of the eastern escarpment nearly all the predictability of monthly precipitation is a result of contingency. However, the relationship of predictability to altitude is weaker on the eastern slopes than it is on the western slopes.

A comparison of the two escarpments shows that predictability of monthly precipitation decreases upslope *whether* the actual amounts of precipitation are falling (as on the eastern escarpment) or rising (as on the western escarpment). In either region irrigation to diminish irregularities in monthly distribution of rainfall is most likely at the higher altitudes of agricultural production.

Temperature

For the most part, annual mean minimum temperature follows the expected pattern on a transect across the central Andes (Figure 2.9): it drops for stations progressively higher on the two escarpments. However, it also appears to fall as one crosses the altiplano from the northeast to the southwest.[11] This decline in minimum temperatures may be more significant than is precipitation for the ecological zonation described by Troll (1968; see discussion above). If so, then it may be more apt to designate these vegetation zones by terms that refer to temperature rather than moisture, perhaps as the cool-dry and cold-dry punas. Viewed across the whole transect, predictability of minimum temperature shows no clear trend and has not been graphed.

A detailed look at the two escarpments gives evidence of an orderly relationship between altitude, temperature, and minimum temperature predictability. On the western escarpment, both mean maximum ($r = 0.74$; $r^2 = 0.55$) and mean minimum ($r = 0.94$; $r^2 = 0.89$) temperatures fall regularly with altitude (Figure 2.10). This relationship is especially strong for the minimum, which has a lapse rate of 5.24°C per 1,000 meters and an explained variance of 0.89. The range between the maximum and minimum temperatures increases as one moves upslope, but this effect would be reduced (and the correlation between mean max °C and altitude enhanced) if we eliminate the coastal stations from the mean maximum temperature data.[12] Both of these sea-level stations are "outliers" in the data set, their maximum temperatures unusually depressed because of proximity to the cold ocean currents.

FIGURE 2.8 EASTERN ESCARPMENT: PREDICTABILITY (P), CONSTANCY (C) AND CONTINGENCY (M) OF MONTHLY PRECIPITATION AS A FUNCTION OF ALTITUDE, WITH SUPERIMPOSED LINEAR REGRESSIONS (SEE TABLE 2.7).

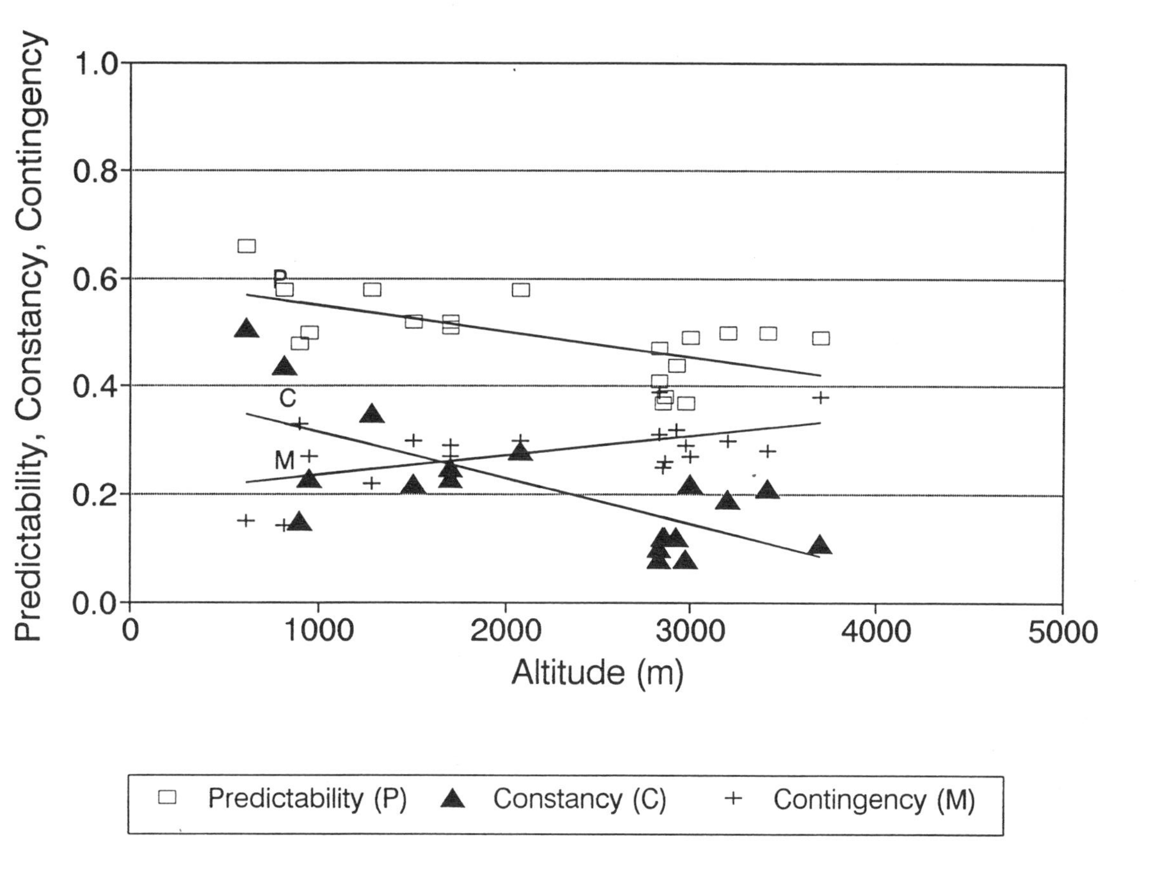

FIGURE 2.9 MEAN MINIMUM TEMPERATURE AT SAMPLE STATIONS AS ARRAYED ON THE NORTHEAST (NE) TO SOUTHWEST (SW) TRANSECT ACROSS THE SOUTHERN PERUVIAN ANDES.

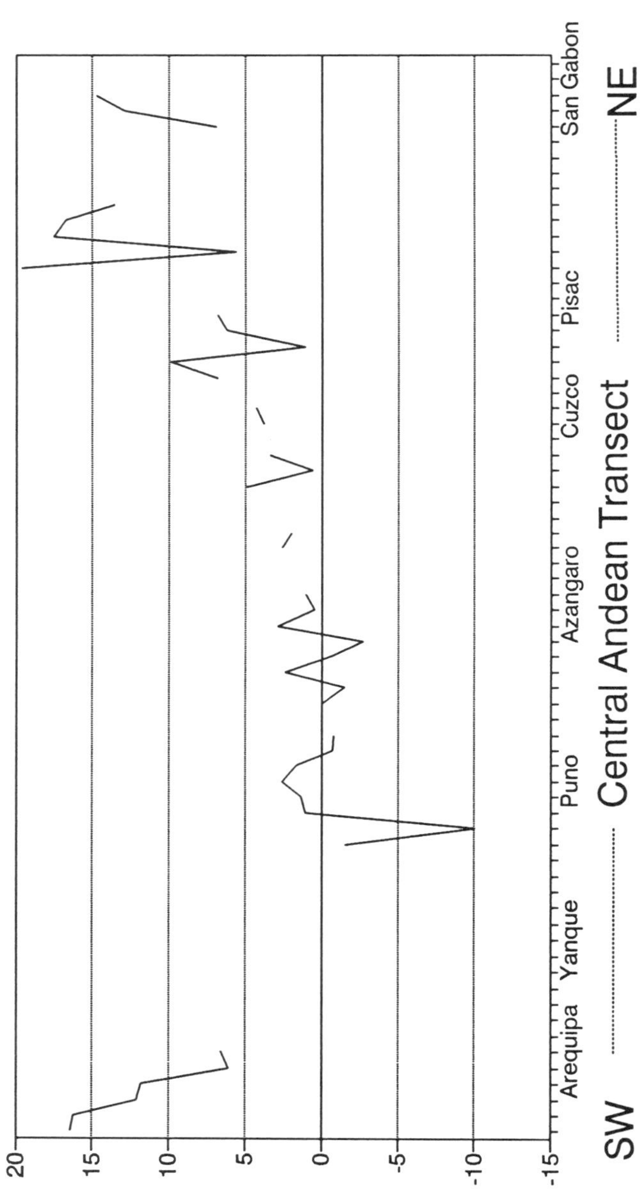

FIGURE 2.10 WESTERN ESCARPMENT: MAXIMUM AND MINIMUM MEAN MONTHLY TEMPERATURES AS A FUNCTION OF ALTITUDE, WITH SUPERIMPOSED LINEAR REGRESSIONS (SEE TABLE 2.7).

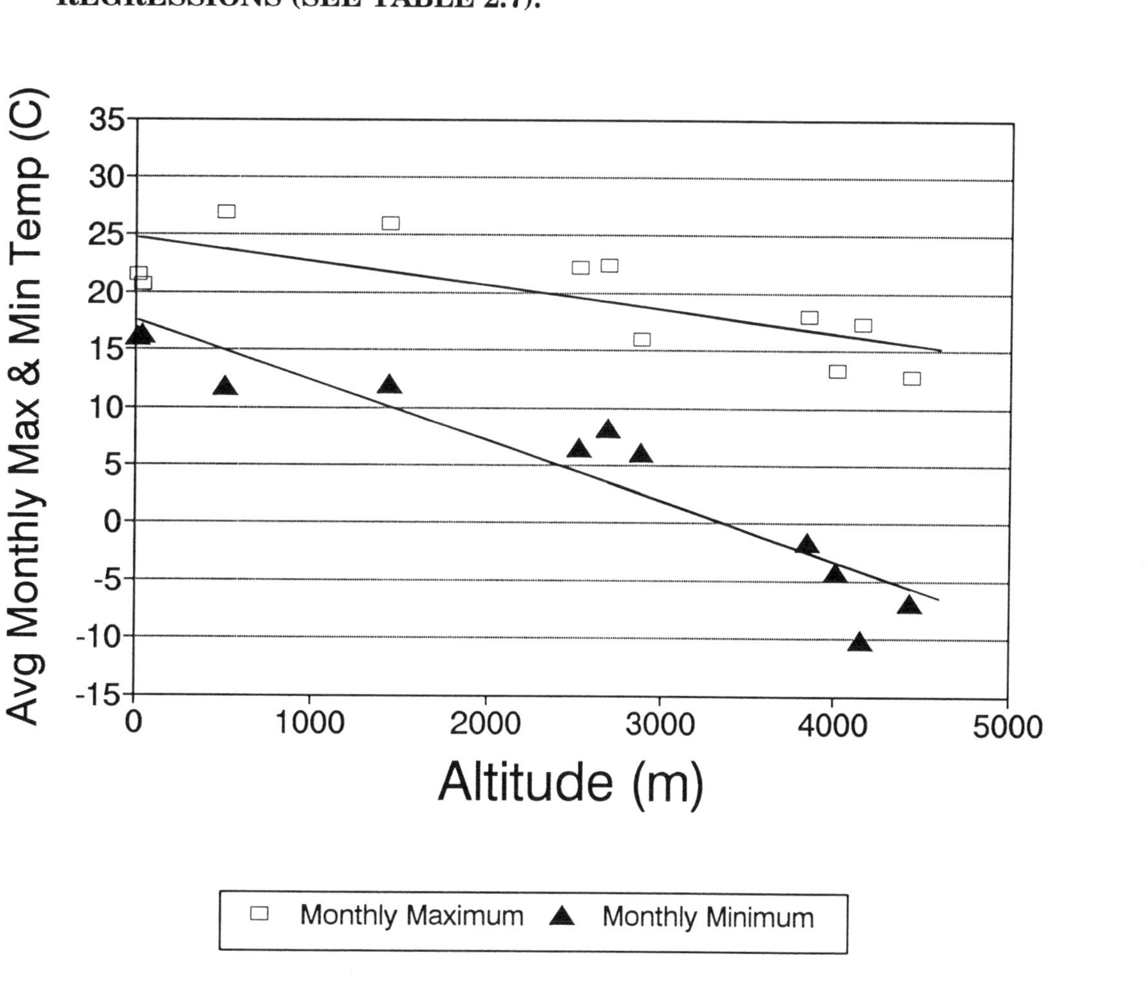

FIGURE 2.11 WESTERN ESCARPMENT: PREDICTABILITY (P), CONSTANCY (C) AND CONTINGENCY (M) OF MONTHLY MINIMUM TEMPERATURE AS A FUNCTION OF ALTITUDE, WITH SUPERIMPOSED LINEAR REGRESSIONS (SEE TABLE 2.7).

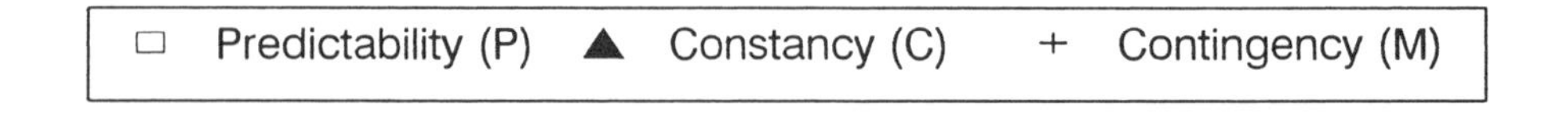

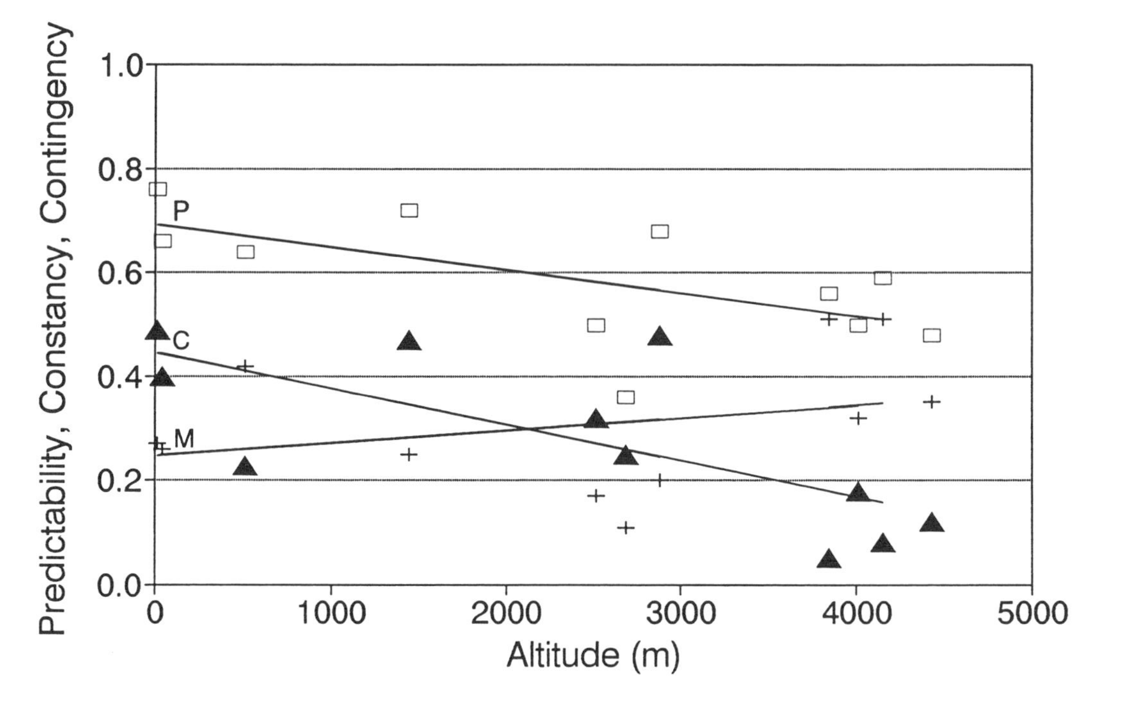

Predictability of minimum temperatures on the western escarpment also falls with altitude ($r = 0.62$; $r^2 = 0.38$), although altitude accounts for less than 40 percent of the variance in this measure (Figure 2.11; Table 2.7). Minimum temperatures are lower and less regular at the higher altitudes.

Much the same pattern of temperature relationships emerges on the eastern escarpment (Figure 2.12). Mean maximum ($r = 0.89$; $r^2 = 0.81$) and mean minimum ($r = 0.98$; $r^2 = 0.96$) temperatures fall with altitude in strong linear relationships. The lapse rates per 1,000 meters are 4.5°C and 5.9°C, respectively. The very high correlation coefficients in these data suggest that the irregular topography of this escarpment affects the relationship between temperature and altitude much less than it does the relationship between precipitation and altitude. The predictability of minimum temperatures diminishes upslope on the eastern escarpment ($r = 0.67$; $r^2 = 0.45$), from slightly less than 0.80 to approximately 0.50 at the highest altitudes (Figure 2.13; Table 2.7).

We can use the regressions from Table 2.7 to compare the eastern and western escarpments for altitude-adjusted minimum temperatures. At 1,000 meters, the western slopes average about 5°C colder than their eastern counter parts, whereas at 4,000 meters the difference is 3°C. Thus, at comparable elevations, minimum temperatures are significantly colder on the western escarpment. This difference may be partially because of the effect of moisture, as frosts generally begin at lower elevations on the slopes of dry mountains (Sarmiento 1986:13).

Onset of Rainy and Thermal Seasons

Figure 2.14 presents the second of the novel measures that I described earlier: onset and conclusion of the thermal and rainy seasons. On the northeastern end of the transect, the thermally propitious season lasts from July through the following June, the full 12 months of the annual cycle. Excepting stations adjacent to Lake Titicaca,[13] from the high northeastern slopes to the high southwestern escarpment the warmer thermal season drops from 10-12 to 4-5 months, mainly because it commences later and later in the year. This parallels the earlier observation of a fall in average minimum temperature along the same portion of the transect. In the middle to lower slopes of the western escarpment there is a rapid transition again to a 12-month thermal warm season.

The rainy season lasts for 12 months only at the lowest stations on the northeastern end of the transect. It diminishes, with the expected irregularities, as one moves up the eastern escarpment. It begins in November and lasts through March on the northeastern half of the Altiplano, and it begins somewhat later in December and lasts through March from the southwest Altiplano to the middle elevations of the western escarpment. It drops abruptly to zero for Arequipa and stations to the southwest. From the perspective of regional agro-ecosystems, it is important that on the eastern escarpment the rainy season is comfortably bracketed by relatively favorable thermal conditions. The coincidence of these two requirements for production, however, grows more restrictive along the transect until the middle levels of the western escarpment, where the lengthy thermal season is not

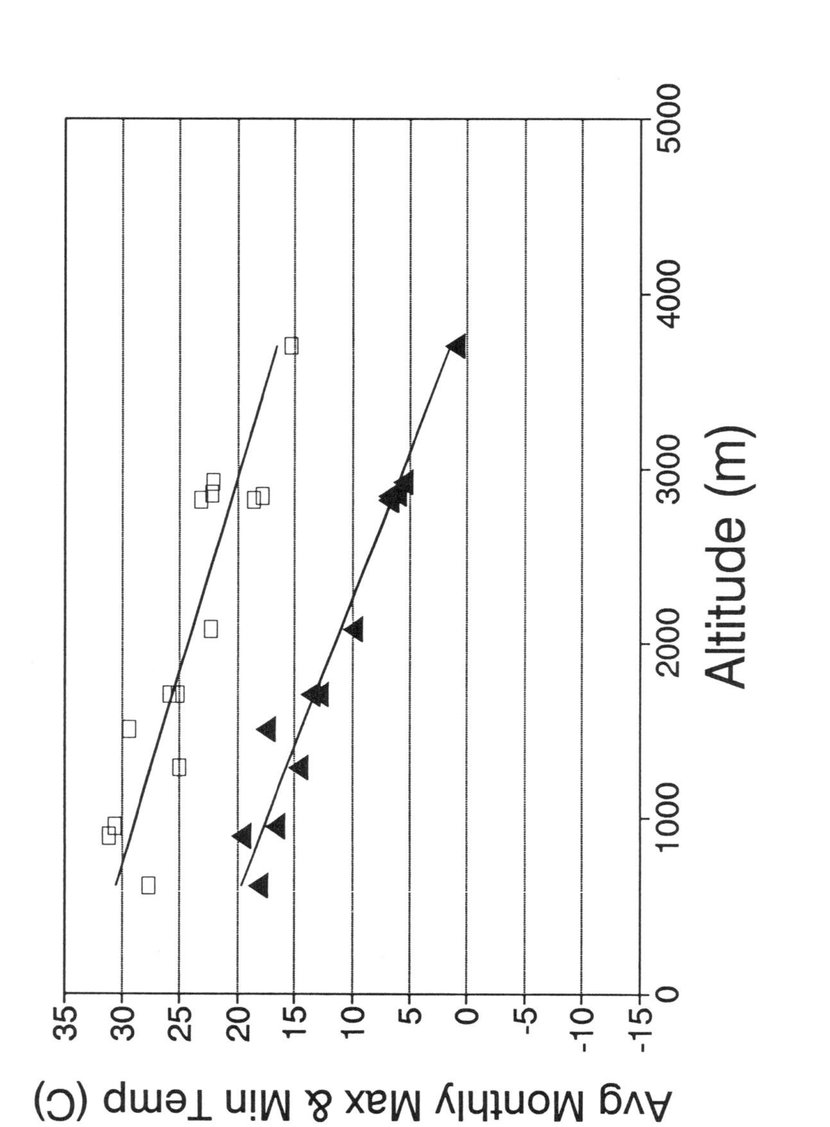

FIGURE 2.12 EASTERN ESCARPMENT: MAXIMUM AND MINIMUM MEAN MONTHLY TEMPERATURES AS A FUNCTION OF ALTITUDE, WITH SUPERIMPOSED LINEAR REGRESSIONS (SEE TABLE 2.7).

FIGURE 2.13 EASTERN ESCARPMENT: PREDICTABILITY (P), CONSTANCY (C) AND CONTINGENCY (M) OF MONTHLY MINIMUM TEMPERATURE AS A FUNCTION OF ALTITUDE, WITH SUPERIMPOSED LINEAR REGRESSIONS (SEE TABLE 2.7).

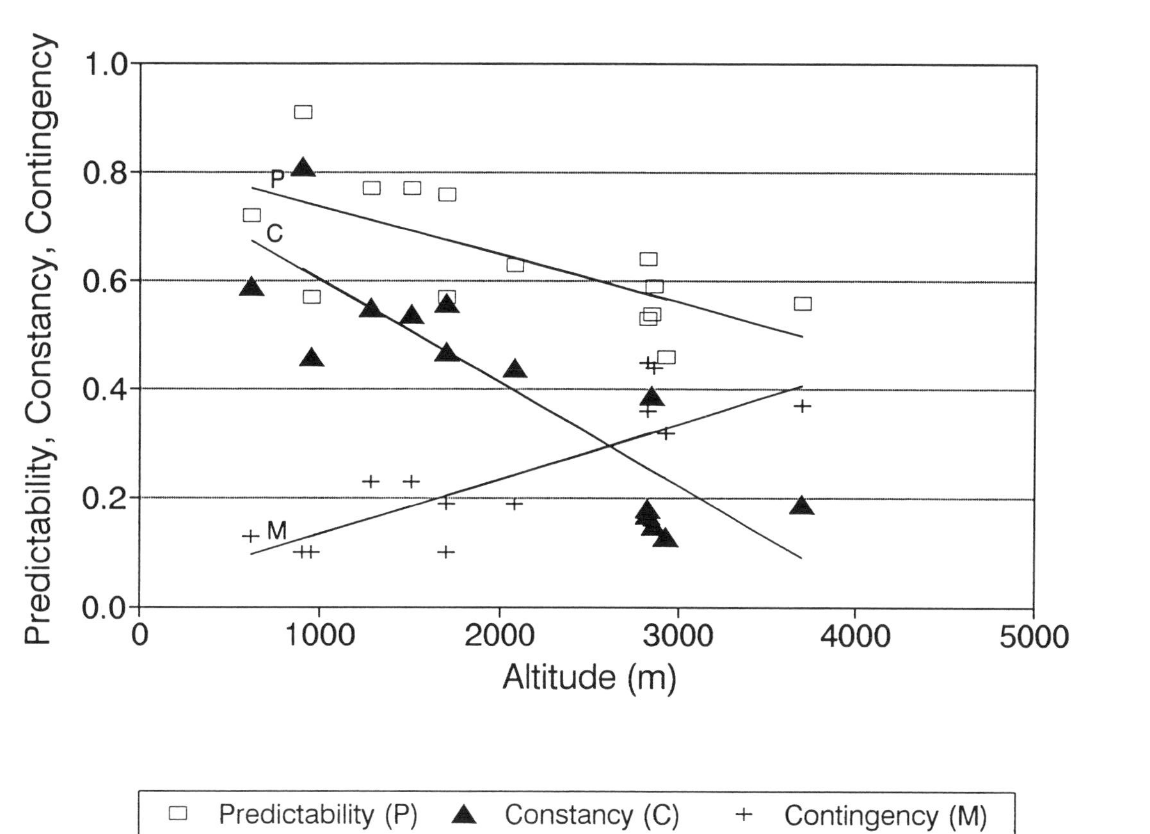

FIGURE 2.14 ONSET AND CONCLUSION OF THE RAINY (DARK STIPPLING) AND THERMAL SEASON (LIGHT STIPPLING) FOR THE SAMPLE STATIONS AS ARRAYED ON THE NORTHEAST (NE) TO SOUTHWEST (SW) TRANSECT ACROSS THE SOUTHERN PERUVIAN ANDES.

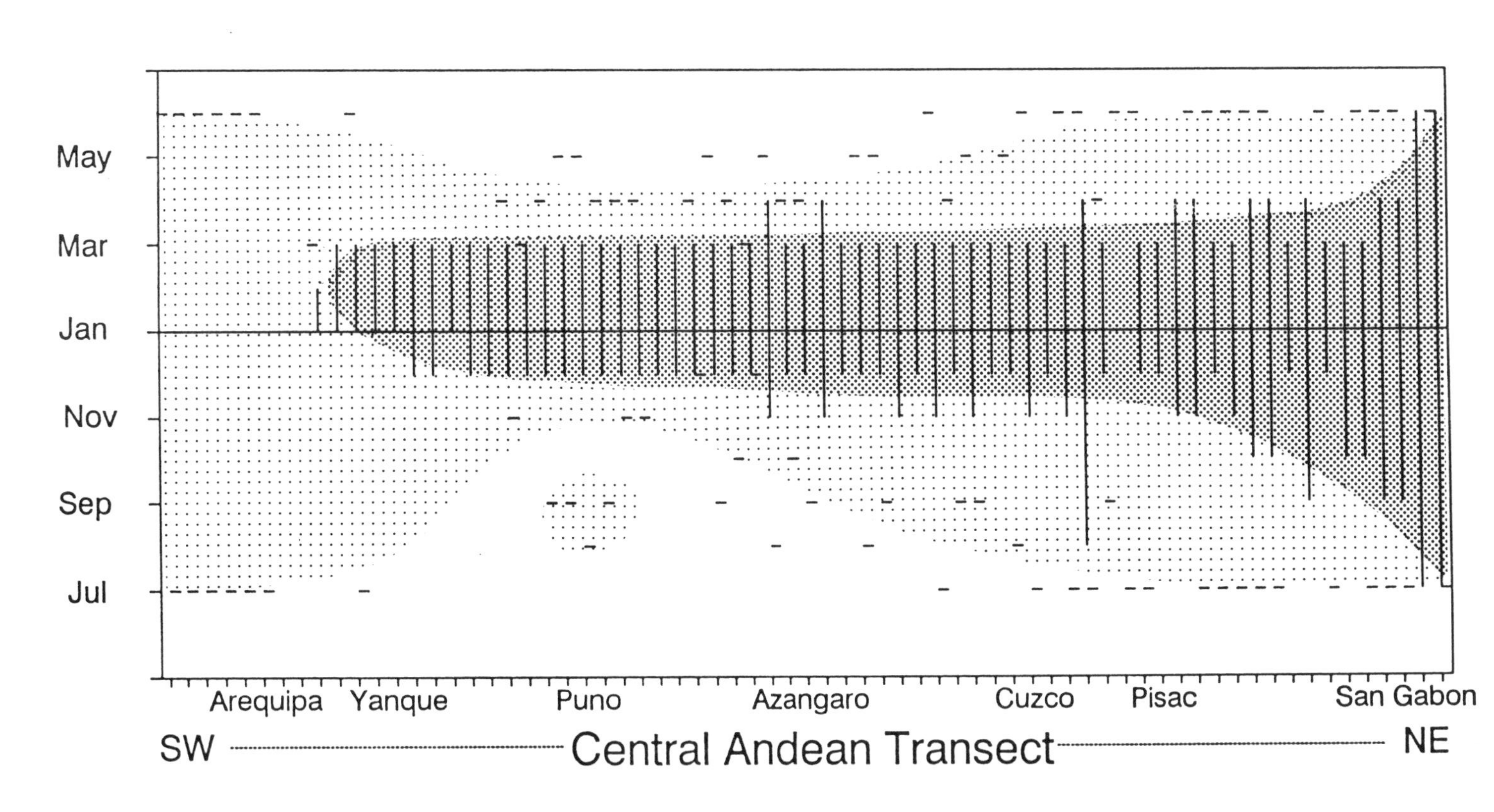

Note: The vertical lines represent the actual data points for duration of the rainy season; the horizontal, dashed lines the actual data points for the beginning and ending of the thermal season. The detached patch of light stippling isolates four stations that are located adjacent to Lake Titicaca.

FIGURE 2.15 WESTERN ESCARPMENT: ONSET AND CONCLUSION OF THE RAINY AND THERMAL SEASONS, AS A FUNCTION OF ALTITUDE.

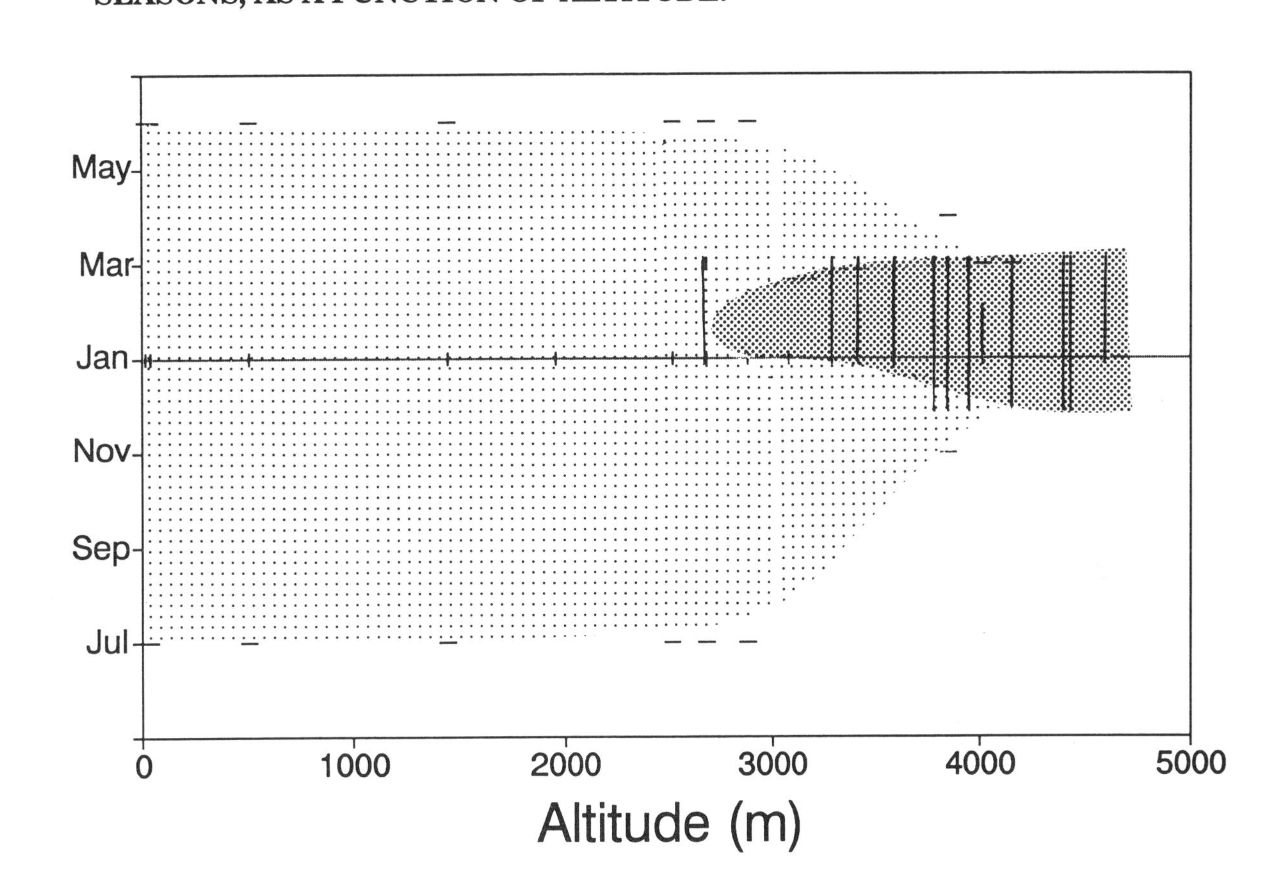

Note: The vertical lines represent that actual data points for duration of the rainy season; the horizontal, dashed lines the actual data points for the beginning and ending of the thermal season.

complemented by a rainy season.

The western escarpment pattern is portrayed in greater detail in Figure 2.15. Note that the length of the thermal season expands from 4 months to nearly 12 months as altitude drops from 4,000 to 3,000 meters. But between these same elevations the effective rainy season drops from 3-4 to 0 months duration. Even without knowledge of past or present agriculture in the Colca Valley, we might predict irrigation in this zone. The abundant water of the highest elevations (Figure 2.5) need only be moved a short distance downslope to areas of favorable thermal conditions where it can augment the quantity and duration and ameliorate the irregularities of the quite limited rains.

SUMMARY

Average annual precipitation is highly variable from site to site on the eastern escarpment and is only partially correlated with altitude. It is nearly constant as one moves from northeast to southwest across the Altiplano. And it is almost completely explained in the statistical sense by altitude on the western escarpment. The amount of precipitation decreases upslope on the northeast side of the Andes. In contrast, it increases upslope on the southwest side. The predictability of monthly rainfall decreases with altitude on both escarpments. Contingency increases with altitude, but it fails to offset the stronger drop in constancy.

Monthly maximum and minimum temperatures diminish with altitude on both escarpments as does the predictability of minimum temperatures. Linear regressions of mean monthly maximum temperatures on altitude are quite similar on the two sides of the mountains, but mean monthly minimum temperatures are consistently higher on the eastern slopes (by 3° to 5°C). Predictability of minimum mean temperatures is higher at low elevations on the eastern escarpment than it is at low elevations on the western escarpment, but the regression lines for this factor are indistinguishable by 3,000 to 4,000 meters.

The *thermal season* constricts dramatically in the Altiplano portion of the transect. The *rainy season* drops from 5 or 6 to 3 or 4 months as one passes from the high slopes of the eastern escarpment across the Altiplano basin to the high slopes of the western escarpment. Favorable moisture and temperature conditions last only for a relatively brief seasonal period at the higher altitudes on the southwest end of the transect.

PROBLEM DISCUSSION

I have used a very focused question (Why is there irrigation in the Colca Valley but not in the Sandia Valley?) to motivate a broader analysis of climate in the southern Peruvian portion of the central Andes. It is appropriate that I return to that question to elucidate some of what has been learned from the data. I will focus on the zones between 3,000 and 4,000 meters on both escarpments.

I earlier hypothesized that irrigation might be used to alter the timing, quantity, predictability, or quality of water available for agriculture. I now can describe these postulated relationships in somewhat greater detail:

Timing: On the higher portions of the eastern escarpment a thermal season of 8-12 months amply brackets a rainy season of 4-6 months. In contrast, the rainy season on the western escarpment is 4 months or much less in the zones that are thermally adequate for agriculture. Irrigation to provide water before (and perhaps after) the normal rainy season would be of much greater benefit in the Colca area than in the Sandia area.

Quantity: According to the regressions established in Table 2.7, at 3,000 meters the Colca Valley gets 290 millimeters of precipitation and the Sandia Valley gets 900 millimeters. At 4,000 meters the difference is much less (518 millimeters in the Colca Valley; 609 millimeters in the Sandia Valley). Roughly speaking, the Colca gets half the incident precipitation of the Sandia Valley and would derive a much greater benefit from irrigation to augment the overall quantity of available moisture. The polynomial relationship of precipitation to altitude on the western escarpment makes the high elevation areas in the upper reaches of the Colca Valley relatively attractive water sources.

Predictability: The two escarpments do not differ greatly in the predictability of minimum temperatures or precipitation at these altitudes. Although month-to-month predictability is quite low in both zones and presumably influences agricultural tactics in a variety of ways (including irrigation management in the Colca Valley), it does not help explain the presence or absence of irrigation in the regional comparison.

Quality: Mean annual minimum temperatures are 3° to 5°C lower on the western escarpment than on the eastern escarpment. Another potential benefit of irrigation, the mitigation by damp soils of localized frost damage, would be of greater importance in the Colca Valley than in the Sandia Valley. Other potential aspects of water quality have not been discussed here.

A Summary of the Evidence

The validity of the climatological proposals developed here to explain the distribution of central Andean irrigation must be measured against empirical evidence of the agro-ecological function of irrigation in the prehistoric and contemporary periods. Unfortunately, most of the anthropological literature on Andean irrigation has focused on its sociopolitical organization and provides little information on its role in production. Quantitative data are sparse (Mitchell 1976). Nonetheless, the comparison given above supports the ecological approach, as do those few studies that provide the necessary information. In the Colca Valley, for example, irrigation "does seem to serve three functions: to extend the cropping season (for some crops); to increase predictability; and, to increase total water available for crops" (William Denevan, personal communication, 6 February, 1990). Treacy (1987:425) notes that Colca Valley farmers use irrigation to extend the cropping season at its initiation (August through November) to ensure that crops mature before they are killed by May and June frosts. For the same region, Guillet notes that irrigation is used in lower elevations to "lengthen the growing season for the long maturing maize" (1987b:82). Farrington states that "irrigation [in highland Peru] offers a valuable supplement to rainfall during the growing season and in places does serve to extend the cultivation year by enabling certain crops to be planted earlier" (1980:298).

Mitchell's (1976 and *intra*) work provides additional support. His description of irrigation in Quinua (located in the inter-Andean valley of Ayacucho) is one of the most complete in terms of the agro-ecological functions of the system. The major part of the food production in this community is based on two agricultural zones, one located in the upper altitudes and one in the lower altitudes of the lower montane savanna. In the upper zone, irrigation is used to extend the growing season immediately before the onset of rains (the September through December period). It is not used at the conclusion of the rainy season, which is limited by frost. The longer season is required to encompass the growing period of crops (especially maize) whose maturation is delayed by altitude. Further, the measured delivery of water before heavy rains allows root systems to develop past the point that they would suffer from rotting in more saturated soils. In the lower zone, irrigation is used to supplement the quantity of precipitation and to cover the intermittent dry periods during the normal period of the rainy season. Here, the overall amount of precipitation is inadequate, although the duration of the rainy season and the maturation period of the crops coincide and match the favorable thermal period. Thus, it is not necessary at this elevation to extend the cultivation interval. Use of water in the two zones is complementary because their agro-ecological requirements are successive in time.

Qualifications

Regional climatological analysis offers insight into one of the factors important in the origins, distribution, and functioning of Andean irrigation systems. Any complete explanation must include such additional considerations as differences in topography, soils, crops, capital, market conditions, and sociopolitical organization. Further, moisture availability to plants can be managed in other ways: by increasing the capacity of soils to retain moisture by adding organic matter, or by selecting crops with greater tolerance for desiccation. In my comparison of the Sandia and Colca valleys, these variables are more or less constant. But elsewhere in the Andes they are not. As Seligmann and Bunker have demonstrated (*intra*), people require both the technical and the political means to develop an irrigation system. Similarly, economic conditions in the contemporary Ayacucho Valley have discouraged the expansion of irrigation in spite of great need for additional agricultural land (Mitchell *intra*).

Even within its own terms, the large-scale of the comparison forces it to overlook significant local heterogeneity. Some of the regional variables, for example, also operate at much smaller scales. For instance, the local thermal conditions, upslope winds (Troll 1968:48), and rain shadows found in some of the deep valleys of the eastern escarpment create conditions similar to those characteristic of the western escarpment. Like the Colca Valley, the warm lower slopes of these valleys typically are xeric and might benefit from irrigation in their intermediate and lower elevations (Bowman 1968[1916]:154). The irrigation system of Quinua (Mitchell *intra*) is an example. The sources of Quinua's irrigation water are the springs and streams of the wet highland zones to the east of the valley.

At this point, climatological propositions concerning the regional patterning and functioning of Andean irrigation systems are somewhat tentative. I hope that raising these issues will motivate Andeanists to gather additional test data.

CONCLUSION

The Andean and general human ecology literature is filled with hypotheses about the effects of environment on the development and form of human adaptive systems. Despite this, ecological analysis in anthropology rarely has provided ecological information in the abundance, detail, and form needed to assess these proposals (Winterhalder 1980). The data often do not exist or are difficult to obtain. The analytical methods of the natural sciences often do not match the analytical needs of the human ecologist, and too often anthropologists have been content to propose but not test ecological explanations. The present essay has attempted to provide meteorological information with sufficient detail to assess the impact of climate on the distribution and function of highland irrigation systems.

Standing at the Carabaya overlook it is easy to appreciate the "noble proportions" (Bowman 1968[1916]) of the Andes and visualize the broad seasonal patterns of precipitation and associated changes of temperature to which they give rise. As Bowman notes, a diverse sample of the earth's climates can be found within a day's walk of a place like this. Significant changes also occur from day to night, week to week, and year to year, challenging the technical and managerial skills of the high altitude peasant who must deal with the resulting drought, flood, and frost. These spatial and temporal variations absorb the attention and organizational energies of Andean peoples. Both must gain the attention of human ecologists if we are to understand the ecological bases of phenomena like the regional distribution and functioning of irrigation systems. The geographical perspective must be matched to an analytical view that encompasses the local details of the weather, especially the short-term vagaries and predictability of precipitation and temperature.

NOTES

Acknowledgments. For assistance, advice, and encouragement, I would like to thank Bradley Bennett, Sally DeGraff, William Denevan, Carol Goland, Margaret Graham, Peggy Gregson, David Guillet, Mary Jane Kellum, Sheryl Gerety, Anne Larme, William P. Mitchell, Jorge Recharte, Paul J. Slabbers, and Karl Zimmerer. Funding was provided by the National Science Foundation (BNS #8313190) and the University of North Carolina (Chapel Hill) University Research Council.

1. The central Andes encompasses the tropical cordilleras of western South America from

southern Ecuador through Peru and into northwestern Bolivia, northern Chile, and Argentina (Brush 1982).

2. This corresponds, by his estimation of lapse rate on the Altiplano, to an average warming of about 2.5°C.

3. "Incidence" here is the probability of occurrence within consecutive ten day intervals, with crop damage assumed if precipitation is less than 0.5 times evapotranspiration, if minimum temperature drops below 0° C, or if hail is recorded.

4. This essay has been extracted from a monograph in preparation ("Climate predictability and patterning in the Central Andes of Peru"), part of a larger research project ("Production, Storage and Exchange in a Terraced Environment on the Eastern Andean Escarpment") funded by the National Science Foundation (BNS #8313190) from July 1984 to December 1987.

5. The key sources are the following: Drewes and Drewes 1957; Grace 1983; Johnson 1976; Winterhalder and Thomas 1982.

6. For precipitation, predictability is calculated using the base-2 log of monthly precipitation. This procedure eliminates the correlation between the mean and variance of rainfall, facilitating comparison among stations with differing average precipitations. The log transform is not used on minimum temperature calculations reported here.

7. The specific figure of 64 millimeters was chosen because it represents a natural break in the log transformed data; see note 6.

8. The position of each station on this transect was determined by measuring its distance from a line drawn parallel to the axis of the central Andes and passing through the northeasternmost station; the sequence is an ordinal ranking. The axis of the southern Peruvian Andes was determined to lie at an angle 64° west of North by visual inspection of a physical map of Peru; the transect was established at an angle of 26° east of North. With only a few exceptions, the transect ordering faithfully lumps together stations assigned to the three biogeographic zones (eastern escarpment, Altiplano, western escarpment).

9. Here is Troll's statement: "Because of this diagonally asymetric arrangement of the zones of humidity, the climato-vegetational zones of moist puna, dry puna, and thorn-and-succulent puna cross the Andes diagonally from NW to SE" (1968:46; see figure 17, p. 37, and figure 20, p. 47).

10. The polynomial model that fit so well on the western escarpment performs poorly on this eastern escarpment data [$n = 19$; without a constant, $r = 0.33$ and $r^2 = 0.11$; with a constant, $r = 0.56$ and $r^2 = 0.31$]. Similarly, a linear model, with the two low-altitude and very high precipitation stations eliminated gives a relatively poor fit [$n = 17$, $r = 0.48$; $r^2 = 0.23$; constant = 17772.9; coefficient = -291.0]. This model appears to be a better predictor at the upper altitude on this escarpment.

11. This may be due to minor differences of elevation. Stations on the northeastern end of the Altiplano portion of the transect are somewhat lower than those on the western end (Figure 2.2). There is a significant relationship between mean minimum temperature and altitude on the Altiplano (Table 2.7).

12. If the two coastal stations are eliminated from the sample, the linear regression of maximum temperature on altitude is much stronger [$n = 9$; $r = 0.91$; $r^2 = 0.83$; constant = 29.9; coefficient = -3.56]. This gives a lapse rate for mean maximum temperature of 3.56°C/1,000 meters.

13. Four stations make up a small isolated cluster on the southwestern end of the Altiplano segment of the transect. Their thermal seasons begin unusually early. They are centered immediately above "Puno" on the x-axis of Figure 2.14. In fact, all are on or near the margin of Lake Titicaca and presumably have their temperatures meliorated (see Figure 2.9) and growing season lengthened due to the proximity and thermal properties of the lake.

REFERENCES CITED

Bowman, Isaiah
1968 [1916] *The Andes of Southern Peru.* New York: Greenwood Press.

Browman, David L.
1987a "Introduction: Risk Management in Andean Arid Lands." In *Arid Land Use Strategies and Risk Management in the Andes*. David Browman, editor, pp. 1-23. Boulder, CO: Westview Press.
Browman, David L., editor
1987b *Arid Land Use Strategies and Risk Management in the Andes*. Boulder, CO: Westview Press.
Brown, Paul F.
1987 "Economy, Ecology and Population: Recent Changes in Peruvian Aymara Land Use Patterns." In *Arid Land Use Strategies and Risk Management in the Andes*. David Browman, editor, pp. 99-120. Boulder, CO: Westview Press.
Brush, Stephen B.
1982 "The Natural and Human Environment of the Central Andes." *Mountain Research and Development* 2:19-38.
Cardich, Augusto
1985 "The Fluctuating Upper Limits of Cultivation in the Central Andes and Their Impact on Peruvian Prehistory." In *Advances in World Archaeology 4*. Fred Wendorf and Angela Close, editors, pp. 293-333. Orlando, FA: Academic Press.
Colwell, Robert K.
1974 "Predictability, Constancy, and Contingency of Periodic Phenomena." *Ecology* 55:1,148-1,153.
Conrad, Geoffrey W.
1981 "Cultural Materialism, Split Inheritance, and the Expansion of Ancient Peruvian Empires." *American Antiquity* 46:3-26.
Denevan, William, Kent Mathewson, and Gregory Knapp, editors
1987 *Pre-Hispanic Agricultural Fields in the Andean Region*. BAR International Series 359(i).
Dew, Edward
1969 *Politics in the Altiplano; the Dynamics of Change in Rural Peru*. Austin: University of Texas Press.
Donkin, Robin A.
1979 *Agricultural Terracing in the Aboriginal New World*. Tucson: University of Arizona Press.
Drewes, Wolfram U., and Arlene T. Drewes
1957 *Climate and Related Phenomena of the Eastern Andean Slopes of Central Peru*. Syracuse, NY: Syracuse University Research Institute.
Farrington, Ian S.
1980 "The Archaeology of Irrigation Canals, with Special Reference to Peru." *World Archaeology* 11:287-305.
Flores Ochoa, Jorge A.
1987 "Cultivation in the Qocha of the South Andean Puna." In *Arid Land Use Strategies and Risk Management in the Andes*. David Browman, editor, pp. 271-296. Boulder, CO: Westview Press.
Goland, Carol
1988 *Prehispanic Occupation of the Eastern Andean Escarpment: A Preliminary Report of the Cuyocuyo Archaeological Survey*. Production, Storage and Exchange (PSE) Working Paper No. 1, Department of Anthropology, University of North Carolina, Chapel Hill.
Grace, B.
1983 *The Climate of the Altiplano, Department of Puno, Peru*. Unpublished MS.
Guillet, David
1986 "Comments." *Current Anthropology* 27:101-106.
1987a "Terracing and Irrigation in the Peruvian Highlands." *Current Anthropology* 28:409-430.
1987b "On the Potential for Intensification of Agropastoralism in the Arid Zones of the Central Andes." In *Arid Land Use Strategies and Risk Management in the Andes*. David Browman, editor, pp. 81-98. Boulder, CO: Westview Press.

Isbell, William H.
1978 "Environmental Perturbations and the Origin of the Andean State." In *Social Archeology: Beyond Subsistence and Dating*. Charles L. Redman, Mary J. Berman, Edward V. Curtin, William T. Langhorne, Jr., Nina M. Versaggi, and Jeffrey C. Wanser, editors, pp. 303-313. New York: Academic Press.

Johnson, A. M.
1976 "The Climate of Peru, Bolivia and Ecuador." In *World Survey of Climatology, Volume 12: Climates of Central and South America*. Werner Schwerdtfeger, editor, pp. 147-202. Amsterdam: Elsevier Scientific Publishing.

Julien, Catherine J.
1983 *Hatunqolla: A View of Inca Rule from the Lake Titicaca Region*. Berkeley: University of California Press.

Kent, Jonathan D.
1987 "Periodic Aridity and Prehispanic Titicaca Basin Settlement Patterns." In *Arid Land Use Strategies and Risk Management in the Andes*. David Browman, editor, pp. 297-314. Boulder, CO: Westview Press.

Lhomme, J. P., and O. E. Rojas
1986 "Análisis de los riesgos de sequía, granizada y helada para la agricultura del Altiplano Boliviano." *Turrialba* 36:219-224.

Malpass, Michael A.
n.d. "Irrigated Versus Non-irrigated Terracing in the Andes: Environmental Considerations" Unpublished MS.

Mitchell, William P.
1976 "Irrigation and Community in the Central Peruvian Highlands." *American Anthropologist* 78:25-44.

Molina, Eduardo G., and Adrienne V. Little
1981 "Geoecology of the Andes: The Natural Science Basis for Research Planning." *Mountain Research and Development* 1:115-144.

Murra, John V.
1972 "El 'control vertical' de un máximo de pisos ecológicos en la economía de la sociedades andinas." In *Visita de la provincia de León de Huánuco en 1562*, Iñigo Ortiz de Zúñiga, pp. 427-476. Huánuco, Peru: Universidad Nacional Hermilio Valdizán.
1984 "Andean Societies." *Annual Review of Anthropology* 13:119-141.

Orlove, Benjamin
1977 "Integration Through Production: The Use of Zonation in Espinar." *American Ethnologist* 4:84-101.
1986 "Barter and Cash Sale on Lake Titicaca: A Test of Competing Approaches." *Current Anthropology* 27:85-106.

Paulsen, Alison C.
1976 "Environment and Empire: Climatic Factors in Prehistoric Andean Culture Change." *World Archaeology* 8:121-132.

Sarmiento, Guillermo
1986 "Ecological Features of Climate in High Tropical Mountains." In *High Altitude Tropical Biogeography*. François Vuilleumier and Maximina Monasterio, editors, pp. 11-45. New York: Oxford University Press.

Schoenwetter, James
1973 "Archaeological Pollen Analysis of Sediment Samples from Asto Village Sites." In *Les Establissements Asto a l'Epoque Prehispanique*. (Tome 15). Daniele Lavallee and Michele Julien, editors, pp. 101-111. Lima: Travaux de l'Institut Francais d'Etudes Andines.

Schwerdtfeger, Werner
1976 "Appendix II–High Thunderstorm Frequency Over the Subtropical Andes During the Summer; Cause and Effects." In *World Survey of Climatology, Volume 2: Climates of Central and South America*. Werner Schwerdtfeger, editor, pp. 192-195. Amsterdam: Elsevier Scientific Publishing.

Stearns, Stephen C.
1981 "On Measuring Fluctuating Environments: Predictability, Constancy and Contingency." *Ecology* 62:185-199.

Treacy, John
1987 "Comment." Current Anthropology 28:425.

Troll, Carl
1960 "The Relationship Between the Climates, Ecology and Plant Geography of the Southern Cold Temperate Zone and of the Tropical High Mountains." *Proceedings of the Royal Society of London, Series B*,152:529-532.
1968 "The Cordilleras of the Tropical Americas: Aspects of Climatic, Phytogeographical and Agrarian Ecology." In *Geo-Ecology of the Mountainous Regions of the Tropical Americas*. Carl Troll, editor, pp. 15-56. Bonn: Ferd Dümmlers Verlag.

Winterhalder, Bruce
1980 "Environmental Analysis in Human Evolution and Adaptation Research." *Human Ecology* 8:135-170.

Winterhalder, Bruce, and R. Brooke Thomas
1982 *Geoecología de la región montañosa del sur de Perú: Una perspectiva de adaptatión humana*. Occasional Paper No. 38. Institute of Arctic and Alpine Research, University of Colorado, Boulder.

CHAPTER THREE

Water and Power: The Role of Irrigation Districts in the Transition from Inca to Spanish Cuzco

Jeanette E. Sherbondy
Washington College

Irrigation was at the heart of Inca culture. The Inca state (A.D. 14th century-1532) relied heavily on maize and invested a great deal of labor in hydraulic works to guarantee the maize harvest (Sherbondy 1982a). Consequently, water matters pervaded all aspects of Inca life: economic, social, political and religious. In this chapter, I explore the relationship between irrigation and Inca social organization and how that relationship helped mold the changes found in the early colonial period.

I focus on the role of irrigation in the organization of the city and valley of Cuzco, the capital of the Inca empire. I am concerned with the late Inca period (just prior to the Spanish conquest of Cuzco in 1533) and the early period of Spanish domination, until approximately 1572, when the last Inca heir to the throne was executed and a stronger colonial regime was implemented. First, I examine the relationship between irrigation and the political organization of Inca Cuzco through an analysis of its system of spatial and social divisions, especially the social units known as *ceques*, *panacas*, and *ayllus*. Then I turn to the early years of Spanish Cuzco, when the *panacas* and *ayllus* were transformed and organized into a system of parishes.

I see the early colonial period as one of many changes and innovations, the result of competing strategies undertaken by factions of Incas, other native ethnic groups, and various groups of Spaniards (soldiers, priests, and Crown officials). It would be an oversimplification to view this period only in terms of Inca continuity and/or the introduction of European changes. Both Andean populations and the Spaniards adapted to new circumstances, so that Andean and Spanish institutions changed together, each acting on the other. Yet most of the altered Spanish institutions retained a recognizable European character and the modified Andean institutions a characteristically Andean one. Indeed, it was only through these adjustments that Andean institutions were able to persist into the colonial period (see also Dover 1992:8). The Andean forms that emerged, therefore, need to be studied with an awareness of the dynamic historical process. They were created through a sequence of choices made as much by Spanish authorities as by the survivors and descendants of the Incas.

This study is not intended to be complete, but rather tries to highlight the scholarly issues and those aspects of them that still need further investigation. The paucity of archeological research on Inca Cuzco and the complete lack of written material from the Inca period are major obstacles to my research. We do not have adequate chronological data on the development of irrigation in Cuzco, but I assume that the Inca utilized and reinterpreted earlier forms of irrigation. The historical data are also problematic, since they derive from Spanish documents in which Inca narratives were retold by Spanish writers. Because the documents contain both cultural and personal biases, Inca culture can be perceived only through layers of distorted perceptions and translations. Some of these historical obstacles, however, have been ameliorated by R.T. Zuidema's structural research on Inca social organization and religion and I use his research as a departure point for my own analysis. (See MacCormack 1991, Rappaport 1990, Salomon 1982, and Zuidema 1964 for further discussion of the problems in using Spanish documents).

I use a multiple methodology to approach Inca irrigation: structural analysis of the sixteenth century works on the Incas, ethnographic inquiries into farming and irrigation matters, contemporary field surveys of Inca canals, and historical analysis based on archival research. Legal documents have been a major source of information on land and water rights.

What emerges from my research is that Inca Cuzco had developed a system that was well adapted to the Andean natural and social environment. It drew on earlier social and technological arrangements that had survived the various Andean conquest states and had proven themselves over the millennia. The Spanish conquest was different in that it imposed foreign political, economic and ideological structures. Nonetheless, even though the Spanish jolted the Andean system, many features of local productive arrangements (especially irrigation and its attendant social organization) survived in modified form, many of these arrangements persisting into the present.

Even though the Inca and Spanish economies were organized differently, they both depended to a considerable extent on local production. For the Incas, Cuzco was one of the valleys best endowed with water and land (see also Mitchell, *intra*), providing them with a strong economic base during the early Inca period. Later, in the imperial period, they no longer relied solely on local production, but supplemented it with tribute from the provinces (Betanzos 1987 [1551]:55-63). When the Incas fell tribute stopped and the population of Cuzco was once again forced to live primarily on local food production. Nonetheless, throughout the early and late periods, Inca Cuzco relied on irrigated cultivation for most, if not all, of its sustenance. In consequence, the political organization of the city and valley of Cuzco was associated with the regulation of the water that provided agricultural prosperity.

When the Spanish colonized Cuzco they introduced many changes, but they utilized Inca customs when they considered it expedient to do so. The Spanish incorporated Cuzco into a commercial empire focused on the production of minerals. Cuzco, located on the route from the mines in Potosí to the port of Callao, was a strategic point for maintaining the mules essential to the transport system. Cuzco produced the alfalfa needed by the animals. The Spanish and the tribute-producing Indians also depended upon the foods grown in Cuzco by the native population (Sherbondy 1979a). From being the major center for the collection of tribute during the late Inca period, therefore, Cuzco became a major supplier of fodder and food for the Spanish conquerors.

Because Spanish success depended upon the efficient use of land, water, and people, they continued to utilize many preexisting agrarian institutions–institutions that had proven successful in the Andean context. In consequence, the Spanish were encouraged to retain some Inca irrigation practices into the colonial period. The native population had also quickly recognized that they needed to defend their land and water rights in Spanish terms, developing a number of strategies to do so in the chaotic period following the fall of Cuzco. Some tried to adopt the Spanish mode individually. Most, however, lacked access to the Spanish legal process, and had to rely on the legal efforts of others. In doing so, they reworked their Andean collective system, emphasizing their status as *ayllu* members, thereby accommodating to the Spanish in a manner that allowed them to exist with a recognizable Andean identity.

This process of incorporating some (but not all) aspects of the Inca system is well illustrated in the transformation of the Inca *ayllus* and *panacas* into the communal *ayllus* and parishes of the Spanish colonial system. The crucial transitional period was the forty years from the Spanish conquest of Cuzco in 1533 to the defeat of the last Inca in 1572, when the viceroy Toledo imposed a new administrative system.

During this transitional period, the Spanish eliminated the upper administrative levels of the Inca (the divisions into *ceques*, *suyus*, and moieties), but retained many aspects of the lower productive levels. The Inca *ayllus* changed in number and sometimes name, but they continued as communal groups controlling land and water. The Inca *panacas* lost their governmental function. Some of the nobility were able to transform *panaca* rights into individual land holdings. Most *panaca* members, however, emphasized the collective character of the *panacas*, transforming them into colonial *ayllus*. The Spanish organized the *ayllus* into parishes but these larger social units were no longer connected to the hydrology of the Valley as the larger units had been in the Inca period.

THE *AYLLU* AND IRRIGATION

Inca *panacas* and Spanish parishes were the chief administrative units in Cuzco that both included lands with their irrigation systems and groups of people that lived there and worshipped at local ritual sites (Sherbondy 1982a, 1987b). The *panaca* was similar to the *ayllu* in that both were kinship groups internally structured as groups descending in a male and a female line for four generations from a male genitor (Zuidema 1977). The founder or head of an *ayllu* may be a mythical ancestor or a living head. The *panaca* was different because it comprised the descendants of a deceased Inca king (with the exception of the king's successor). *Panacas* and *ayllus* also had corporate rights to specific lands and waters (Zuidema 1964, 1973; Sherbondy 1979b, 1982a, 1987b). The territory of the *ayllu* was sometimes not contiguous, but was intercepted by the lands of others in the form known as an archipelago (Murra 1972), although this pattern may have been the result of the disruption of *ayllu* lands. Ideally Andean *ayllus* controlled a variety of ecological niches: pasture lands for the herds, lands for potato and the other Andean tubers, well-watered and well-drained lands in a warm zone for maize production, and if possible, fields for coca, hot pepper, cotton and other hot climate cultigens. The territory of an *ayllu* provided a basis for self sufficiency at the very least, and at best, surplus for exchange.

There is a third dimension of the *ayllu* that should be included in its definition because it was the glue that held the people and their lands and waters together in an Andean fashion: the *ayllu* was also a spiritual community that held its origins to be in its lands and waters. The members of the *ayllu* worshipped the lakes, rocks, springs, hills, and mountains of the *ayllu*, a worship that was the most important focus of their religious life. Catherine Allen (1988) has masterfully described this spiritual dimension of a modern *ayllu* in the Cuzco area (see also Bastien 1985). Today, this profoundly spiritual relationship with the land is constantly ridiculed by the mestizo majority, but in the Inca period it was a fundamental part of the belief system of both elites and the peasantry.

The origin myths of the *ayllus* help us understand the relationship of the people of an *ayllu* with their lands and waters (Sherbondy 1982c). The ancestors are intimately linked with the earth and with the sources of water on the *ayllu's* land. Many Andean peoples believed that their ancestors were created at Lake Titicaca, where they entered the earth and traveled along subterranean waterways until they arrived at the site that they were to claim for their descendants, their *ayllu*. They emerged from inside the earth through the openings that were water sources for the most part (lakes, springs, rivers, but also caves) and then founded their *ayllus* (see for example, Betanzos 1987 [1551]:11-12).

These origin myths functioned to legitimate the *ayllu's* rights to its lands and waters (Sherbondy 1982a:22-24). For this purpose the most important element of these myths was the statement that the founding ancestors emerged at specific water sources that were also used for the irrigation systems built by the members of the *ayllu*. Linking ancestors with water rights is essential. Rights to lands depend on rights to waters in large part because there is a strong sense that lands belong to the irrigation systems that water them.

People made offerings and prayers to the sources of the water to ensure their goodwill and supply (see also Paerregaard, *intra*). The act of participating in these rituals reaffirmed their membership in an *ayllu* and established their rights to water. Often group labor projects were linked to these rituals. The rituals preceded and accompanied group labor projects of building canals and dams, cleaning already existing canals, or repairing parts of an irrigation system. Only legitimate users of these hydraulic works could participate in the work because the work had spiritual as well as legal significance (Sherbondy 1982a:22-24). These yearly rituals and group labor events involved all the members of an *ayllu*. Whenever these customs disintegrated, the *ayllu* also lost its cohesiveness.

In summary, the irrigation systems were an essential part of an *ayllu's* economic base, providing the infrastructure for agriculture and livestock raising. The work projects on irrigation canals united *ayllu* members for a common good, while the worship of the water sources and ancestors that had emerged from them unified the people of the *ayllu* ideologically, creating a common kinship group and spiritual community.

The *ayllu* was also the basic building block for larger political structures, such as villages and urban settlements that consisted of several *ayllus*. A minimum of two *ayllus* provided two halves, a fundamental Andean organizational principle. Two common patterns were villages of either two halves of two *ayllus* each or of two halves of three *ayllus* each. Similar principles were used to organize Cuzco, a large city and the capital of the Inca empire during the fourteenth to sixteenth centuries A.D. (Zuidema 1964). The *ayllu*, transformed into the royal *panaca*, constituted the basic administrative unit of the Inca capital. These Cuzco *panacas* were further organized into four *suyus* and two moieties, providing the basic organization of the city of Cuzco and of the empire.

IRRIGATION IN INCA CUZCO: THE ORGANIZATIONAL PRINCIPLES

By the late fifteenth century Cuzco was a large city of thousands of inhabitants. Population growth by natural increase, conquest, and alliances required a complex organization of the imperial capital. We know of one massive reorganization of Cuzco in the century before the Spanish invasion that is attributed to the ruler Pachacuti Inca (who may or may not have been an historical figure). Zuidema (1990:18-19) argues convincingly that this reorganization reflects the kind of redistribution of lands and water that must have occurred with the succession of each Inca king. The details were recorded by *quipu*, a mnemonic device that consisted of knots on cords, and then recounted to a Spanish informant who wrote them down. A partial account is found in the papers of an early governor of Cuzco, the Licenciado Polo de Ondegardo (1917 [1571]) and a complete version is in the chronicle of Bernabé Cobo (1956 [1653]:169-186) who may have based it on a complete list made by Polo (Rowe 1979; Zuidema 1964).

The description consists of a list of lines, or *ceques*, and the sacred sites or *huacas* that were located on them (see Figure 3.1). These lines were imaginary, comparable to the imaginary lines of parallels and meridians that we use for mapping space. All the *ceques* emanated from one central point like spokes of a wheel. This central point was Coricancha, the Temple of the Sun and the major temple of the ancestors of the Incas. The *ceques* divided up the entire valley into areas shaped like pieces of a pie. There were 41 *ceques* in this system and 328 sacred sites or *huacas* unevenly distributed along them. The care and worship of the *huacas* on certain *ceques* were assigned to specifically named *ayllus* and *panacas*. These *panacas* and *ayllus* received lands for cultivation. The *panacas* were also responsible for the maintenance of a section of the Huatanay River and the heights along side it. These lands, called "*chapa*" (Zuidema 1990a and 1990b), were distributed to the *panacas* to construct storehouses for the tribute from provincial lords.[1]

The *panacas* as royal kin groups had the highest social rank and political power. The founding Inca king of each *panaca* provided the rank of the whole *panaca* according to his genealogical distance from the current ruler. The *panaca* cared for the mummy of the founder and also cared for the offical *huacas* located along the *ceques* assigned to that *panaca*. The attendant rituals were more than religious duties. They were also state obligations. The *huacas* marked the location of the *ceques*, which in turn delineated the spatial divisions of Cuzco. Therefore, the *panacas* maintained the political boundaries established by the state.

An understanding of this system of *ceques* is necessary in order to understand how the Incas thought about and managed space, peoples, and resources in the area around their capital city. Inca informants to the Spanish claimed that this model for urban administration was replicated throughout the empire. Water sources and irrigation played a key role in the logic of the system.

Since most of the *huacas* were features of the natural landscape, it has been possible to identify many of them and locate most of the *ceques*. Sixteen of the 41 *ceques* were assigned to specific *panacas* and *ayllus* (Cobo 1956[1653]). Each one of these groups was associated with a particular location or direction within Cuzco. Furthermore, I (Sherbondy 1982a) found that the water sources of the major irrigation canals in the Cuzco valley were *huacas* located on the *ceques* assigned to the *panacas* (see Figure 3.2). The less important hydraulic systems were fed at *huacas* on *ceques* assigned only to *ayllus*. In this fashion the Incas assigned the care of the canal water sources to various corporate groups in a hierarchical order. The *panacas* and *ayllus* also contolled the canals that flowed from these water sources.

Another organizational principle was the moiety. The *panacas* and kings of Cuzco were divided into two groups: those of Hanan (or Upper) Cuzco and those of Hurin (or Lower) Cuzco. The kings of Hanan Cuzco enjoyed the higher rank. This moiety designation referred not only to social organization but to the geographical division of Cuzco (see Map 3.1), which has a topographical basis (see also Mitchell 1976). Hanan Cuzco consisted of the headwaters of the Huatanay River, the river that drains the Cuzco Valley. The two major tributaries of that river flow through

FIGURE 3.1 DIAGRAM OF THE CEQUE SYSTEM

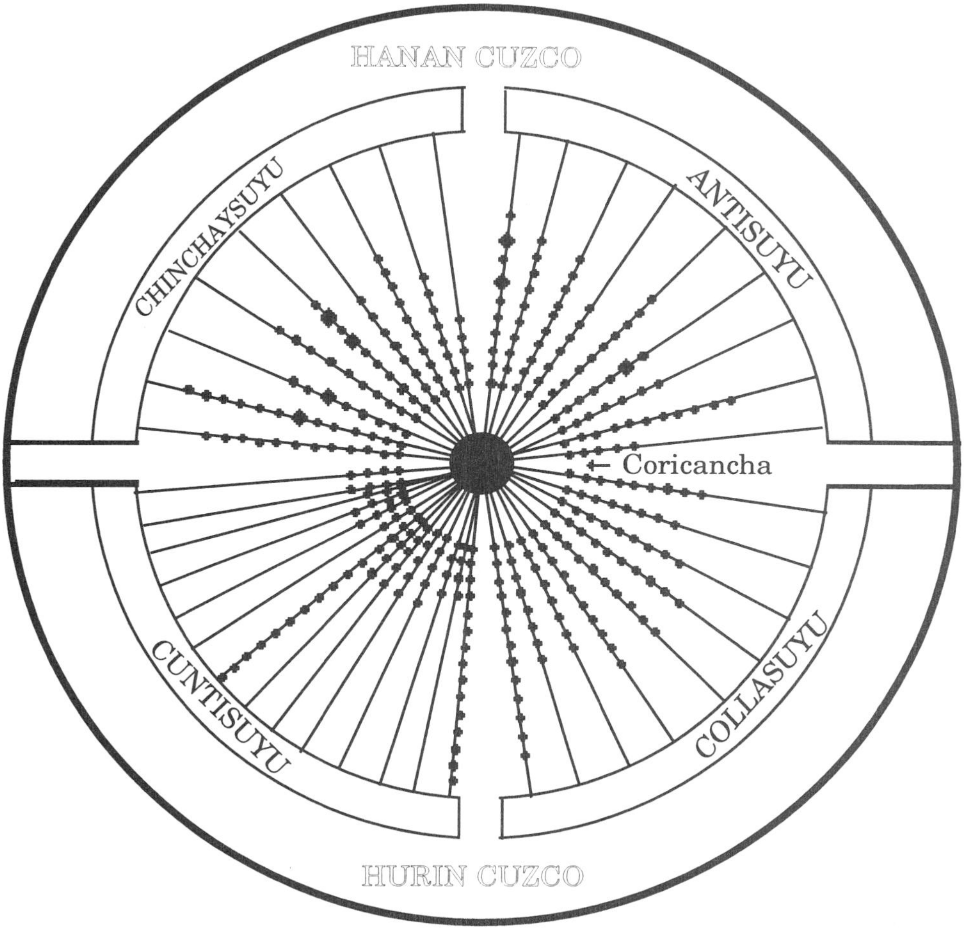

HANAN CUZCO
CHINCHAYSUYU
ANTISUYU
Coricancha
CUNTISUYU
COLLASUYU
HURIN CUZCO

The *ceque* system was conceptualized as lines (*ceques*) emanating from the center (Coricancha) with *huacas*, depicted here as dots, located on them. The numbers of *huacas* on each *ceque* varied. The *ceques* were grouped by *suyu* and variety.

FIGURE 3.2 DIAGRAM OF THE SECTORS BELONGING TO PANACAS IN HANAN CUZCO WITH THEIR MAJOR CANALS, RANKED FROM ① (HIGHEST) TO ⑤ (LOWEST).

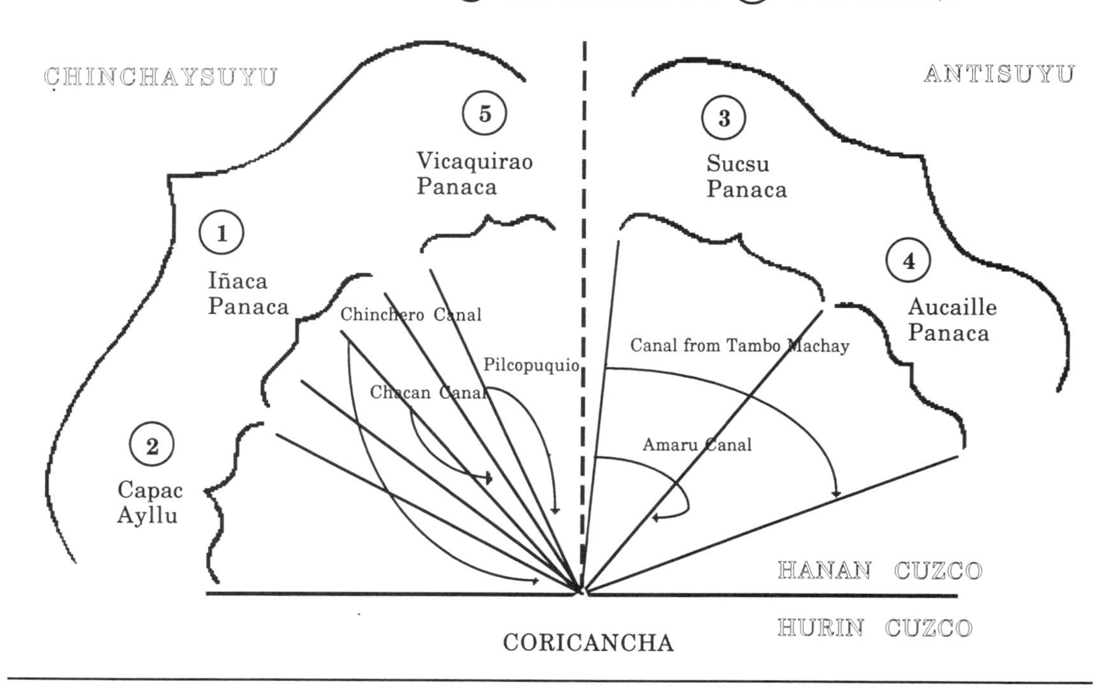

Hanan Cuzco. They join at Coricancha (the Temple of the Sun) to form the Huatanay River, a boundary that also marks the beginning of Hurin Cuzco.

The use of drainage topography to form moiety divisions is another example of how the Inca used water to create sociopolitical organizations. Through this means the Cuzco valley was divided hydrologically and politically into a higher ranking half and a lower ranking half. In Hanan Cuzco five *panacas* controlled the rivers, springs, and canals of that half of the Cuzco valley; in Hurin Cuzco, four *panacas* did the same (see Figure 3.3).[2] Likewise, there were officially five *ayllus* of Hanan Cuzco and five of Hurin Cuzco.

There was one more major organizational principle that the Incas used: the *suyu*, based on a division of the valley into four quarters, two within each moiety. The borders of the four *suyus* were extended to the outermost limits of the Inca empire. *Suyus* became major subdivisions of the empire: the name of the Inca empire, Tahuantinsuyu, literally meant "the parts that in their fourness make up a whole" (Mannheim 1991:18)

The term "*suyu*" also implied a division of tasks, a division of group labor (González Holguín 1952 [1608]:333). Since the Inca empire taxed its subjects in labor (Murra 1980), the *suyu*, *ceques*, *panacas*, and *ayllus* helped organize labor tribute both outside and inside the Cuzco valley (Zuidema and Poole 1982). A large

MAP 3.1 INCA CUZCO

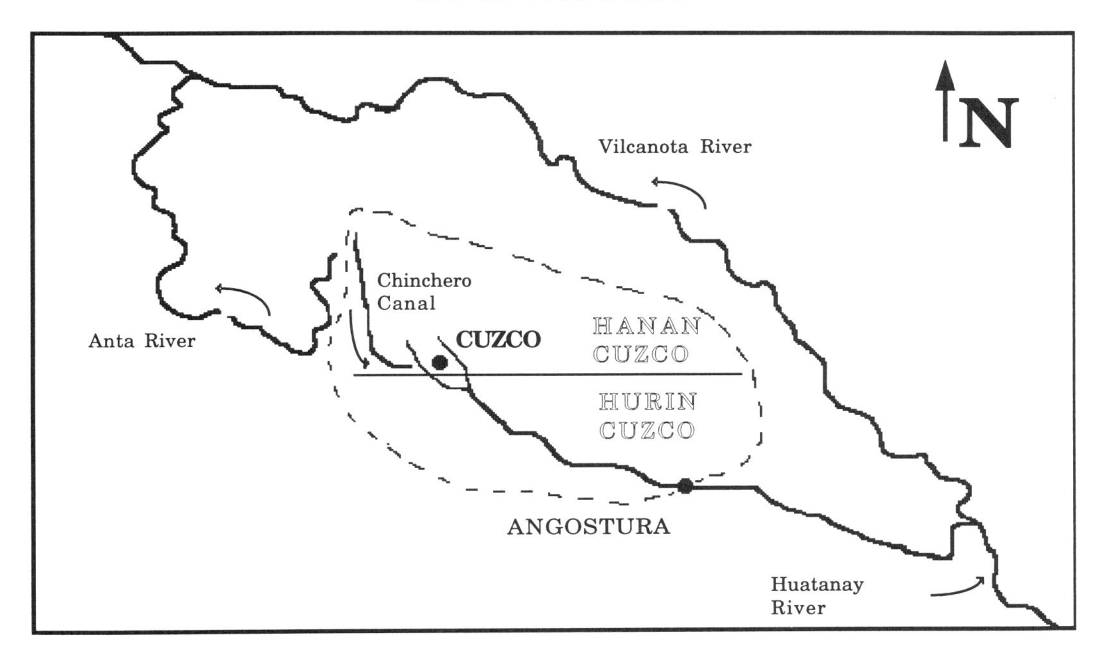

Inca Cuzco: Area that was defined by *ceques* is designated by dotted line. It was internally divided into Hanan Cuzco (upper) and Hurin Cuzco (lower).

---

part of the labor due the state was agricultural: planting, cultivating, and harvesting lands as well as building and maintaining works related to agriculture such as terraces and irrigation works. The assignment of irrigation networks and lands to specific *panacas* and *ayllus* in Cuzco also entailed responsibility for their repair as well as use. Such care was both an obligation and a privilege. It is likely, therefore, that the *ceques* served an additional function of organizing the labor tasks of the *panaca* and *ayllus* (Sherbondy 1982a:92-95).

In sum, the Inca organizational principles that formed territorial and social groups were based on the natural hydrology of the Cuzco valley and associated irrigation works. In this way, the Inca state organized Cuzco into social groups that also organized labor, production, irrigation, and rituals. Significant points in the natural and hydrological landscape were formalized as *huacas*. Over one third of the 328 *huacas* have been clearly identified as water sources or features associated with water sources, canals, and other hydraulic works. The area defined by these *huacas* and *ceques* was further assigned to moieties (Hanan Cuzco and Hurin Cuzco) and *suyus* (Chinchaysuyu, Artisuyu, Collasuyu, and Cuntisuyu). Each *suyu* was divided into three sectors and these were likewise divided into three (see Figure 3.4).

FIGURE 3.3 DIAGRAM OF THE SECTORS OF THE SUYUS IN CHARGE OF THE PANACAS AND THE AYLLUS

FIGURE 3.3 DIAGRAM OF THE SECTORS OF THE *SUYUS* IN CHARGE OF THE *PANACAS* AND THE *AYLLUS*

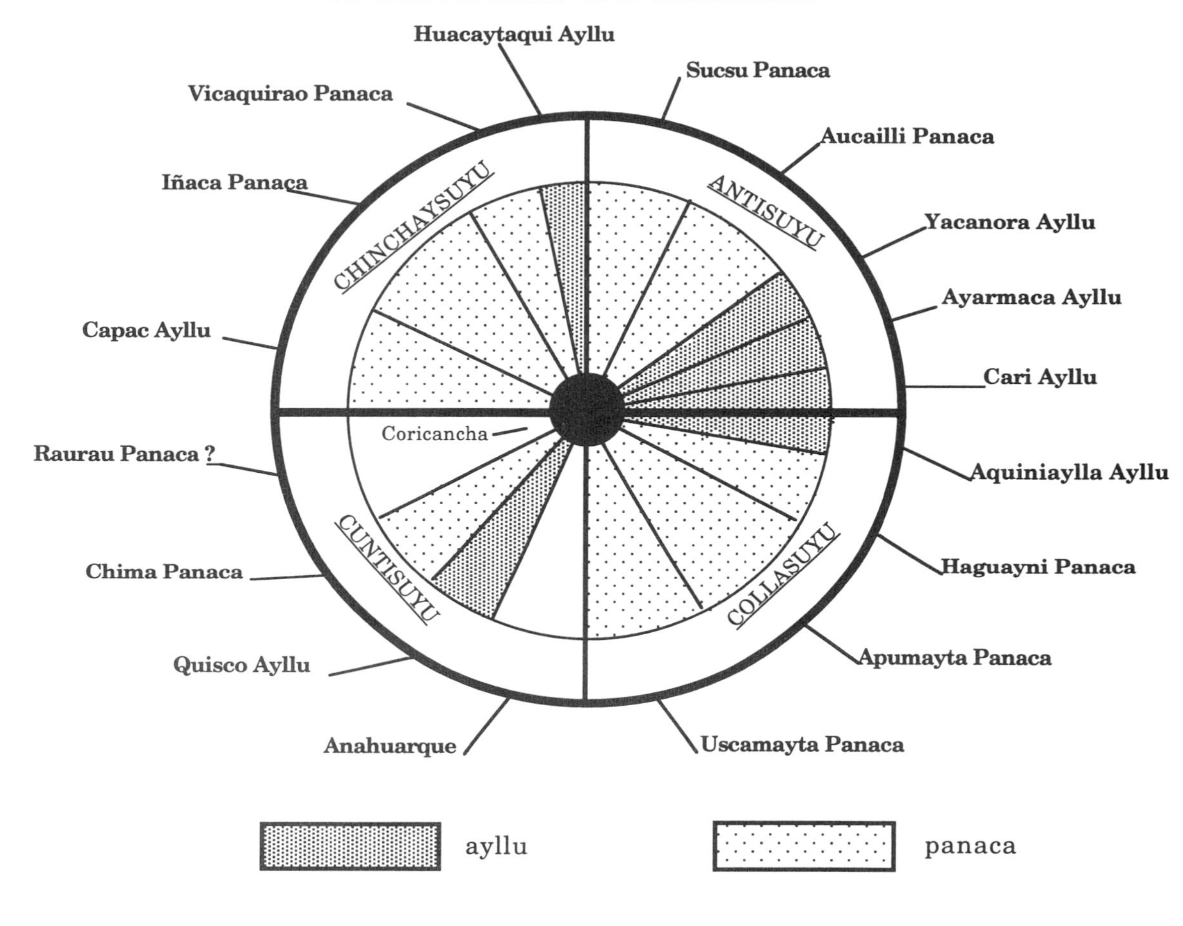

They were the essence of the radial division of Cuzco. All of the above forms of subdividing a multi-ayllu community have been documented in the historical past as well as in the present (Albó 1972; Isbell 1971, 1978; Mitchell 1976; Sherbondy 1982a; Zuidema 1964).

IRRIGATION IN INCA CUZCO: THE PRACTICE

How did the *ceque* organization work out on the ground in Inca Cuzco? It was applied to the upper part of the Huatanay River drainage basin, an area approximately 18 kilometers in length and 6 kilometers in width. This area contained the urban

FIGURE 3.4 DIAGRAM OF GROUPS OF 3 *CEQUES* (COLLANA, PAYAN, AND CAYAO) IN CHINCHAYSUYU AND PROPOSED TERRITORIAL DIVISIONS OF THE PANACAS AND AYLLUS.

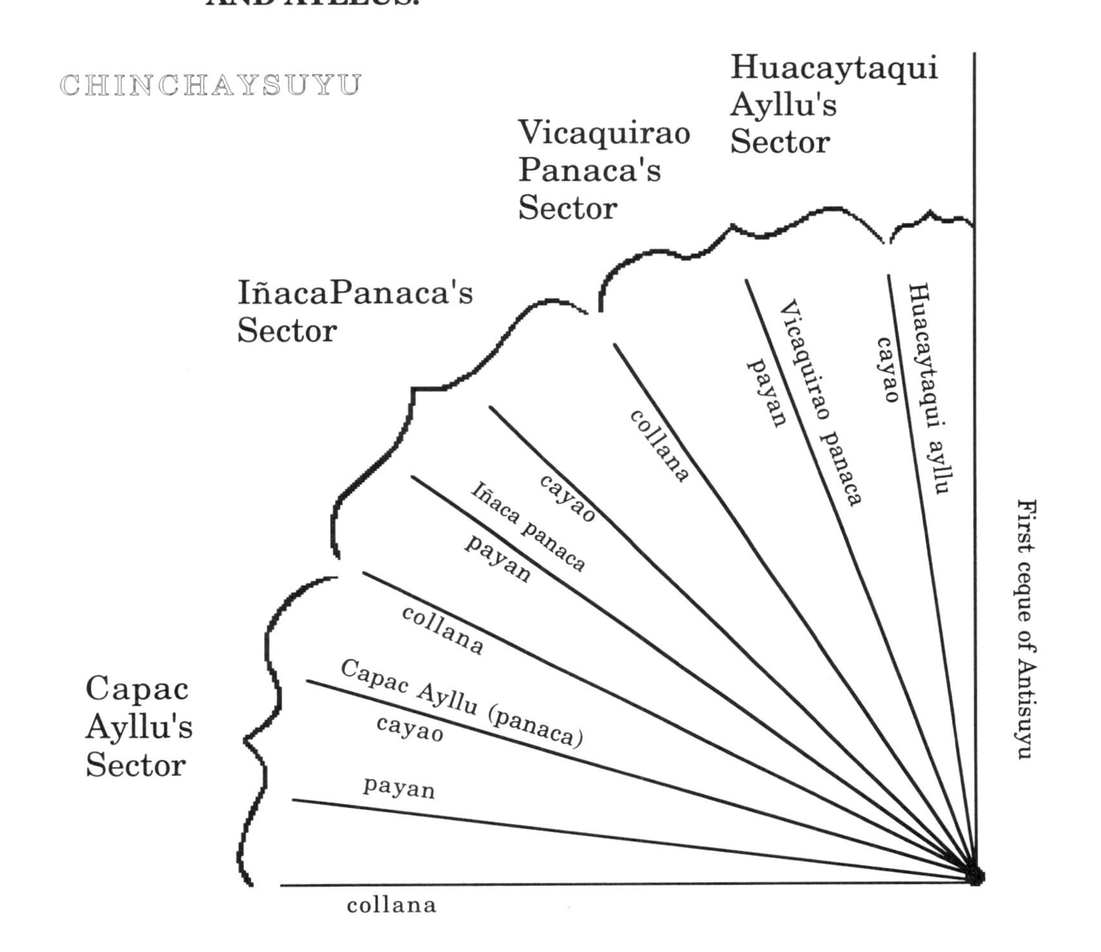

settlement of Cuzco as well as settlements scattered throughout the valley as far as the first gap through which the Huatanay flows. The *ceque* system, thus included the natural drainage basin and a few artificial extensions, notably in Chinchaysuyu, in which the boundaries bulge out to the northwest to include the sources and initial sections of a very long canal (about 18 kilometers), now known as the Chinchero canal, that brought water from springs in the Anta valley to the Huatanay valley. Ideally it ordered a circular area, but as it adjusted to topographic reality, it assumed a more rectangular and irregular shape.

Each *ceque* division contained both urban settlement and rural areas. The center of the system was urban Cuzco, the political and ritual center of the empire and the location of most palaces and temples. Other dispersed settlements scattered throughout the agricultural lands, pastures, and quarries of the valley, delimited by the *ceque* system, gave it a mixed urban-rural character.

Various types of evidence indicate that the population of the Cuzco Valley was substantial. The earliest figure for the population of the city of Cuzco by the Spanish observer, Cristóbal de Molina, "el Almagrista" (1968:73) was 40,000 *vecinos* (citizens) implying heads of households or 150,000-200,000 inhabitants, a sizeable population (Cook 1981:212-219). Zuidema (1990a:7), however, suggests this figure was not a real count but the "official" figure that indicated the administrative rank of an imperial province. Molina estimated the population of the area outside the city of Cuzco, within a radius of 10-12 leagues, at 200,000 additional Indians and commented that it was the most populous area of the Inca empire. Part of this estimate would have included the Huatanay valley from Cuzco to its juncture with the Vilcanota river. The Cuzco valley as delimited by the *ceque* system had a radius of only two leagues, a small portion of this area. Sancho (1962 [before 1550]), Pizarro's secretary and one of the first Spaniards to enter Cuzco, estimated that there were more than 100,000 houses in the Cuzco valley, but these included not only permanent residences but temporary ones for recreation as well as storehouses. Even though a simple multiplier can not be used, the data indicate that the population of the Cuzco valley must have been considerably larger than 100,000 inhabitants.

The irrigation canals of Hanan Cuzco mostly originated there and watered fields within its boundaries. While the canal network of Hurin Cuzco is less well known, the evidence we have indicates that it too was largely contained within its moiety although some canals in Hurin Cuzco may have had sources in Hanan Cuzco. The Huatanay river itself united the two moieties in a dependency relationship: the river in lower Cuzco was formed by the flow of the water from the tributaries in upper Cuzco. This dependency was translated into a political ranking in which the population of Hanan Cuzco enjoyed superior rank over that of Hurin Cuzco.

Geological faulting has produced still another divergence between Hanan Cuzco and Hurin Cuzco: the two sides of the valley reflect different geological formations and different soils. Hanan Cuzco has calcareous rock formations from the Middle Cretaceous whereas Hurin Cuzco has sandy and clayey soils from an Upper Cretaceous formation (Marocco 1978). The soils of Hurin Cuzco are not less fertile, but are more unstable, so that canals built on them would have required far more maintenance than those in Hanan Cuzco. This situation probably accounts for the abandonment of many of the Hurin Cuzco canals in the early colonial period (Astete 1984). Some early Spanish observers commented that the best water sources were to be found in the upper moiety, an opinion still held by many farmers in the valley today. The early abandonment of many of the Hurin canals and the subsequent loss of their water sources may have influenced these observations, but the Hanan water sources may also have been superior in volume, reliability, and chemical content during the Inca period.

The large *suyu* divisions similarly reflected hydrological realities. In Hanan Cuzco, Chinchaysuyu (the northwest quarter) contained the more important natural streams and the most important canals. It also enjoyed higher status than Antisuyu (the northeast quarter). The Saphi River and the Tullumayo, the two major tributaries of the Huatanay river, were contained in Chinchaysuyu. The major irrigation canal systems were also there. The Chacan canals, which derived their water from the Saphi river, figured prominently in Cuzco origin myths and in Inca rituals. The first king of Hanan Cuzco, Inca Roca, is said to have built these canals after discovering their water sources. According to the myth, Cuzco suffered from a lack of irrigation water until the canals were built, so that the development of Inca Cuzco was attributed to their construction.

The springs that feed the Saphi river are located on Huayna Corcor mountain (also known as Sencca mountain). Springs on the other side of this same mountain fed the other major canal of Chinchaysuyu, the Chinchero canal. The Chacan and Chinchero canals were the two most important canals for Inca Cuzco and both were located within the highest ranking *suyu*, Chinchaysuyu. Hydrologically and ritually, therefore, Chinchaysuyu was built around the mountain and the irrigation canals dependent on it (see Map 3.2). These canals also had the most direct importance for Cuzco because they provided water for urban Cuzco itself.

The canals of Antisuyu also had important ritual and symbolic significance, even if not as important as those in Chinchaysuyu. The Antisuyu canals irrigated the lands east of Cuzco. The Amaru canal, closest to the city, was said to have had miraculous properties. According to one Inca myth, during a drought that devastated Cuzco only the terraced fields that were watered by the Amaru canal flourished, saving the population from starvation (Sherbondy 1982a:67).

The canal from Tambo Machay, also known as the canal belonging to Sucso and Aucaille (two former *panacas*, now *ayllus*) was also considered a major contributor to the prosperity of Inca Cuzco from the time of Inca Roca. A myth tells that Inca Roca married Mama Micay, whose first act as queen of Inca Cuzco was to send irrigation water to drought stricken Cuzco. This myth tells us that she incorporated the major source of water for the irrigation canals of her chiefdom, Lake Coricocha, into Inca territory (Sherbondy 1982a:140). According to the story, underground channels from Lake Coricocha provided water to the *huaca* of Tambo Machay, the source of water for the Sucsu and Aucaille canal. The people who continue to live in the region of the former Antisuyu still believe that their irrigation water comes from Lake Coricocha. They tell us that the Incas built subterranean channels to bring the water to Cuzco and they point to the Inca stone structures that channel several of the springs that feed the canals. However, there is no evidence that there are channels from Lake Coricocha (Sherbondy 1982a:140-144, 1982b).

This contemporary belief is a transformation of the Inca myth that had the function of validating Inca rights to waters that were considered the property of Mama Micay's people. The marriage union of Mama Micay, who controlled the land and waters of what came to be Antisuyu, with Inca Roca, who was credited with the discovery of the water sources and development of the irrigation canals of what came to be Chinchaysuyu, established clear water rights for Hanan Cuzco.

A comparable myth for Hurin Cuzco tells us that the ruler Sinchi Roca, who mirrors Inca Roca in name and action, drained the area of urban Cuzco and made

MAP 3.2 THE MYTHICAL SOURCES OF WATER FOR CHINCHAYSUYU (HUAYNA CORCOR MOUNTAIN) AND ANTISUYU (CORICOHCA LAKE).

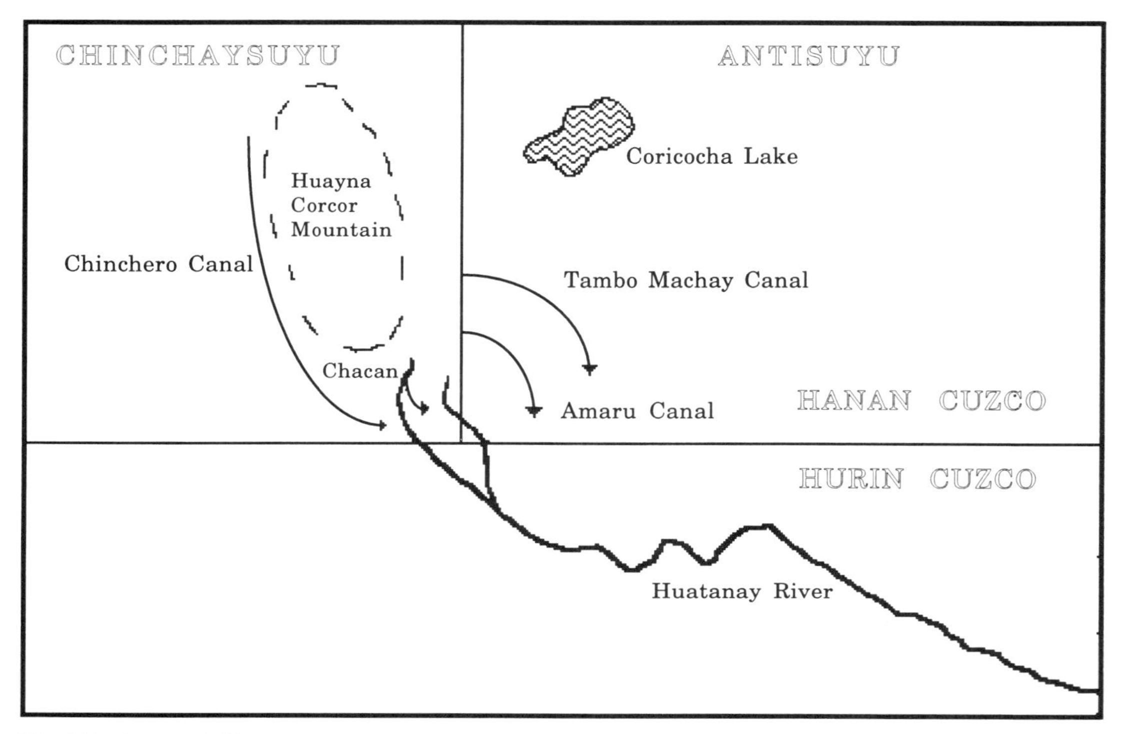

The Chinchero and Chacan canals as well as the Saphi River are believed to come from the waters of Huayna Corcor Mountain. The Tambo Machay and Amaru canals are believed to come from the waters of Coricocha Lake.

---

it possible to found the city there.

Water is essential for life and survival. Adequate technologies, however, are often needed to channel water for useful ends. Drainage systems direct water away from an area and irrigation canals conduct them to an area. Both of the above Inca myths credit these early kings with hydraulic works essential for the subsequent populating of Cuzco. By emphasizing the original construction of these hydraulic works, the myths utilize Andean legal discourse to establish the legitimacy of Inca ownership of these works and surrounding land, for the Inca people were considered the descendants of Sinchi Roca and Inca Roca.

Most of the Huatanay river flows through Collasuyu, but a small but very significant portion flows through Cuntisuyu (Sherbondy 1987a:127-129). This segment begins at the confluence of the two major tributaries of the Huatanay, the Saphi and Tullumayo Rivers which originate in Hanan Cuzco (see Map 3.3). It

MAP 3.3 CUNTISUYU AS THE *CHAUPI* OR MIDDLE SECTION OF HUATANAY RIVER

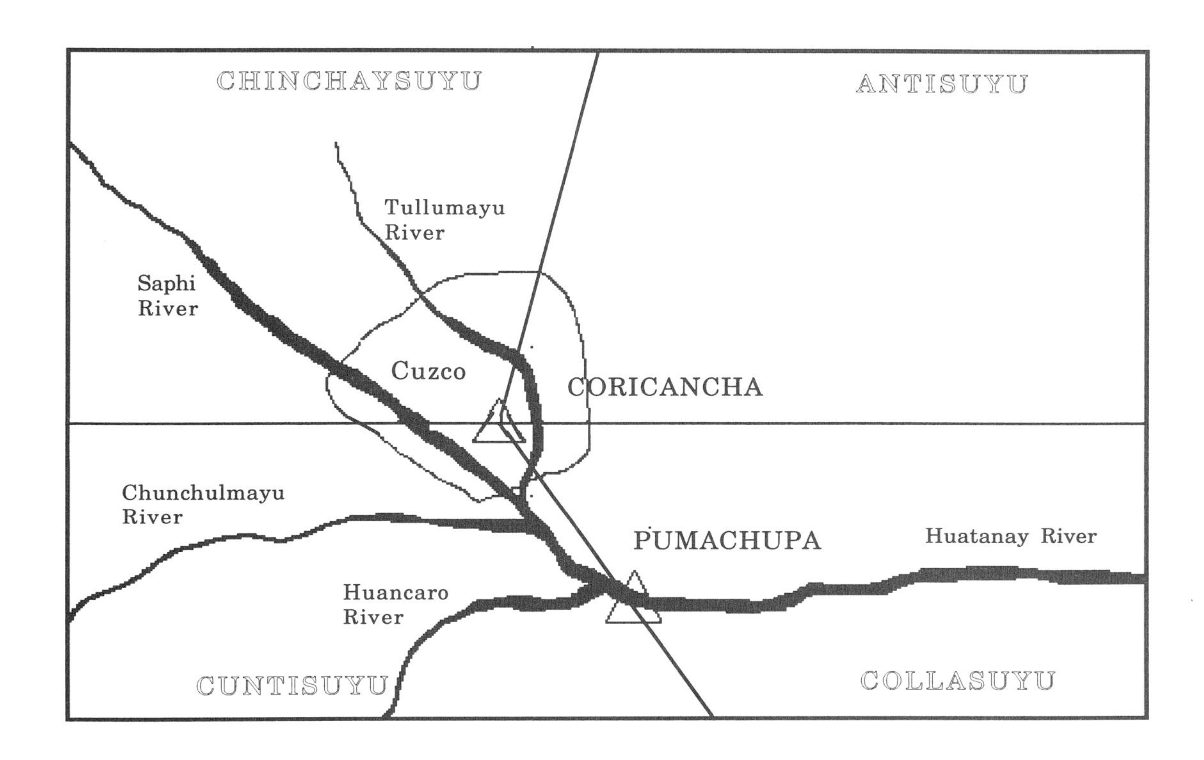

continues past the union of the Huatanay with the Chunchulmayo River (flowing from the west) and ends at the juncture with the Huancaro River (also entering from the west). This Cuntisuyu branch of the Huatanay River was ritually significant for the Incas, as indicated by the *huaca* Pumachupa. This *huaca*, very important in Inca rituals, meant "puma's tail," a reference to Cuzco as a symbolic puma (Zuidema 1983a; Barnes and Slive 1990). The body of the puma was the major area of urban Cuzco that included the major temples and palaces and plazas and the tail was the Huatanay river. The *huaca* Pumachupa was the site of rituals that involved the river. Pumachupa also marked the border between Cuntisuyu and Collasuyu within Hurin Cuzco.

This section of the Huatanay river in Cuntisuyu was thought of as *chaupi*, that is an essential central section that mediates between two other zones. The *chaupi* region resolves the differences between the two zones by transforming one into the other. This is an apt description of the area where several tributaries are transformed into the Huatanay river.

The Chinchero canal and the Chacan canals of Chinchaysuyu mentioned above were the two most important irrigation systems of Cuzco (Sherbondy 1982a:30-48). The Chinchero canal began on Iñaca *panaca's ceque*, flowed across another *ceque* (that Iñaca *panaca* undoubtedly also cared for) where it picked up another water source (marked as a *huaca*). On Capac *ayllu's ceque* it obtained an important additional source of water. It continued across the next *ceque* (associated also with Capac *ayllu*), where it might have picked up another water source, and finally to the last *ceque* the Chinchaysuyu, also associated with Capac *ayllu*. Here it followed the *ceque* into Cuzco (Sherbondy 1982a). I have deduced that Capac *ayllu* and Iñaca *panaca* shared the ownership of the canal. These were the two highest ranking *panacas*, those of the last two Inca kings to have their *panacas* incorporated into the *ceque* system. Iñaca *panaca* was that of Pachacuti Inca, who according to tradition organized the *ceque* system as we know it; Capac *ayllu* was the *panaca* of his successor, Tupa Inca Yupanqui. Since Pachacuti Inca had reorganized Cuzco and the *ceque* system, he made his *panaca* the highest ranking one and provided for his son's *panaca* to be the next most prestigious.

This relative ranking is reflected in the way in which the Chinchero canal is distributed between the two *panacas*. Pachacuti Inca's *panaca* (Iñaca *panaca*) controlled the upper part outside the drainage basin of Cuzco and Tupa Inca Yupanqui's *panaca* (Capac *ayllu*) controlled the lower part inside it. Iñaca *panaca* thus controlled the upper part of the canal including excellent springs located on the slopes of Huayna Corcor mountain. Capac *ayllu* controlled Ticatica, a very important spring in the Cuzco valley. This spring, well known for its excellent quality and quantity of water, substantially increased the volume of the canal. Iñaca *panaca* also owned within its boundaries the Chacan canals attributed to Inca Roca. These canals had the highest symbolic and ritual significance and were the most important sources of water for the area closest to urban Hanan Cuzco.

The water sources of the Chinchero and Chacan canals (the springs, rivers, and their mountain home) were formally incorporated into the *ceque* system as *huacas*. Parts of the *ceques* or certain *huacas* on them also functioned to delimit the boundary between Capac *ayllu's* section of the Chinchero canal and Iñaca *panaca's* section as well as to mark the boundaries of the district irrigated by the Chinchero canal and of the district irrigated by the Chacan canals.

In Chinchaysuyu, Uicaquirao *panaca* pertained to Inca Roca. Since he was the first king of Hanan Cuzco, his *panaca* was assigned the lowest rank among the kings of Hanan Cuzco. This *panaca* was assigned to the *ceque* that included the water source Uiroypaccha, located within urban Cuzco and probably used to irrigate terraces inside the city and to provide water for drinking and cooking. It was a prestigious fountain but did not have the high economic value of the canals of Chinchaysuyu that irrigated large extensions of agricultural land.

Huacaytaqui *ayllu* is the only *ayllu* connected to Chinchaysuyu in the context of the *ceque* system. It is associated with Pilco Puquio, which provided the water for the stream now known as Qenqo Mayu and which flows through Cuzco from the ritual site of Qenqo (Sherbondy 1987a:141). This irrigation district lies on the eastern fringe of Chinchaysuyu, on the border with Antisuyu.

The major irrigation systems of Antisuyu flow in an easterly direction away from urban Cuzco, in contrast to the waters of Chinchaysuyu which all flow into the city itself. The largest canal system of Antisuyu was the Tambo Machay canal, which irrigated the lands of Sucsu *panaca* and Aucaille *panaca*. Like the Chinchero canal, the upper section of the Tambo Machay canal was cared for by one *panaca* (Sucsu *panaca*) and the lower section by another (Aucaille *panaca*). The Andean logic of upper/lower was applied, assigning a higher ranking value to Sucsu *panaca* and a lower ranking value to Aucaille *panaca*. This ranking corresponded to the relative rank of the founders of the two *panacas*: Yahuar Huacac, the founder of Aucaille *panaca*, was the father of Viracocha Inca, the founder of Sucsu *panaca*. In addition, all the sources of water were located at the upper end of the canal. The higher ranking Sucsu *panaca* controlled the water source and the lower ranking Aucaille *panaca* was dependent on the flow of water from the upper part.

The highly valued Amaru canal was also located within the sector of the highest ranking *panaca* of Antisuyu, Sucsu *panaca*. This canal enjoyed higher status than the Tambo Machay canal by being located closer to urban Cuzco. The Inca believed that Lake Coricocha provided water to the Amaru canal, the Tambo Machay canal and most of the springs in Antisuyu. The Amaru canal belonged to Tupac Amaru Inca, the brother of the Inca king (Tupa Inca Yupanqui) and second in power (Segunda Persona). The king was in charge of military affairs and his brother was responsible for the priestly structure of the Inca state and agrarian affairs, especially irrigation (Sherbondy 1979b, 1982a:64-68). Tupac Amaru Inca was closely associated with Sucsu *panaca* (Zuidema 1983b).

The lower boundary of the regions irrigated by the Tambo Machay canal was a *ceque* assigned to Yacanora *ayllu*. Although this *ayllu* had its own source of water, it also depended on a branch of the Tambo Machay canal. This *ayllu* was lower in status than Sucsu *panaca* and Aucaille *panaca* because of its lower topographical relationship to them and its dependence on the higher parts of the Tambo Machay canal network.

Three of the *ceques* in Antisuyu were controlled by *ayllus* rather than *panacas*: Yacanora *ayllu*, Ayarmaca *ayllu*, and Cari *ayllu*. These three *ceques* follow one on the other as if they were a single line (Sherbondy 1979:53-55). Yacanora *ayllu* cared for the *huacas* closest to Cuzco, Ayarmaca *ayllu* those farther out, and Cari *ayllu* those farthest from the center. These assignments are consistent with an internal ranking among the three *ayllus* in which the one closest to the urban center enjoyed the highest rank. Some of the *huacas* on these *ceques* are springs that flow southerly towards the Huatanay river, suggesting that these *ayllus* might have irrigated the lower slopes and the Huatanay river plain (Sherbondy 1982a). Fields in these locations are more vulnerable to frost than those on the more protected higher slopes of the Tambo Machay canal. For this reason and because it was lower in altitude, the territory associated with these *ayllus* enjoyed lower prestige than that associated with the Tambo Machay canal.

From the above analysis, it is clear that the canals of Hanan Cuzco were mapped onto the *ceque* system. It is reasonable to hypothesize, therefore, that the water sources and canal districts of Hurin Cuzco were similarly encoded. The distribution of *ceques* among the *panacas* and *ayllus* of Hurin Cuzco must have similarly indicated ownership of the canals.

It would be tempting to conclude that the *ceques* simply delineated the *panacas* and *ayllus* much like the division of *ayllu* lands in contemporary San Andrés de Machaca, Bolivia (Albó 1972). The picture in Inca Cuzco, however, was more complex and I have reassessed my earlier position that the *ceques* were a mental template dividing *panaca* lands (Sherbondy 1986, 1987a). The *ceques* functioned more like parallels and meridians, serving as spatial reference points, rather than as simple boundaries between lands.

The case of Yacanora *ayllu*, for example, suggests that one *ayllu* may even have cared for the *ceque* associated with another *ayllu*. The data are unclear, but it appears that Yacanora *ayllu* cared for the springs on a *ceque* below its own lands, a *ceque* that cuts across the lower reaches of the lands of Sucsu *panaca* and Aucaille *panaca*. The lands of Yacanora *ayllu* are irrigated today by a spring fed canal that informants say comes from the Tambo Machay canal via a subterranean link. If this were the case in the past–and the evidence is uncertain–then Yacanora *ayllu* cared for a spring on a *ceque* associated with an irrigation system that it did not use. Nonetheless, Yacanora appears to be an exception to the general pattern in which *ceques* mapped the lands and irrigation systems of associated *panacas* and *ayllus* (see also Zuidema 1990b).

FROM *PANACAS* TO PARISHES

After the conquest of the Inca state in 1532 and the fall of Cuzco in 1533, the Spanish built on what remained of the Inca system after the native population had endured several years of stress and morbidity caused by infectious diseases and war (not only the war of conquest, but the subsequent internecine conflicts of both Incas and Spaniards). These traumas disrupted the Inca system, but did not destroy it completely. The conquerors took direct control of the highest levels of political organization, eliminating the *ceque* system, the divisions into moieties and *suyus*, and the political functions of the *panacas*. The Spanish nonetheless maintained some of the lower levels of organization within, of course, the new sociopolitical context.

Many of the *panacas* and *ayllus* became local administrative units that, like those in the Inca period before them, were strongly centered on irrigation. These colonial *ayllus* consisted of groups of people that used land and irrigation systems in common and worshipped at local religious centers on those lands. The Spanish facilitated the formation of these groups in order to maintain production. Natives also learned to utilize the Spanish tenure system to secure rights to land and water by forming communal groups of land and water users, often combining former *ayllus* and *panacas* of the Inca period into new groupings.

Spanish Control Over Lands

The founding of the Spanish municipality in March of 1534 transferred the management of lands and waters from Inca to Spanish administration. At that time Francisco Pizarro distributed city house plots to his men and he made the first distribution of Indian labor (*reparto de encomiendas*). This initial distribution of land and labor was followed by several others (in 1536, 1540, 1548, and 1550). The native population viewed these *encomiendas* as comparable to the Inca *chapas*, which was the Quechua term they used to translate the Spanish *encomienda* (González Holguín 1989 [1608]:96). Religious institutions also received grants of land and services, such as the grants to the monastery of La Merced in 1539-40.

The municipal council (*cabildo*) was the governing and judicial body for Cuzco. As in Castilian municipalities, it made the decisions about the management and development of the agricultural lands in its jurisdiction (Nader 1990:ix). The municipal council made decisions about water rights and appointed officials for the regulation of water. In 1548, for example, the council assigned the city of Cuzco domain over the fountain in the former Coricancha, rather than allotting it to the Dominican monastery that occupied the site.

In 1574, the office of the native attorney (*juez de naturales*) was created and it began to appoint water officials to regulate the canals that had both Spanish and native users. In that year, the native attorney appointed a sheriff (*alguacil repartador*) to distribute water in the Pacpachire canal to alleviate the conflicts among three groups of users: a pre-Inca community (Cachona), the indigenous servants of a Spaniard (D. Diego Maldonado), and the monastery of La Merced.

In 1600, the *cabildo* formulated the first irrigation regulations for Cuzco, revising them in 1659 (Villanueva and Sherbondy 1979). The *cabildo* based the water regulations on custom, consulting the elder natives to establish the traditional schedule for each canal. Preference was given to the native cultivation of maize and other food crops. When alfalfa (introduced by the Spanish) competed with native maize for water, the council determined that water could not be taken away from the fields it had customarily irrigated. The distribution judge appointed by the *cabildo* (*juez de la repartición de aguas*) was charged with enforcing these regulations.

The preference given to the irrigation of maize was considered essential to guarantee the production of food. Because the Spanish did not want to become food growers, which was not a lucrative pursuit, they recognized that the most efficient system was to have the indigenous population feed itself and produce a surplus for tribute. Both Spaniards and natives, therefore, were motivated to maintain the traditional communal control of canals and lands.

Strategies On The Part Of The Indians

The natives that survived the conquest realized that they needed to get Spanish recognition of their claims to land and water. They employed at least two strategies.

One strategy was to claim rights to lands as inherited private property. The other was for natives to claim land based on pre-Hispanic communal rights.

To claim land as private property, the claimant had to establish a genealogical legitimacy acceptable to the Spanish, an approach used by the Inca nobility who were in favor with the Spaniards. Since they were politically valuable to the Spanish, they had a better chance of retaining some of their land and water rights as individuals. Thus, the puppet rulers, such as Paullu Inca, were able to leave their holdings to their descendants. Other influential Inca nobles were similarly able to establish individual land rights.

In the communal strategy, *ayllus* and *panacas* organized in order to maintain their communal land and water rights. The *panacas* often presented themselves as *ayllus* to do so. *Ayllu* land rights were generally tied to water rights, which provided the infrastructure for production. Thus, the *ayllus* continued their ties to irrigation and water sources, even after the *ceque* system had disappeared.

The Parishes

The Spanish first relied primarily on their relationship with the former Inca lords and other leaders to indirectly control the *ayllus*. They later replaced this informal system with parishes. The Spanish Crown was motivated to establish the parishes, not only to ensure greater political control over the natives, but to secure the collection of tribute and therefore its income (Aparicio Vega 1963).

The first Indian parish was Los Reyes, later named Belén, founded in 1550 in the neighborhood of Cayaocache, located southwest in the urban nucleus of Cuzco, where the majority of the Inca population lived. In 1559, the viceroy asked Juan Polo de Ondegardo, the governor (*corregidor*) of Cuzco, to establish additional parishes (see Map 3.4). Polo set up four parishes, bringing the total to five. Four parishes were located in Cuzco (Belén, Santa Ana, San Cristóbal, and San Blas) and one outside Cuzco in the valley (San Lázaro, later renamed San Sebastián). In 1570 the Viceroy Toledo increased the number of parishes to eight in order to incorporate into parishes those natives who had not been paying tribute.

Toledo also acted to curtail the rights of nobles in order to end Inca sources of power. In 1572, for example, he confiscated the lands of D. Carlos Inca in Colcampata to build a fortification there. He also introduced foreign populations into Cuzco, such as the anti-Inca Cañaris and the Chachapoyas, to whom he gave land in San Cristóbal parish (Hemming 1970:451).

The parishes were constructed more for geopolitical and labor considerations, than for their role in land and water rights. Consequently, the Spanish did not utilize precise Inca administrative divisions to construct the parishes, so that the parishes no longer reflected the topography and hydrology of the valley in the same way that Incaic organization had.

Unfortunately, we do not know the exact boundaries of the parishes, nor the complete list of which *ayllus* were included in which parish. The earliest parish records I have seen are from 1573, after Viceroy Toledo had already added three more parishes to the original five. Nonetheless, it is useful to speculate about the

MAP 3.4 PARISH CHURCHES OF CUZCO (1559) COMPARED TO *SUYU* BOUNDARIES

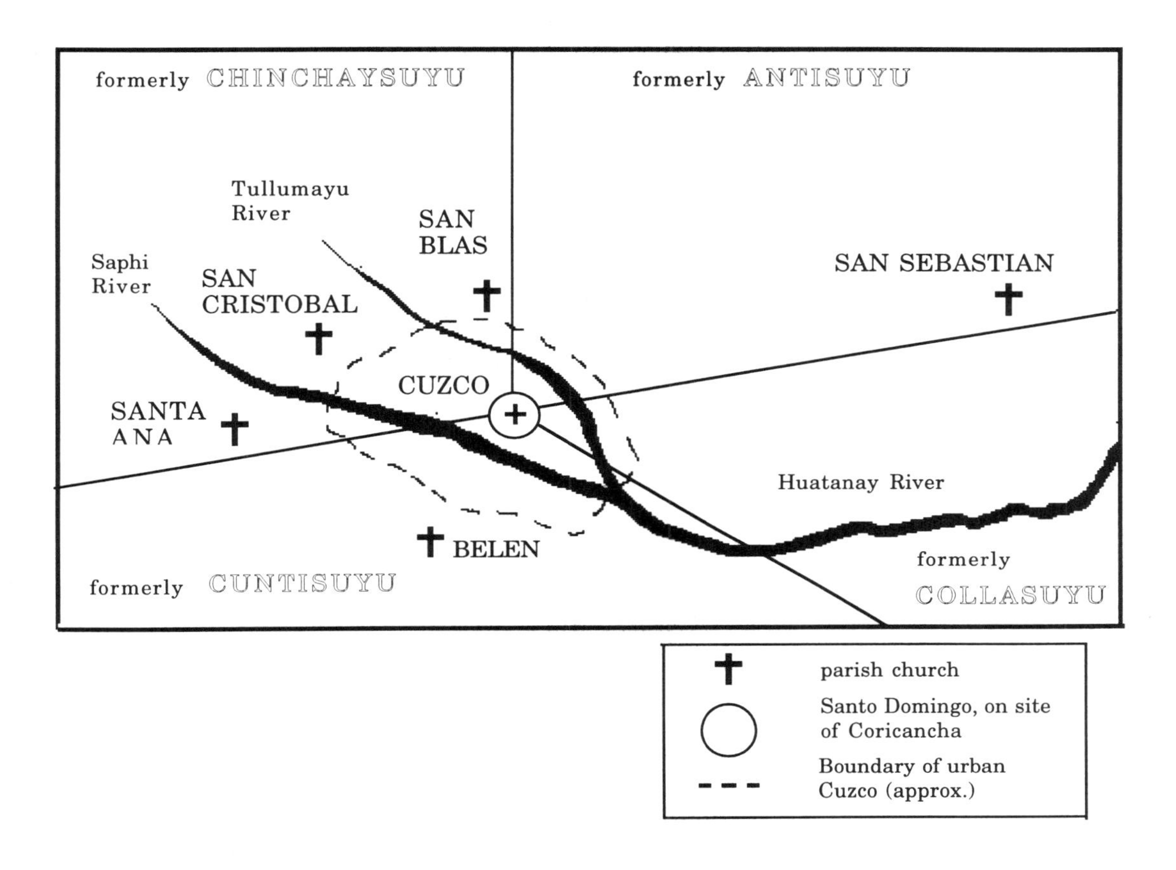

possible organization of the parishes, to assist the direction of future research. The Spanish certainly did not use the Inca division into the two moieties of Hanan Cuzco and Hurin Cuzco to construct the parishes. This dual division, the most important organizational unit of the pre-Columbian period, was based on topography and hydrology, and there is no evidence that the parishes conformed to this plan. The evidence we do have, moreover, indicates that the Spanish did not clearly understand the relationship between the Inca moiety divisions and Inca land and water rights.

There is also no strong evidence that the Spaniards utilized the Inca division of four quarters (the *suyus*) in creating the parishes (see Map 3.4). The parishes definitely did not divide the native population in the same way that the Inca *suyus* had. The parishes, moreover, grouped several *ayllus* together in unfamiliar ways. Some *ayllus* protested that *ayllus* that did not belong together were put together, while others that should have been put together were separated.

Nonetheless, the general Incaic principle of quadripartition may have been used in the initial design of the parishes, although the evidence here is unclear and contradictory. There is a close correspondence in organizational principle between four *suyus* and the four parishes in the city of Cuzco (the fifth parish was located outside the city). This correspondence in quadripartite structure leads one to speculate that Polo de Ondegardo may have used the quadripartite principle in creating the parishes. He was very knowledgeable of Incaic organization and we do know that he had utilized his investigation of Inca customs in the service of Spain (Sherbondy 1982b, 1992). Spanish policy, moreover, encouraged the superimposing of their institutions on already existing ones. Polo, therefore, may have used the familiar quadripartite principle to facilitate the harvesting of parish labor for municipal work projects, while at the same time scrambling the *ayllus* in the parishes to break up any inter-*ayllu* alliances that might have existed.

Thus, the city's population may have been divided into four parishes and the rest of the Valley aggregated into a fifth parish (that of San Sebastián). If that were the case, it may have been Polo's intention to incorporate Antisuyu into the parish of San Sebastián, to divide the population of Chinchaysuyu between the parishes of San Blas and San Cristóbal, to combine the populations of Collasuyu and Cuntisuyu into the parish of Belén, and to set up the parish of Santa Ana for the Chachapoyas allies of the Spaniards who had been living in Cuzco prior to the Spanish invasion. If this was Polo's reasoning, he may have modified the *suyu* system to respond in part to changes in the population of the valley, such as the incorporation of the Chachapoyas.

It is also possible that Polo made a conscious effort to scramble the earlier *suyu* divisions in order to dilute their political power. Evidence for this hypothesis is found in the location of the parish of San Sebastián. This parish, physically located on the border between the old Antisuyu (Hanan Cuzco) and the old Collasuyu (Hurin Cuzco), incorporated *panacas* and *ayllus* from both moieties, notably Sucso and Aucaille *panacas* of Hanan Cuzco (Antisuyu) and Chima and Raurau *panacas* of Hurin Cuzco (Cuntisuyu). The parish church was constructed on the plaza of Colcapampa, a *huaca* that contained a rock revered by the Incas.[3] Additional evidence that the Spaniards may have deliberately blurred the *suyu* divisions is found in the structure of the parish of San Blas. This parish also straddled two *suyus*: Chinchaysuyu and Antisuyu. As in San Sebastián, the San Blas church was located near the *suyu* border. In this case it was built in Chinchaysuyu but on the border of Antisuyu.

The next lower level of Inca organization was based on the tripartite division of each quarter into sectors (indicated by *ceques*) that included the waters and lands of the *panacas* and *ayllus*.[4] It is on this level that I find the most important continuities as well as some significant changes. At this level the political organization of Inca Cuzco closely corresponded to the ecological demands of irrigation, unlike the imperial level, which was more removed from irrigation matters. Each sector was centered on an irrigation system and the lands it watered, that is, an irrigation district. As a result, the sectors were irregularly shaped by the demands of topography and water flow, even though they were theoretically geometrical. Some sectors were much larger than others.

The *panaca* and *ayllu*, social groups with communal (corporate) rights to certain lands and waters, remained essentially intact after the Spanish conquest. There were political and economic reasons for this. Spanish officials governed indirectly through the traditional leaders of the *panacas* and *ayllus*. The *panacas* and *ayllus*, the basic productive units, provided tribute to the Spanish Crown and food for the entire community. The Spanish leadership, especially Polo de Ondegardo, understood that to maintain production of maize, potatoes and the other Andean foods they had to facilitate native irrigation. In consequence, Spanish administrators kept (with minimal modification) the Inca irrigation calendar (Villanueva and Sherbondy 1979:3ff). In the case of a conflict between the irrigation of Old World crops such as alfalfa and New World maize, moreover, the Spanish gave preference to maize. The survival of Inca Cuzco as well as of Spanish Cuzco depended on maize production and that in turn was dependent on the careful management of irrigation.

As a result, native farmers continued to turn to their *panaca* and *ayllu* for their primary identity. They kept their chiefs, who continued to distribute irrigation water, rotate lands, and organize other communal concerns involving their lands and waters, albeit with some new functions.

Let us look at the *panacas* and *ayllus* in the colonial period beginning with the ones that most successfully kept their identity, cohesiveness, and lands. The most notable case is that of the two *panacas* of Antisuyu: Sucsu *panaca* and Aucaille *panaca*. These two groups survived by joining forces and frequently acting as one *ayllu*: Sucsu-Aucaille *ayllu*. Today they are an officially recognized indigenous community that still controls the canal from Tambo Machay. We can trace their efforts to defend their land and water rights in the colonial period through contemporary conflicts with nearby landowners and communities. Indeed, Sucsu-Aucaille has survived as an *ayllu* largely because of the solidarity created by its successful efforts to defend water rights. Their canal, the major source of water for production, requires daily cooperation for its use and maintenance. Individuals who do not fulfill their obligations to the community, such as making a contribution to the Feast of the Holy Cross, are still punished by being denied access to this important resource.

The *panacas* of Chinchaysuyu were Capac *ayllu*, Iñaca *panaca*, and Uicaquirao *panaca*. In the colonial period Capac *ayllu* and Hatun *ayllu* were joined as the powerful Capac-Hatun *ayllu* in San Blas parish in an area that had formerly been part of Antisuyu. By the early eighteenth century this *ayllu* was aggressively encroaching on the Tambo Machay canal and lands of Sucsu-Aucaille *ayllu* in the former Antisuyu.

It is intriguing to speculate why Capac-Hatun *ayllu* was located in San Blas parish in the colonial period. The Incaic Capac *ayllu* had actually been located in what became the parish of Santa Ana in the former Chinchaysuyu.[5] In the colonial period, however, the parish of Santa Ana consisted primarily of Chachapoyas (non-Incas). One explanation for the shift from Capac *ayllu* to the Chachapoyas may be that during the Inca period Capac *ayllu* was displaced by the Inca king Huayna Capac to make room for the Chachapoyas, who were his favored escort. It is likely that Huayna Capac would have given the Chachapoyas lands in Cuzco even though they did not qualify as a *panaca*. Why were they not included in the *ceque* system as an *ayllu*? Is it that the lands coded into the *ceque* system had not been redistributed since Tupa Inca Yupanqui's reign? The *ceque* system does not include the *panaca* of his successor, Huayna Capac. Or were the Chachapoyas given usufructory rights to one of the *panacas'* lands, perhaps those of Capac *ayllu*? In the early colonial period Capac *ayllu* and the Chachapoyas were distinct *ayllus* in the parish of Santa Ana (in the former Chinchaysuyu). Therefore, it may be possible that the Chachapoyas had rights to some of Capac *ayllu's* lands in the Inca period and that Capac *ayllu*, displaced in part by the Chachapoyas, pushed eastward into San Blas parish (in Antisuyu) to cultivate lands irrigated by the Tambo Machay canal. We cannot assume that change only occurred after the Spanish invasion. However, the lack of written records from the Inca period, makes it more difficult to study the changes at that time.

Another hypothesis suggests that Capac *ayllu* and Hatun *ayllu* lost the lands irrigated by the Chinchero canal in Santa Ana parish because of Spanish usurpation of that canal. Capac and Hatun *ayllu* then joined and were either given or claimed lands in the Tambo Machay canal district. This hypothesis depends on Hatun *ayllu* being another name for Iñaca *panaca*, as Sarmiento (1947:220) suggests. We could hypothesize that since Capac *ayllu* and Iñaca *panaca* had been closely associated and shared the Chinchero canal in the Inca period, they continued to act as one *ayllu* in the colonial period the way that Sucso and Aucaille *panacas* did in asserting mutual ownership of the Tambo Machay canal. However, the Tambo Machay canal survived the conquest, but the Chinchero canal was destroyed during it, perhaps purposely by the Incas during the seige of Cuzco in order to create a muddy battlefield and mire down the Spaniards' horses. At any rate, it had to be rebuilt beginning in the 1550s (Sherbondy 1982a:48-49), reconstructed by the Spanish municipal government for the primary purpose of providing water for the urban needs of Cuzco. Irrigation was never mentioned in the documents and it appears that the Spaniards controlled the canal, not the *ayllus*.

The hypothetical loss of their canal and lands in the Chinchero canal district in the colonial period may explain why Capac-Hatun *ayllu* tried successfully to gain water and land rights in the parish of San Blas, coming into conflict with Sucsu *ayllu* and Aucaille *ayllu* in the process.

The territory of Iñaca *panaca* corresponded to what became the parish of San Cristóbal, located between the parish of Santa Ana and the parish of San Blas. Nonetheless, although the Spanish parish and the Inca irrigation district coincided territorially, the Inca social group (Iñaca *panaca*) seems to have been displaced by other *ayllus* in the Spanish period. Iñaca *panaca*, for example, does not appear on the registers for San Cristóbal, even though Capac-Hatun *ayllu* does. Indeed, evidence of Iñaca *panaca* in the colonial period is scarce, although some of their lands were located far from urban Cuzco in San Jerónimo parish, in the lower part of the valley. Spaniards also occupied many of the lands that would have been part of Iñaca *panaca's* territory. Although only two of the three major Chacan canals continued to be used in the sixteenth century, dwindling to only one at a later date, the Spaniards valued Chacan water highly (Villanueva and Sherbondy 1979:27-30).

The fifth parish, that of Belén, was located in the old territory of Cuntisuyu in

Hurin Cuzco. This parish, known as the parish of the Indians, grouped many *panacas* and *ayllus* together, serving as a catchall for *ayllus* that did not have a firm base elsewhere. It included *ayllus* that had formerly been part of Hanan Cuzco as well as of Hurin Cuzco. Perhaps the *ayllus* not powerful enough to negotiate with the Spaniards were reduced to Belén. We do know that the lord of Hurin Cuzco at the time of the Spanish conquest was very young and that he called upon a lord of Hanan Cuzco to help him defend his lands from being taken by Spaniards (Sherbondy 1982b).

The weak leadership of Hurin Cuzco at the time of the conquest and the abandonment of the canals of Hurin Cuzco shortly after the conquest were two major factors that debilitated the position of the *panacas* and *ayllus* of Hurin Cuzco. The *panacas* and *ayllus* that survived were those that kept control over their canals and therefore over their lands. In so doing, they created and sustained a strong organization based on the need to maintain and defend the canals.

Shortly after the conquest, the *panacas* dropped the designation of "*panaca*" and became simply *ayllus*, even though Spanish and natives continued to maintain distinctions between Inca nobles and commoners. This loss of usage helps illuminate the differences between *panacas* and *ayllus*. The term "*ayllu*" (unlike "*panaca*") is still current because the *ayllus* are corporate groups that control lands, water, and associated rituals. The *panacas* not only did that, but they were also state institutions. The members of the *panacas* were of noble descent and they held the most important political, social, economic, and ritual offices. The Spanish usurped these state functions of the *panacas* and assigned the highest ranking administrative posts to Spaniards. Spanish individuals and institutions (eg. monasteries, convents) appropriated many of the economic resources of the *panacas* (land and water). State ritual was entrusted to the Catholic church, creating a dual religious system. The Christian tier held political and economic power, while the native tier was disenfranchized politically and economically. In the Inca empire elites and commoners shared a common belief system, even though the state used religion to control its subjects (see Silverblatt 1987). In the process of these changes, the *panacas* either disappeared or continued as *ayllus*, productive and religious institutions without formal state functions.

Aside from Christianizing the Andean people, the parishes collected tribute for the Crown and provided labor for city works. The native nobility, the members of the former *panacas*, were not subject to tribute, as were the *ayllus*. Each parish included several tribute paying *ayllus*. European economic organization was very different from the nonmonetary trade and redistributive economy of the Incas. In the Inca empire tribute was paid in labor on Inca state land. The state provided the seed for these lands and they served food and drink to the tribute payers while they were working. In contrast, the Spanish system required tribute payment in money and goods as well as labor. The *ayllus* had to raise the tribute payments by using their own lands, waters, seeds, and livestock, a much more exploitative system.

Nonetheless, for the economic stablilty of Cuzco, it was expedient for the Spanish to keep the irrigation districts intact as much as possible so that the *ayllus* could maintain the canal network and produce food crops. During the colonial period, the

Spanish commonly lamented that canals and the knowledge of water sources were lost, often permanently (see also Seligmann and Bunker, *intra*). They admired the fine hydraulic works that they had seen when they arrived in Cuzco, and admired the smooth functioning of the Inca systems. Many of these irrigation systems crumbled soon after the conquest and only a few were rebuilt. The ones that survived did so as a result of the efforts of the *ayllus* and re-structured *panacas*. The overall organization of the Cuzco valley, however, was lost. It was this organization that had permitted the construction and maintenance of such large hydraulic works as the stone channeling of an entire river and the coordination of drainage and irrigation projects involving the cooperation of all the *panacas* and *ayllus*. The destruction of this function of the *panacas* occurred with the defeat of the Inca state and was only imperfectly replaced by Spanish administration.

CONCLUSIONS

During the first 50 years of Spanish occupation of Cuzco, control over lands and canals provided continuity to the *panacas* and *ayllus*. Spanish administration protected the *panacas* and *ayllus* as corporate groups to some degree by organizing them into parishes that recognized their traditional rights to lands and waters and their traditional roles in maintaining the irrigation systems. The Spanish established parishes partly to undermine the state and cosmological roles of the *panacas* and *ayllus*.

The policy of protecting and recognizing preexisting rights and customs regarding irrigation helped preserve part of the Inca irrigation system in Cuzco for the benefit of the Spanish colony. In other parts of the Andes as well, the survival of pre-Hispanic irrigation traditions has been beneficial for the conservation of waters and soils (see Guillet 1989). The Andean *ayllu* with its special relationship to its natural environment is often looked on as a backward, detrimental force within the Peruvian nation. In reality, it is one of the forces that has maintained the social and economic integrity of the Peruvian peasantry, helping provide it with the irrigated infrastructure necessary for survival.

NOTES

Acknowledgments. The field and ethnohistorical research on which this article is based were conducted in Peru in 1975-76, 1977, 1985-86 with support provided by two Fulbright Hayes fellowships, funding from the Graduate College of the University of Illinois, and a Wenner-Gren grant. The Joint Committe on Latin America of the Social Science Research Council funded the ethnohistorical research in 1983 at the Lilly Library, Indiana University, Bloomington.

1. The *panaca* was described by some of the later chroniclers (cf. Sarmiento de Gamboa 1947:134, Zuidema 1964) as the *ayllu* that descended from each Inca king, with the exception of his successor, who founded his own *panaca*. Zuidema (1990a:14-33), however, using data from the earliest chroniclers, has argued that at the time of his succession, each Inca king redistributed the ten subdivisions of Cuzco to nobility so that the *panacas* did not develop one by one upon the death of each

Inca but always existed as ten units of Cuzco.

2. Raurau*panaca* is not mentioned in the description of duties to *ceques* (Cobo 1956 [1653]). This may have been an error or that *panaca* may have had a different set of responsibilities (Sherbondy 1987a:150-151).

3. This rock is probably the same one now known as Ecce Homo, which is housed in its own Christian chapel.

4. Another hypothesis is that the lands of the *panacas* were based on the *chapas*, lands distributed to the *panacas* for the purpose of storing tribute brought to Cuzco by provincial lords (Zuidema 1990a, 1990b). Roughly, these *chapas* were strips of land that ran up slope from the Huatanay River on which the *panacas* built storehouses. These *chapas* divisions also had a hydrological basis because they were separated from each other by ravines that contained the streams that flowed into the river.

5. Capac-Hatun *ayllu* may have had lands in Antisuyu before the Spanish conquest since one early source states that the founder of that *ayllu* was Amaru Tupac Inca whose canal and lands were located in what became San Blas parish (Sherbondy 1982a:64-69).

REFERENCES CITED

Albó, Javier

1972 "Dinámica en la estructura intercomunitaria de Jesús de Machaca." *América Indígena* 32:773-816.

Allen, Catherine J.

1988 *The Hold Life Has: Coca and Cultural Identity in an Andean Community*. Washington: Smithsonian Institution Press.

Aparicio Vega, Manuel Jesús

1963 "Documentos sobre el Virrey Toledo." *Revista del Archivo Histórico del Cusco* 11:119-128.

Astete Victoria, José Fernando

1984 *Los sistemas hidráulicos del valle del Cusco (prehispánico)*. Unpublished thesis, Universidad Nacional San Antonio Abad del Cusco.

Barnes, Monica and Daniel J. Slive

1990 "The Cuzco Puma: Ynga City Plan or European Conceit?" Paper presented at the 23rd Chacmool Conference. Calgary: University of Calgary.

Bastien, Joseph

1985 *Mountain of the Condor: Metaphor and Ritual in an Andean Ayllu*. Prospect Heights, IL: Waveland Press.

Betanzos, Juan de

1987 [1551] *Suma y narración de los Incas*. Madrid: Atlas

Cobo, Bernabé

1956 [1653] *Historia del Nuevo Mundo*. Biblioteca de Autores Españoles, 92. Madrid: Atlas.

Cook, Noble David

1981 *Demographic Collapse: Indian Peru, 1520-1620*. Cambridge: Cambridge University Press.

Dover, Robert V.H.

1992 "Introduction." In *Andean Cosmologies Through Time; Persistence and Emergence*. Robert V.H. Dover, Katherine E. Seibold, and John H. McDowell, editors, pp. 1-16. Bloomington: University of Indiana Press.

González Holguín, Diego

1952 [1608] *Vocabulario de la lengua general de todo el Perú llamado lengua qquichua o del Inca*. Porras Barrenechea, editor. Lima: Instituto de Historia, Universidad Nacional Mayor de San Marcos.

Guillet, David

1989 "The Struggle for Autonomy: Irrigation and Power in Highland Peru." In *Human Systems Ecology*, Sheldon Smith and Ed Reeves, editors, pp. 41-57. Boulder: Westview Press.

Hemming, John
1970 *The Conquest of the Incas*. San Diego: Harcourt Brace Jovanovich.
Isbell, Billie Jean
1971 "No servimos más." *Revista del Museo Nacional* 37:285-298(Lima).
1978 *To Defend Ourselves*. Austin: University of Texas Press.
MacCormack, Sabine
1991 *Religion in the Andes*. Princeton: Princeton University Press.
Mannheim, Bruce
1991 *The Language of the Inka Since the European Invasion*. Austin: University of Texas Press.
Mitchell, William P.
1976 "Irrigation and Community in the Central Peruvian Highlands." *American Anthropologist* 78:25-44.
Molina, Cristóbal de ("el Almagrista," now identified as A. de Segovia)
1968 [1533] "Relación de muchas cosas acaecidas en el Perú." Madrid: Biblioteca de Autores Españoles 209: 57-95.
Murra, John Victor
1972 "El 'control vertical' de un máximo de pisos ecológicos en la economía de las sociedades andinas." In *Visita de la Provincia de León de Huánuco en 1562,* by Iñigo Ortiz de Zúñiga, volume 2, pp. 429-476. Huánuco: Universidad Nacional Hermilio Valdizán.
1980 *The Economic Organization of the Inka State*. Greenwich, Connecticut: JAI Press, Inc.
Nader, Helen
1990 *Liberty in Absolutist Spain*. Baltimore: Johns Hopkins University Press.
Ortiz de Zúñiga, Iñigo
1972 [1562] *Visita de la Provincia de León de Huánuco*. Huánuco, Peru: Universidad Hermilio Valdizán.
Polo de Ondegardo, Juan
1917 [1571] "Relación del linaje de los Incas y cómo extendieron ellos sus conquistas." In *Informaciones acerca de la religión y gobierno de los Incas*, pp. 45-94. Colección de Libros y Documentos referentes a la Historia del Perú, Volume 4. Lima: Sanmarti.
Rappaport, Joanne
1990 *The Politics of Memory*. Cambridge: Cambridge University Press.
Rostworowski de Díez Canseco, María
1964 "Nuevos aportes para el estudio de la medición de tierras en el virreynato e incario." *Revista del Archivo Nacional del Perú* 28.
Rowe, John Howland
1979 "An Account of the Shrines of Ancient Cuzco." *Ñawpa Pacha* 17:1-80.
Salomon, Frank L.
1982 "Chronicles of the Impossible: Notes on Three Peruvian Indigenous Historians." In *From Oral To Written Expression*. Rolena Adorno, editor, pp. 9-40. Syracuse: Foreign and Comparative Studies, Latin American Series, no. 4.
Sancho, Pedro
1962 [c.1550] *Relación de la conquista del Perú*. Madrid: Bibliotheca Tenenitla, Libros Españoles e Hispanoamericanos, 2. Ediciones José Porrúa Turanzas.
Sherbondy, Jeanette E.
1979a "Estudio Preliminar." In *Cuzco: Aguas y poder*. Horacio Villanueva and Jeanette Sherbondy, pp. v-xix. Cuzco: Centro de Estudios Rurales Andinos Bartolomé de las Casas.
1979b "Les réseaux d'irrigation dans la geographie politique de Cusco." *Journal de la Societé des Américanists* 66:45-66.
1982a *The canal systems of Hanan Cuzco*. Ph.D. dissertation. University of Illinois, Ann Arbor: University Microfilms International, 8218563.
1982b "Lands and Waters of Hurin Cuzco." Paper presented to the American Society for Ethnohistory, Nashville, Tennessee.
1982c "El regadío, los lagos y los mitos de origen." *Allpanchis* 17:3-32.
1986 "Los ceques: código de canales en el Cusco incaico." *Allpanchis* 27:39-74.

1987a "Organización hidráulica y poder en el Cuzco de los incas." *Revista española de antropología americana* 17:117-153.

1987b "The Incaic Organization of Terraced Irrigation in Cuzco, Peru." In *Pre-Hispanic Agricultural Fields in the Andean Region*, BAR International Series 359(1):365-371.

1992 "Water Ideology in Inca Ethnogenesis." In *Andean Cosmologies Through Time; Persistence and Emergence*. Robert V.H. Dover, Katherine E. Seibold, and John H. McDowell, editors, pp. 46-66. Bloomington: University of Indiana Press.

Sarmiento de Gamboa, Pedro

1947 [1572] *Historia de los Incas*. 3rd. edition. Buenos Aires: Emecé.

Silverblatt, Irene

1987 *Moon, Sun, and Witches: Gender Ideologies and Class in Inca and Colonial Peru*. Princeton, New Jersey: Princeton University Press.

Villanueva U., Horacio and Jeanette E. Sherbondy

1979 "Cuzco: aguas y poder." *Archivos de historia rural andina*, 1. Cuzco: Centro de Estudios Rurales Andinos Bartolomé de las Casas.

Zuidema, R. T.

1964 *The Ceque System of Cuzco: The Social Organization of the Capital of the Inca*. Leiden: E.J. Brill.

1973 "La quadrature du cercle dans l'ancien Pérou." *Signes et Langages des Amériques* 3 (1-2):147-165. Québec.

1977 "The Inca Kinship System: A New Theoretical View." In *Andean Kinship and Marriage*, Ralph Bolton and Enrique Mayer, editors, pp. 240-281. Washington, DC: American Anthropological Association.

1983a "The Lion in the City: Royal Symbols of Transition in Cuzco." *Journal of Latin American Lore* 9(1):39-100.

1983b "Hierarchy and Space in Incaic Social Organization." *Ethnohistory* 30(2):49-75.

1990a *Inca Civilization in Cuzco*. Austin: University of Texas Press.

1990b "Ceques and Chapas: An Andean Pattern of Land Partition in the Modern Valley of Cuzco." In *Circumpacifica, Festschrift für Thomas S. Barthel*. Bruno Illius and Mathias Laubscher, editors, pp. 627-643. Frankfurt: Peter Lang.

Zuidema, R.T. and Deborah A. Poole

1982 "Los límites de los cuatro suyus inaicos en el Cusco." *Bulletin de l'Institut Français des Études Andines* 11(1-2):83-89.

CHAPTER FOUR

Teaching Water: Hydraulic Management and Terracing in Coporaque, the Colca Valley, Peru

John M. Treacy
George Washington University

INTRODUCTION

Every year in August, the people of the village of Coporaque hold a Festival of Water (*Fiesta de Aguas*) to celebrate the imminent return of the rains for the upcoming planting season. The jubilant climax of the Festival is a ceremonial exercise in hydromancy, or water divination. Villagers gather on the dry, recently cleaned bed of the major reservoir, awaiting the release of water held in check at the intake from Coporaque's main water source, the Sawara River. Upon the water's release, people race the incoming flow as it begins to fill the reservoir. If the stream of water outpaces its joyous escort, the year's supply will be good.

The Festival of Water, above all, highlights the pivotal role of irrigation in the lives of Coporaqueños. Water is a life-giving, nourishing substance that all farmers need in the semiarid Colca Valley. Without supplementary irrigation to extend the growing season and to save crops from periodic droughts, farming in the valley would be hazardous and untenable.

The Festival also symbolizes mastery over water because farmers have brought it great distances over rugged terrain to water their fields. Villagers use the Quechua expression *yaku yachachiy* (teaching water) in reference to water and irrigation methods. *Yachachiy* commonly translates as "to teach," but it also means "to discipline" or "to direct" (Lira 1941); thus the expression conveys the thought that water is an uncontrolled liquid that requires direction. An irrigator with a shovel guiding water in a field teaches water to move in desired directions over uneven soil surfaces. In addition, there is another, older translation of *yachachiy* as "to prepare" or "to make fit" (González Holguín 1952). Farmers in Coporaque harken to this meaning when they speak of water that has been taught (*yaku yachachisqa*) so that it flows relatively unaided. In this sense, the prime instruments of instruction are landform modifications ranging from canals and terraces to soil ridges and swales on individual fields.

The notion of teaching water, among other methods of water control, sheds light upon indigenous hydrological management methods and attitudes toward technology and water. The major theme I discuss here is canal irrigation and agricultural landforms. Although scholars have described the close relationships between terracing and irrigation (Donkin 1979; Guillet 1987; Denevan 1980), conceptualizations of terraces have tended to be narrow and dominated by the general image of them as walled fields upon steep slopes (cf. Cook [1916] "Staircase Farms of the Ancients"). A physical geography of canals and terraces in Coporaque, however, demonstrates that there are organic links between water and agricultural landforms and management, and that the idea of agricultural terracing requires a broader contextual analysis.

Terracing is more than building and farming walled fields; it is but one exercise in hydraulic management wherein the landscape is reengineered into a hydraulic template through which water may be directed and controlled. I argue that the notion of agricultural terracing should be closely linked to water management; in addition, the concept of terracing should be amplified to encompass a wide range of field types.

THE COLCA VALLEY AND THE VILLAGE OF COPORAQUE

The Colca Valley is located on the western flanks of the southern Andes approximately 100 kilometers north of the city of Arequipa. The agricultural area of the valley (zones between 3,200 meters and 3,800 meters) corresponds to the semiarid Montane Steppe climate of Holdridge (ONERN 1973) or the dry *puna* climate of Troll (1968). Rainfall averages 433 millimeters yearly, but variability is great: some years have rainfall bonanzas approaching 700 millimeters, while others are very dry with less than 200 millimeters. The rainy season occurs between November and March (providing 75 percent of annual precipitation), while May through August is both the dry and frost season. These elements of uncertainty and seasonality require farmers to push the start of the growing season back into September by using irrigation, thereby allowing them to harvest their crops before the frosts in May (Mitchell 1976).

The village of Coporaque, one of thirteen in the agricultural area of the valley, is located on the north (right) bank of the Colca River twelve kilometers east of the village of Lari (see Guillet 1987). The population of Coporaque was 1,163 in 1981 (IICA-UNSA 1985:18) and the maximum amount of land cultivable today with irrigation is 520 hectares. Primary crops planted in the village are alfalfa, barley, maize, broad beans (*habas*), wheat, quinoa, and potatoes. Farmers irrigate all crops, except in those rare instances when abundant rains permit seedings of crops such as barley or potatoes in small, unirrigated fields.

WATER SOURCES FOR THE COPORAQUE IRRIGATION SYSTEM

Coporaque draws the bulk of its water (up to 80 percent in the rainy season) from streams flowing from the peak Huillcaya (5,275 meters), and a host of other peaks generally referred to collectively as Mismi. The rolling, grassy plains of the sub-alpine *puna* (from 3,900 to near 4,800 meters in the Colca region) and the flanks of mountains are watersheds for rain which adds water to streams in the wet season. Major streams of the puna have been refashioned by dredging or by course diversion to speed water toward fields far below. Upland lakes are additional sources of water. Flow-through (seepage) forms springs which add to streamflow. Smaller springs may dry up seasonally. Some may disappear altogether. However, new springs have been known to appear suddenly, prompting authorities to try to splice them into the irrigation system.

Minor river valleys that dissect shoulder slopes immediately above cultivated areas are conduits for water that are sometimes linked to the irrigation system by crossover canals. Some ravines were channelized in pre-Hispanic times by the construction of interior stone walls. The water is led to the terraced mountain foot slopes and finally to the alluvial plains at the bottom of the valley, where the wholly artificial hydrology of canals gradually replaces the predominantly natural hydrology of streams.

WATER MANAGEMENT SYSTEMS IN COPORAQUE

The components of the irrigation infrastructure of Coporaque are listed in Table 4.1. Judging from the descriptions of canal irrigation in nearby Yanque (Valderrama and Escalante 1988) and Lari (Guillet 1987), Coporaque's technological repertoire of canals, reservoirs, and terraces is probably representative of the types of systems in the Colca Valley as a whole.

Primary Feeder Canals

Primary feeder canals take water from streams, springs, and lakes and transport it to secondary canals in the cultivation area. The Waynaqorea and Aquenta streams form the heart of the primary feeder canal network at the headwaters of the Sawara River (see Map 4.1, boxed numbers 2 and 3). However, through countless cleanings, dredgings, and occasional reroutings to cut through

TABLE 4.1 A TYPOLOGY OF HYDRAULIC AND WATER MANAGEMENT FEATURES IN COPORAQUE

Feeder Canals:	Channeled Streams Crossover Canals (*Wijchus*)* Valley-Side Canals (*Acequias, Yarqas*)
Control Valves (*Wijchus*)	
Reservoirs (*Estanques, Cochas*)	
Distribution Canals:	Valley-Side Canals Pampa Canals (*Acequias, Yarqas*) Drop Canals (*Kalchas*)
Terraces:	Bench Terraces (*Andenes*) Broadfield Terraces (variant of bench terraces) Valley Floor Terraces
Field Hydraulic Features:	Intakes or Offtakes (*Boquerones*) Backwall Canals (*Ocoñas*) Water Drops (*Pajchas*) Scratch Canals (*Killas*) Impounding Features (*Chakas*)

Source: Field observations and interviews, Coporaque, 1985 1986.

*Local names in italics.

meander loops, Coporaque's farmers have modified these streams into hybrid channeled streams, in part natural flows and in part cultural artifacts. The Aquenta stream, flowing east from Huillcaya, is the larger branch measuring three to five meters in width as it turns southward and winds through undulating topography. The Waynaqorea is a stream channeled along major stretches to redirect its course away from the Qallachapi River on the east and toward the Aquenta on the west. The diverted stream, averaging around sixty centimeters in width and thirteen centimeters in depth, gradually becomes an artificial canal which crosses a small ridge and empties into the Aquenta to increase the latter stream's discharge. South of the point of union of the two courses is a new, artificially enlarged river, termed the Sawara (Map 4.1, boxed number 4.) The Sawara proceeds down its course acquiring additional water from the Chachayllo, Yanchawi, and other minor valley streams which enter in a dendritic fashion from the west.

Other streams and feeder canals enter the agricultural area of Coporaque east and west of the Sawara. The wholly artificial canal from the small lake (*bofedal*) at Qachulle feeds fields in the eastern portion of Coporaque (Map 4.1, boxed number 8), while the Chilliwitira River waters fields on the west via the Wallallik'ucho canal (Map 4.1, boxed number 10).

The short Yanchawi and Laqraque canals south and east of the Aquenta are crossover canals (Map 4.1, boxed number 9). These canals circumvent natural dendritic patterns by taking water from one stream and conducting it across an interfluvial ridge to augment another stream. The Yanchawi takes water from the Yanchawi stream at midcourse, directing it into the Chilliwitira to boost that stream's flow, and the Laqraque canal does the same from the Laqraque stream. Another major crossover canal from the Waynaqorea carries water about 6 kilometers across Pampa Finaya to replenish a large seasonal lake used by cattle while at pasture there (Map 4.1, boxed number 6). An abandoned extension of the canal that leads to the Qachulle canal indicates that at one time this canal also boosted the waters of Qachulle (Map 4.1, boxed number 8). Another abandoned set of crossover canals furnished water for the now abandoned terraces at Ch'ilaqota (Map 4.1, boxed number 11).

There are two contrasting strategies for harnessing upland waters. The first is to take advantage of natural streamflows as much as possible and then to merge them, a hydrological method known in Quechua as *iskaymanta ch'ullullaman yaku*, or "merging two sources of water." The expression is based upon *ch'ullay*: making a single thing out of two like things, or melding like things (González Holguín 1952). This strategy of merging streams, as the Coporaqueños have done with the joining of the Aquenta and Waynaqorea streams, requires less engineering effort than constructing multiple canals from multiple streams.

The other strategy is to bypass natural dendritic patterns completely by using crossover canals to force water to flow from one watershed to another. Crossover canals are expedient because they transfer water to channels needing more discharge. Also, despite the apparent topographical independence of the source streams providing water for Coporaque, crossover canals enabled ancient irrigation authorities to consider the hydrological system as a unified one by shunting water from one sector to another.

Control Valves and Offtakes (Wijchus)

During the rainy season, rivers and canals may swell with water that could scour canals and cause overflow damage. To prevent this, Coporaqueños modulate canal flow by means of control valves and branch canals that divert the excess water out of the system. Termed *wijchus* (from *wijchuy*: to throw aside, to cast down, to open and let fall [Lira 1941]), they are built along canals, often as simple stone gates (or valves) above steep ravines into which water is cast away. The crossover canal on Pampa Finaya also functions as a control valve during the height of the rainy season when too much water fills the Sawara system.

Feeder canals invariably divide into small canals leading directly to reservoirs

MAP 4.1 LEGEND AND KEY TO NUMBERS

Legend

Permanent snowpack

Stream course

Canal

Key to Map Numbers

1. Carhuasanta intervalley canal from the Carhuasanta River, around the Mismi massif, to the Qallachapi River.
2. Waynaqorea channelized stream.
3. Aquenta channelized stream from Huillcaya.
4. Confluence of the Aquenta and Waynaqorea streams and beginning of the Sawara River.
5. Pampa Finaya canal (offtake of the Waynaqorea stream).
6. Pampa Finaya canal-Qachulle crossover canal.
7. Qachulle canal from the Qallachapi River.
8. Qachulle canal from the *bofedal* at Qachulle.
9. Yanchawi and Laqraqe canal crossover canal complex.
10. Walallik'ucho valley-side feeder control.
11. Ch'ilaqota crossover canal complex.

MAP 4.1 MAIN FEEDER CANAL NETWORK OF COPORAQUE

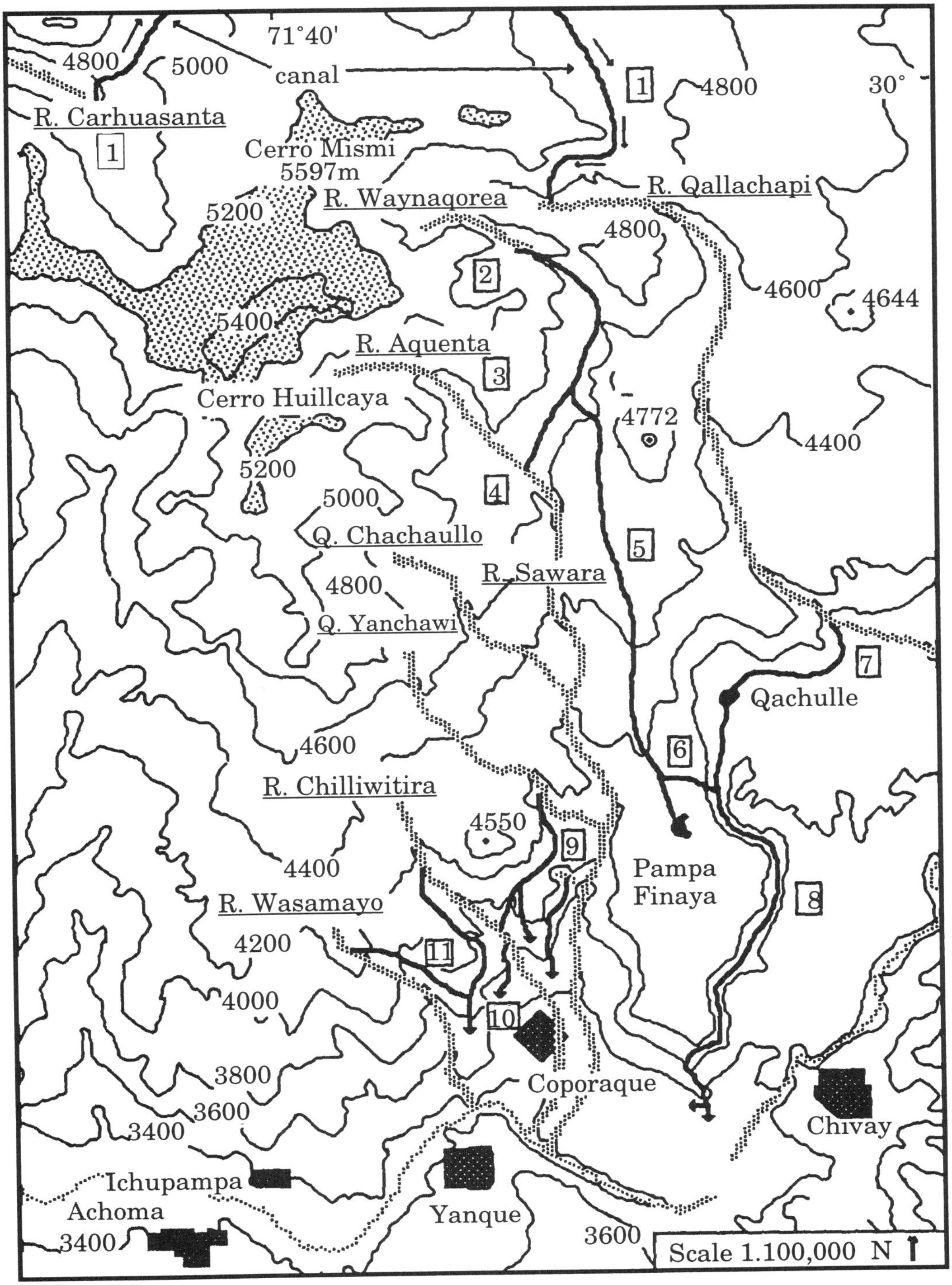

and fields, or to secondary canals within field areas. The offtakes leading water from the main canals into the smaller ones are also called *wijchus*. The local Quechua-Spanish expression for dividing canals is *iskayman partikun yarqa*, or "canal parting into two." It is the reverse of the funneling concept of *ch'ullay* and signifies the spreading of canals towards field areas.

Main feeder canals and their attendant offtakes are associated with irrigation sectors (*sectores*) associated with water from each of the three main sources. Each sector has one or more water reservoirs, some having up to 10,000 cubic meters capacity, to store night time stream flow.

Secondary Distribution Canals

Distribution canals are totally human-made water courses in the agricultural area. There are three principle forms based upon topographical position: valley-side canals, pampa canals, and drop canals. Some canals combine elements of the three types. Controlling water velocity in a canal is largely a function of controlling slope; for valley-side canals, builders maintained gentle slopes over long distances. Older villagers recall that their parents built or rebuilt canals using shovels and trial-and-error methods to control slope (in other words, "teach water"). They dredged a short segment, released water, evaluated its velocity adding corrections if necessary, and proceeded segment by segment to complete the canal. Canals in other parts of Peru are made similarly (Charles Stanish 1987: personal communication), and it is probable that ancient canals were made in the same fashion. The average slope for valley-side canals, measured in the field, is just under three percent, the maximum recommended slope according to hydraulic engineers in Peru today (Juan Arias 1986: personal communication).

Drop canals run vertically down steep slopes, bringing water rapidly from streams or canals. Called *kalchas* (probably from *kalchay*: to cut), they are often lined with stone and have stone berms. Occasional protruding stone steps in drop canal channels create small hydraulic jumps to slow the charging flow of water. Drop canals are also built between terrace rows fifty to sixty meters apart to deliver water from valley-side canals.

Pampa canals follow sinuous paths on uneven terrain and straight paths on level terrain to water fields on the relatively level areas of Coporaque's wide alluvial plains. The majority of the distribution canals in Coporaque are pampa types, some with valley-side or drop segments.

Hydrological studies of the Coporaque canal system show that the mean discharge rate for main distribution canals is 0.068 cubic meters per second, a low rate compared to irrigation systems in the valley of Arequipa. Cross-sectional areas (internal diameters) of canals are extremely uniform in size, averaging between 0.212 and 0.252 square meters (n = 95), as are cross-sectional profiles (internal canal shape). Canals tend to be triangular (deep) instead of parabolic (wide). Triangular channels enhance velocity more than parabolic channels of the same capacity because there is less friction between the water and the inside perimeters of the canal.

Fields and Attendant Hydraulic Features

Most of the irrigated fields in Coporaque have been artificially leveled to permit careful control of irrigation water. Leveled fields with high walls (up to three meters) on steep slopes are bench terraces; their diagnostic characteristics include artificial cobble fill behind the retaining walls, and the cut-and-bench method used to make them (Figure 4.1). The technique involves excavating to a relatively solid surface beneath the topsoil to seat stone retaining walls while backfilling the terraces with soil from above. Bench terraces are often bordered by thick walls (up to two meters) termed *sukwas* that are used for paths connecting the different terraces or as platforms for drop canals.

Broadfield terraces are a form of bench terrace with wide platforms and low walls, commonly built on footslopes or upon moderately sloping terrain (Figure 4.2). Broadfield terraces may or may not have artificial cobble horizons. In Coporaque there are 270 hectares of bench and broadfield terraces, of which 115 appear to have been abandoned since the early Colonial period.

Many fields on the wide alluvial pampas of Coporaque are leveled by gradual slope processes or by continuous cultivation and soil transfer downslope. Called valley floor terraces, these fields are very broad with walls as low as 30 to 40 centimeters (Figure 4.3). Since they are scattered throughout the cultivation area it is difficult to calculate their precise areal extent, but I estimate that they cover 241 hectares.

The remainder of Coporaque's irrigated field types are house plots (*huertas*) and gently sloping or essentially level unterraced fields on the pampas.

Almost every field under cultivation in Coporaque is linked to the canal system by an adjacent intake from a distribution canal (*boquerón*), or more commonly, from secondary or tertiary canals normally able to accommodate the flow of a distribution canal. Each field also has a series of canal or canal-like features that help irrigators control water.

Bench terraces have the most complex irrigation features (Figure 4.4). Farmers almost invariably own terraces in rows one on top of the other so that they may be irrigated as a unit. A farmer begins irrigating by taking water from a drop canal or valley-side canal, directing the flow into the terrace's backwall canal (*ocoña*). Unreconstructed abandoned terraces in Coporaque have backwall canals made of stone, but today most are earthen, formed by a deep, slow pass with the traction plow, usually the finishing touch applied to a field after planting. By quickly opening and closing the canal's earthen wall with a shovel at lateral intervals, the water that spills forth can be patiently coaxed over the soil. The first irrigations following the seasonal fallowing period and before planting are the most difficult because "uninstructed" water flows quickly and with little control over hardened earth. Irrigating requires the skill to teach water to trickle over the field in order to thoroughly soak the bare soil. After planting and weeding, the intricate mounded and ridged microtopography of the soil becomes a template that helps channel water to plants.

Every terrace has minor slope idiosyncrasies demanding special irrigation

FIGURE 4.1 SCHEMATIC VIEW OF BENCH TERRACES AND THEIR CHARACTERISTICS

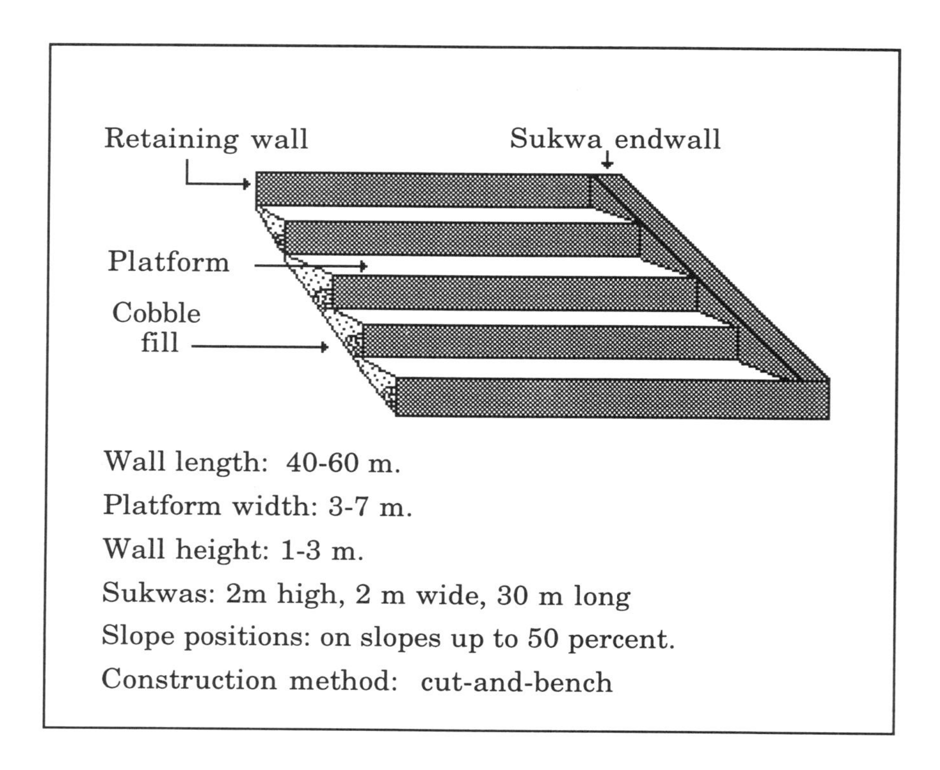

techniques. If a terrace slopes to one side, a farmer may proceed to irrigate slowly lengthwise down the terrace from one end to the next following the slope. If the terrace slopes forward, two or more people may use two drop canals and start at both ends and work towards the middle, although farmers often avoid pooling water at the midsection of a terrace. When the terrace is practically flooded, the farmer closes the terrace's water intake, and proceeds to another terrace.

Some irrigators start at the bottom of a row of terraces and move upslope, bounding up built-in wall stair steps (*takilpus*) from one terrace to the next. The advantage of this method, according to some farmers, is that by working upslope, an irrigator can glance down from above from time to time to see if the terraces are thoroughly irrigated, and make touchups if needed.

Conversely, one may start at the top terrace and work down, passing water from terrace to terrace by gravity drops (*pajchas*) located along the front wall edge. These are flat stones placed in a V-shaped configuration, or rectangular stone chutes. With gravity drops, fields are commonly irrigated one half at a time, alternating between twin drops spaced evenly above a field, or directing the flow from a centrally located

FIGURE 4.2 SCHEMATIC VIEW OF BROADFIELD TERRACES AND THEIR CHARACTERISTICS.

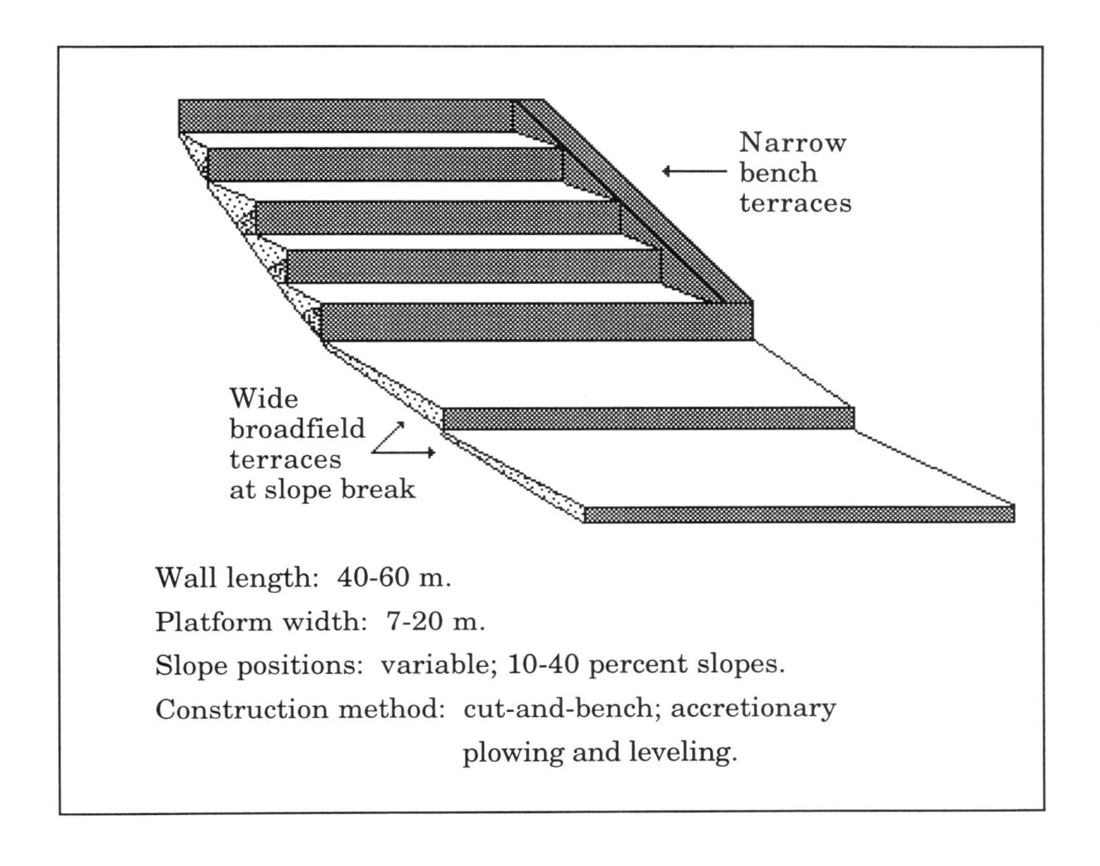

drop. Some farmers do not favor drops, claiming that falling water carries off soil, crushes plants on terraces below, and can in time cause walls to collapse. Some terraces have internal drains (see Figure 4.4) to eliminate internal water buildup.

Pampa fields are irrigated by back wall or back edge canals and by long scratch canals (*killas*) located every 3 to 6 meters upon field surfaces. An irrigator manages the scratch canals in much the same fashion as an earthen backwall canal, piercing their edges now and then to spread water evenly over the field.

HYDRAULIC CONTROL AND AGRICULTURAL LANDSCAPES

Water by nature is an untamed and unpredictable substance that presents irrigators with hydraulic challenges. Canal builders are interested in water velocity

FIGURE 4.3 SCHEMATIC VIEW OF VALLEY FLOOR TERRACES AND THEIR CHARACETERISTICS

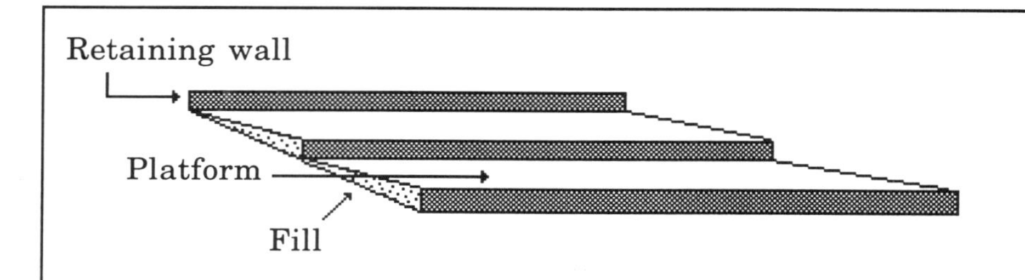

Wall length: up to 60 m.
Platform width: 20-30 m.
Wall height: 0.20-1 m.
Slope positions: on slopes from 5-20 percent.
Construction method: accretionary plowing and leveling, slope processes

---

and how to improve it. High velocity implies abundance, as the water race during the Festival of Water indicates. Rapid canal flow means irrigators may work faster and irrigate more fields in a day, not a trivial matter in Coporaque when people often must wait for hours until others finish irrigating. During the cold, dry season from May to August, when farmers irrigate alfalfa, canals often freeze and cease to deliver water. Since water runs only when the ice melts in the afternoon sun, irrigators want it to speed downslope as rapidly as possible.

On the other hand, too much velocity is a hazard since fast moving water damages canals and washes away soils and manure fertilizer. A steeply sloping field exacerbates high overland flow velocity. To manage this water well irrigators must scurry about with shovels in hand to ridge and furrow soil quickly, so that the advancing water moves in a disciplined fashion to moisten all parts of the field. This form of "teaching water" demands skill and is tedious. The best method of handling water is to reduce velocity to a gradual trickle, and that is done by reducing slope. In other words, it is accomplished by terracing.

Flattening fields to reduce slope and control water is not the sole reason for terracing. Terraces also enhance water infiltration and retention and control or eliminate erosion. But the essence of both water control and terracing is field flattening, and this I suggest is the key to understanding the relentlessly thorough terracing of the Colca Valley.

FIGURE 4.4 BUILT-IN FEATURES OF BENCH TERRACES

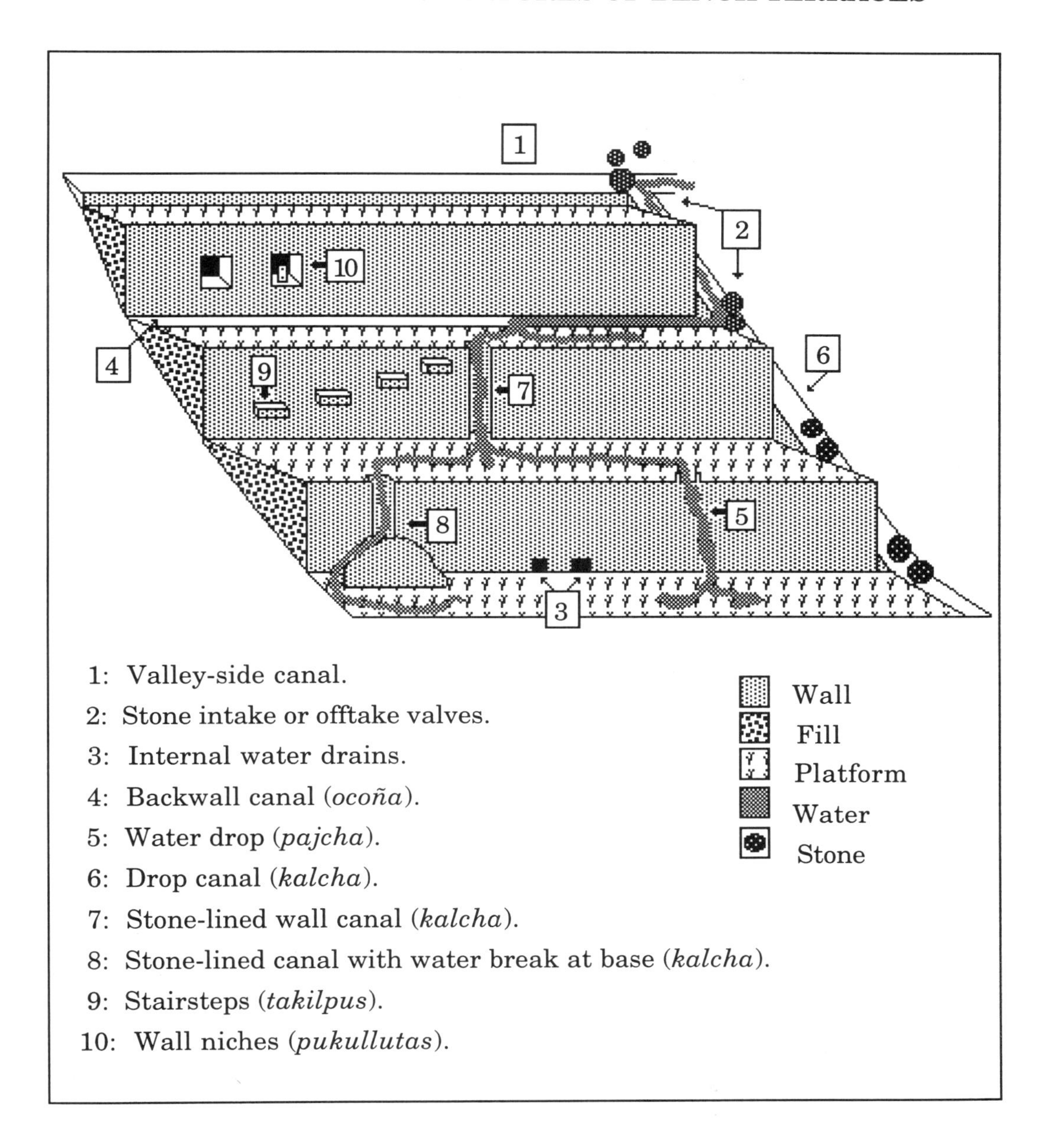

We may refer to the definition of "agricultural terrace" advanced by Spencer and Hale (1961:3): "any artificially flattened surface upon which crops are grown subsequent to the flattening, no matter how small, how crude, or how purposeful." By this definition, almost all of the irrigated fields in the Colca Valley are terraces. This includes valley floor terraces with minimum-sized walls on almost level terrain. These fields have been graded to gentle slopes in order to direct water gradually over their surfaces. Valley side and valley floor terraces are morphologically different in wall height, platform size, and internal structure (valley floor terraces do not routinely have cobble fills), but they function similarly.

The relationship between field flattening and irrigation becomes clearer if the irrigated fields of the Colca Valley are compared to an older series of abandoned and unirrigated fields above the limits of irrigation canals (Treacy 1989:122-127). These fields are almost invariably sloping (up to 50 percent), and have low retaining walls and elongated, walled platforms for capturing runoff. The form corresponds to the sloping dry-field terrace (Spencer and Hale 1961:9) built for runoff farming. Similar sloping field terraces (sometimes called *pata pata* fields) are common in the rainfed lands of the department of Cuzco. Other sloping fields ("slumping" fields or "hanging" fields) are used for potatoes in rainfed regions since slopes encourage drainage (excess water rots tubers).

Even though some valley floor terraces are dismantled so that they may be more easily plowed, many farmers today in the Colca Valley continue to level fields carefully in order to enhance irrigation efficiency, and in some cases they relevel steep bench terraces by building smaller terraces on them.

CONCLUSIONS AND IMPLICATIONS OF HYDRAULIC LANDSCAPE STUDIES

The metaphor of "teaching water" is useful for characterizing the hydraulic landscape of Coporaque because it provides a comprehensive context in which to examine agricultural technologies and water management. Water is untutored and elusive, requiring training to make it behave correctly. The ability to manage water is a necessity for irrigators, and the basic skills for irrigating and handling flows are well known, even though some farmers are better at managing water than others. This generalized ability to manage water helps clarify the issue of whether canal builders in the Andes were the equivalents of hydraulic engineers. The data from Coporaque suggests that canal building was managed by local people who used patient "teaching" methods. Even if they had been directed by someone skilled at teaching flows, he probably would not have been a specialist (see Seligmann and Bunker, *intra*, for a contrasting view).

Rapid water is a hazard which makes thorough irrigation difficult. Farmers respond to this problem by leveling fields. The link between flattening fields and controlling water broadens our definition of agricultural terracing to embrace a wide spectrum of field types. Terraced fields comprise a morphological continuum from simple, low earthen bunds to steeply walled bench terraces. The exclusive focus upon bench terraces has caused us to ignore the flattened fields that are not appreciably walled but are undoubtedly terraced. It is true that many bench terraces in the Colca Valley are unique and are differentiated terminologically (farmers call them *andenes*). This terminological distinction, however, may be in part a function of the crops planted in the different fields. *Andenes* are commonly associated with maize, a prized crop in the valley, and therefore the local definition may be based upon crop suitability as well as upon form. In thinking only of bench terraces, observers interested in terrace restoration miss the point that bench terraces are only one kind of terrace variant.

Finally, there may be social analogs to hydrological control patterns that may illuminate irrigation management policies. The following is an example from Coporaque. On the day when the Aquenta and Waynaqorea channelized streams are scheduled for cleaning during the Water Festival week, the entire village is at work, and the event merits comment since it reveals a key dimension of Coporaque's moiety structure. On or about August 8, the townspeople of Coporaque ascend in unison to the streams' respective headwaters. Splitting into moiety halves, Urinsaya members clean the Aquenta leg while Anansaya members clean the Waynaqorea. They descend, working gradually, and meet to rest and celebrate at the point of union of the two canal branches. Then, as one, the townspeople proceed to "clean" the Sawara River by hoisting boulders from its bed. In other Andean communities, canal cleaning is associated with fertility, renewal, and union. In Coporaque, the event signifies the unity of social halves by emulating the hydraulic principle of *ch'ullay*, in which two sources of water are melded and subsumed into a unitary flow. Canal cleaning is the socially symbolic counterpart signifying the essential unity of people dependent upon water.

A principle which governs the social organization of irrigation in Coporaque is equity, that is, the natural right to water that all village residents enjoy. The uniform size of distribution canals suggests that all fields have access to similar amounts of water, thus guaranteeing equity and a sense of hydraulic equality among farmers. Water equality is in part a reflection of the belief that water is an indispensable lifeblood to which all fields and farmers have right of equal access.

NOTES

Acknowledgments: Funding for field work in Peru was provided by a Fulbright-Hays Doctoral Dissertation Abroad Grant during 1985-1986. Above all, I am grateful to Richard A. Waugh for helping to design and conduct the hydrological fieldwork and for the statistical analyses. David Guillet, William P. Mitchell, and William M. Denevan kindly read drafts of the paper and offered useful suggestions. [Note: the editors of the volume are grateful to William M. Denevan for assuming editorial responsibility for John Treacy's paper.]

REFERENCES CITED

Cook, Orator F.
1916 "Staircase Farms of the Ancients." *National Geographic Magazine* 29:474-534.

Denevan, William M.
1980 "Tipología de configuraciones agrícolas prehispánicas." América Indígena 4:619-52.
Denevan, William M., editor
1986 *The Cultural Ecology, Archaeology, and History of Terracing and Terrace Abandonment in the Colca Valley of Southern Peru*. Technical Report to the National Science Foundation and National Geographic Society, Volume 1. Department of Geography, University of Wisconsin-Madison.
Donkin, Robin A.
1979 *Agricultural Terracing in the Aboriginal New World*. Tucson: University of Arizona Press.
González Holguín, Diego
1952 [1608] *Vocabulario de la lengua general de todo el Perú lamada lengua Qquechua o del Inca*. Lima: Instituto de Historia, Universidad Nacional Mayor de San Marcos.
Guillet, David
1987 "Terracing and Irrigation in the Peruvian highlands." *American Anthropologist* 28:409-30.
IGN
1985 *Carta Nacional 1:100,000*. Lima: Instituto Geográfico Nacional.
IICA-UNSA
1985 *Diagnóstico agro-socio-económico del Distrito de Coporaque, Valle del Colca*. Lima: Instituto Interamericano de Cooperación para la Agricultura, Universidad Nacional San Augustín, PISCA, Arequipa.
Lira, J.
1941 *Diccionario Kkechua-Español*. Tucuman: Universidad de Tucuman.
Mitchell, William P.
1976 "Irrigation and Community in the Central Peruvian Highlands." *American Anthropologist* 78:25-44.
ONERN
1973 *Inventario, Evaluación, y Uso Racional de los Recursos Naturales de la Costa: Cuenca del Río Camaná-Majes*, 2 vols. Lima: Oficina Nacional de Evaluación de Recursos Naturales.
Spencer, J.E. and G.A. Hale
1961 "The Origin, Nature, and Distribution of Agricultural Terracing." *Pacific Viewpoint* 2:1-40.
Treacy, John M.
1987 "An Ecological Model for Estimating Prehistoric Population at Coporaque, Colca Valley, Peru." In *Pre-Hispanic Agricultural Fields in the Andean Region*, William M. Denevan, Kent Mathewson, and Gregory Knapp, editors, pp. 147-62. Oxford: British Archaeological Reports (BAR) International Series 359.
1989 *The Fields of Coporaque: Agricultural Terracing and Water Management in the Colca Valley, Arequipa, Peru*. Ph.D Dissertation. Madison: University of Wisconsin.
Troll, Carl
1968 "The Cordilleras of the Tropical Americas." In *Geo-Ecology of the Mountainous Regions of the Tropical Americas*. Carl Troll, editor, pp. 15-56. Bonn: Fer. Dümmers Verlag.
Valderrama, Ricardo and Carmen Escalante
1988 *Del Tata Malku a la Mama Pacha; riego, sociedad y ritos en los Andes peruanos*. Lima: Centro de Estudios y Promoción del Desarrollo (DESCO).

CHAPTER FIVE

Transforming Colquepata Wetlands: Landscapes of Knowledge and Practice in Andean Agriculture

Karl S. Zimmerer
University of Wisconsin-Madison

INTRODUCTION

Wetland habitats have presented a Janus-like countenance to Andean agriculturalists: the wetlands offer the potential advantages of critical subirrigation and fertile soil yet at the same time require major environmental modifications before they can be used.[1] To overcome the agricultural obstacles of wetland environments, inhabitants of the Andes once cultivated extensive areas of elevated fields in the major highland basins and river valleys between regions that are currently found in northern Bolivia and central Colombia.[2] Most elevated fields were abandoned before or shortly after the Spanish invasion in the sixteenth century, thus escaping the attention of the earliest chroniclers. Although the extent of current wetland production is far less than that of the pre-European period, montane bogs continue to provide many sites for small-scale cultivation in the equatorial Andes. The importance of wetlands in local food production along with other associated uses merit the close examination of present-day agriculture in bog habitats.

This chapter examines the recent modification of bog environments by peasant cultivators in the Colquepata region of southern Peru. It focuses on the relationship between agricultural development on the one hand and the acquisition of knowledge and the social organization of production on the other. The twin themes of knowledge and social organization are key features of farming systems that, despite their importance, have seldom been considered jointly (Turner and Brush 1987). The social construction of so-called practical knowledge–defined as "knowledge that is learned by watching and doing" (Thrift 1985: 373)–shapes agricultural change and land-use activities. The current examination of agricultural practices, sociocultural organization, and physical environment elucidates contrasts between contemporary bog cultivation and prehistoric counterparts. Field attributes (form, function, environment) of bog cultivation are used to facilitate comparisons with other wetland agricultural systems.

Cultivators develop stocks of practical knowledge necessary for wetland agriculture by drawing on communication and work experiences in a variety of sociocultural contexts. However, most "resource management" and "cultural ecological" studies of wetland field systems have not yet examined these sociocultural dimensions of knowledge formation, lacunae found in research in highland Guatemala (Mathewson 1984), Mexico and Central America (Wilken 1987) and Papua New Guinea (Waddell 1972). These studies have relied solely on an adaptation perspective to account for variation in land-use practices. Although this perspective explains many differences in land-use techniques, it would be strengthened substantially through assessment of the sociocultural context that helps structure the acquisition of the necessary practical knowledge.

Social life structures the way in which people experience and communicate the practical knowledge of land use (see also Seligmann and Bunker *intra*). Studies of farming in several central Andean peasant communities have explored the social relations of agricultural production, such as reciprocal labor exchange, but they have not examined how these social relations facilitate or impede the adoption of new cultivation techniques (Brush 1977; Fonseca 1972; Mayer 1979; Orlove 1977). This essay explores these relationships, even if in only a preliminary fashion.

The data on wetland agriculture in this paper also supply useful preliminary comparisons with Andean irrigated farming systems. Wetland production not only shares ecological properties with irrigated farming, but it also has a similar small-scale level of social organization. Mitchell (1976), for example, reports that the social coordination of water control stimulates the uniformity of certain irrigation practices among cultivators (i.e., the timing of irrigation). This chapter considers whether contemporary wetland agriculture likewise shows socially fashioned patterns of variation in cultivation techniques.

Farming techniques in Andean peasant agriculture vary significantly (Brush 1977; Gade 1975; Knapp 1984), and this variation both in time and across space indicates that "traditional land-use techniques . . . [are] not simple, extensive, or unchanging" but instead are complex, locally variable, and changing (Denevan 1980: 217). To explain such differences, the studies of Andean agriculture adopt views similar to those of Allan Johnson (1972), who asserts that variation in practices among individual agriculturalists is because of individual experimentation. Overemphasis on the individual cultivator, however, risks removing agriculturalists from the social relations and cultural milieu in which they carry out their farm production. To show the everyday contexts of experience and communication as well as the biophysical elements important in agriculture, I follow Ian Farrington (1985:5) in using the term *landscape*.

The first section of this chapter ("Landscapes of Agricultural Experience") examines the social context of agriculture in the study area. The section begins with an assessment of the timing and location of recent wetland farming in the southern Peruvian highlands. Ecological and social divisions of agricultural experience among cultivators are discussed. The second section ("Landscapes of Labor and Knowledge") discusses knowledge and practices involved in the transformation of bog environments into fields. It relies on a case study and frequent comparisons to non-wetland cultivation systems in the nearby uplands. In the third section ("The Transformed Landscape"), the ecological characteristics of wetland agriculture are evaluated. The concluding discussion illustrates how variation in these characteristics reflects the continual interpenetration of economic and social practices.

THE STUDY AREA: LANDSCAPES OF AGRICULTURAL EXPERIENCE

During the late 1960s and 1970s agriculturalists inhabiting at least twelve peasant communities in the district of Colquepata in Paucartambo Province (eastern Cuzco) transformed the agricultural landscape through the local innovation and expansion of a wetland cultivation system (Map 5.1). Peasant farmers in the region converted waterlogged bogs at intermediate elevations (3,500-3,900 meters) into agricultural fields. To use the bogs as agricultural sites, Colquepata cultivators annually ditch and mound a field system made up of elevated beds for planting and interlocking drainage canals. The two field elements are arranged in a herringbone pattern to facilitate the removal of excess water from the planting surfaces (Figure 5.1).

Present-day wetland fields in Colquepata differ from earlier wetland cultivation in the Andes in many respects, many of which are field layout patterns. Two major differences here are the size and the patterning of planting surfaces. The platforms of prehistoric wetland fields were from one to more than twenty meters in width and laid out in a variety of patterns, including checkerboard, ladder, linear, irregular embanked, and serpentine (Denevan 1970; Parsons and Denevan 1967; Smith et al. 1968). In contrast, contemporary fields in Colquepata contain planting ridges that are narrow (less than .5 meters) and oriented in a herringbone configuration.

The design of wetland fields in Colquepata District differs also from contemporary analogues in the Andes. The most widespread agricultural landform currently used for high-elevation tuber production on rainfed upland sites in the central and northern Andes (southern Bolivia to central Colombia) is the narrow ridge known in Quechua as a *wachu* (Gade 1975; Knapp 1984; West 1959).[3] Mounded in parallel rows, this upland field type is considered a "ridged field" (Denevan and Turner

MAP 5.1 THE AREA OF WETLAND RIDGED-FIELD AGRICULTURE IN COLQUEPATA DISTRICT

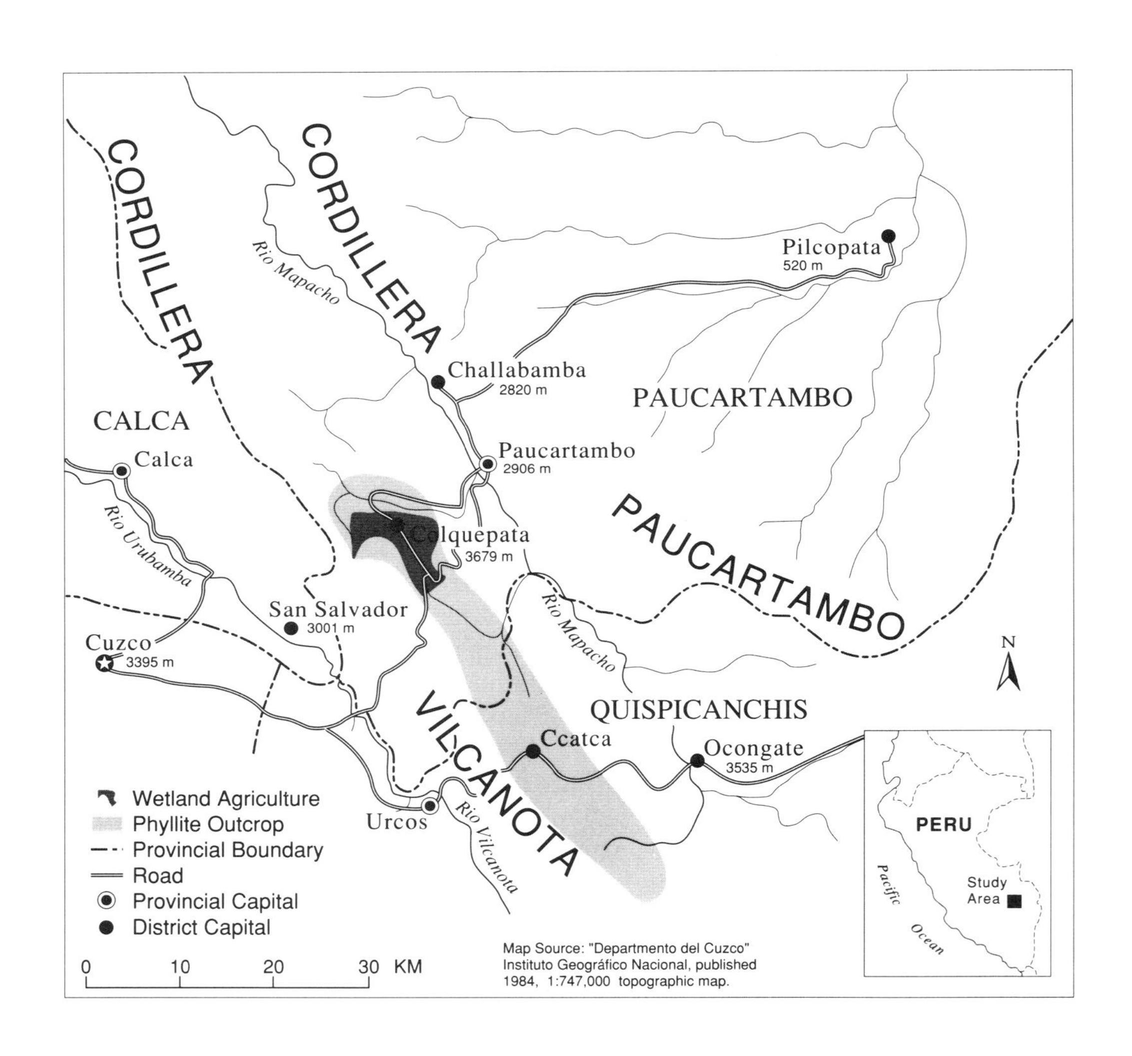

FIGURE 5.1 THE LAYOUT OF DRAINAGE FEATURES IN THE WETLAND RIDGED FIELDS OF COLQUEPATA

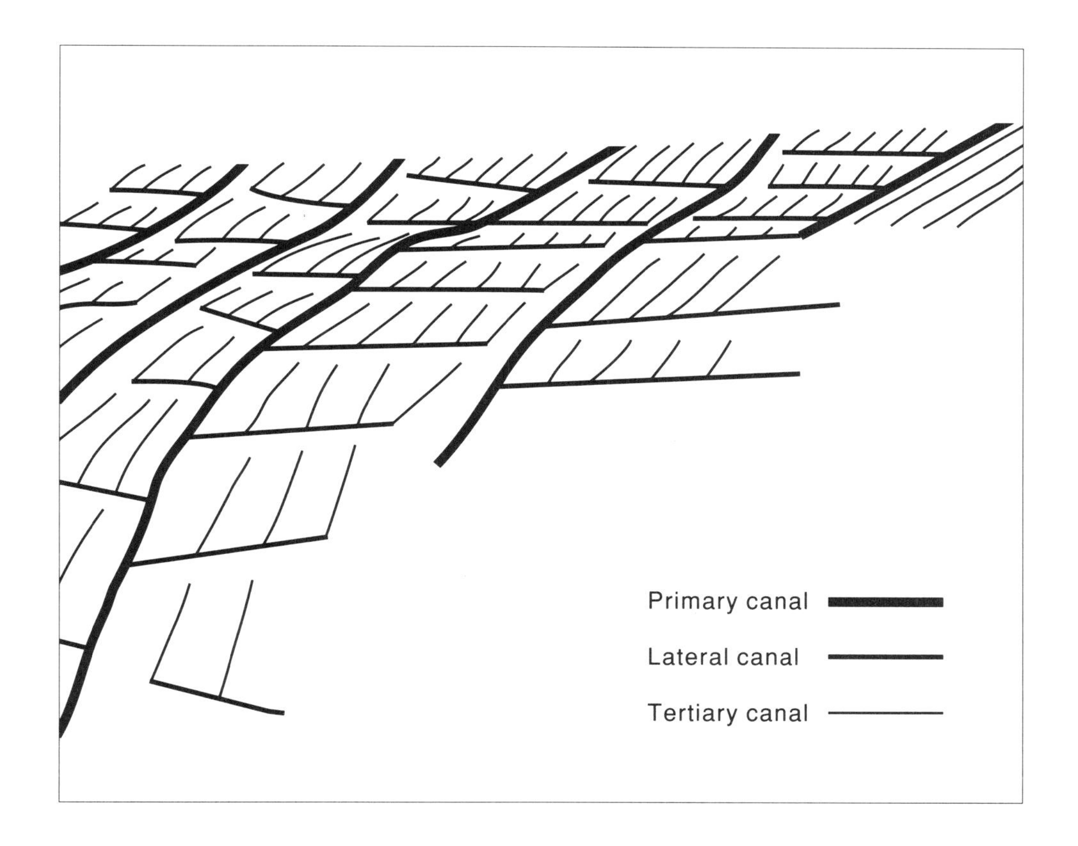

1974:25). Although the wetland fields of Colquepata rely on a similar narrow ridge for planting, they also include interlocking drainage canals and thus differ from the ridged fields found in upland areas. The wetland version of the ridged field, therefore, is subsequently referred to as a "wetland ridged field."

Although the current wetland ridged fields in Colquepata are different from prehistoric and contemporary counterparts, they are not novel, for their design was borrowed from agricultural layouts employed in nearby high-elevation bogs. In the high elevations (3,900-4,150 meters) bordering Colquepata, and similarly across the contiguous upland of Pampallaqta in Calca (Sánchez 1983), the wetland ridged fields dot the Cordillera Vilcanota. Distinctiveness of the present-day wetland ridged fields in Colquepata arises from their location in the previously uncultivated bogs (each referred to as a *wayllar* by residents) that fill topographic swales at intermediate elevations.[4] Today, most intermediate-elevation bog sites in the Colquepata region support agricultural fields, forming an important economic resource for several thousand Quechua-speaking peasants.

Colquepata cultivators use the wetland ridged fields to produce an early planting (*maway*) of potatoes. This planting is one of three potato "crops," each with a distinct growing calendar, that also include a middle planting (*chawpi maway*, the "middle early-planting") and a main planting (*hatun tarpuy*, the "big planting") (Table 5.1). Cultivators sow the early planting during the height of the dry season (July) and are able to harvest it several months before the main crop. This early planting, unlike later ones, requires moisture from sources other than precipitation. Its location in topographic swales, which accumulate and retain abundant soil moisture, assures year-round subirrigation. (In neighboring regions of Paucartambo Province, cultivators rely on cloud-forest environments, irrigation, and floodplain habitats for the early planting of potatoes.)

A combination of price and production incentives motivated the development of wetland agriculture in Colquepata. In the regional market center of Cuzco (1981 population circa 300,000), the price for potatoes harvested in January had regularly been almost double that which prevailed during harvest of the main planting in June. "Pro-peasant" policy measures implemented under the Revolutionary Military Government and its leader General Juan Alvarado Velasco (1969-1976) reinforced these market forces. Agricultural credit, improved potato varieties, and subsidized prices for agrochemical inputs, especially fertilizer, helped ensure the economic viability of wetland cultivation (see Zimmerer 1991 for a discussion of how Colquepata peasants captured a disproportionate share of production subsidies). Early-planting production in swales first appeared in Colquepata during the mid-1960s. Fueled by the timely convergence of production and market incentives, it expanded rapidly during the late 1960s and the 1970s.

Economic incentives and government subsidies triggered the expansion of wetland cultivation, but they do not account completely for its spatial distribution. Social relations between cultivators and village elite, transportation costs, and local environmental conditions further favored the development of wetland ridged-field agriculture in Colquepata rather than in surrounding areas. Social relations conducive to commercial production of the early crop in Colquepata bogs emerged

TABLE 5.1 THE MAJOR POTATO PLANTINGS OF COLQUEPATA (TEMPORAL ORGANIZATION)

	Early Crop Cycle	Middle Crop Cycle	Main Crop Cycle
Local Name	*maway*	*chaupi maway*	*hatun tarpuy*
Approximate Planting Date	July 15	September 1	October 1
Approximate Harvest Date	January 15	March 1	June 1
Primary Habitat	swale bogs	swale bogs	high-elevation grassland
Primary Elevation Range (meters)	3500-3800	3800-3950	3900-4100
Secondary Habitat	irrigation	none	slopes/ inner-canyon
Secondary Elevation Range (meters)	3500-3700		3500-3900/ 3150-3500

through centuries of conflict between Quechua peasants (*runa*) and the nonindigenous local elite (*mistis*). Following the disintegration of forced resettlement (*reducción*) in the village of Colquepata during the late sixteenth century (Allen 1988), the Quechua-speaking peasants and independent communities of the surrounding rural area resisted repeated attempts by local elites to capture their land and labor (Zimmerer 1991). Estate owners only gained a minor and periodically precarious foothold in Colquepata during the colonial and republican periods.

Today, the nonindigenous local elites exercise significantly less social power in Colquepata than in other regions of Paucartambo Province. In the nearby Mapacho Valley, for instance, the former owners of large estates (*haciendas*) that had dominated virtually all land and labor until the Agrarian Reform in 1969 have retained considerable socioeconomic power even after the appropriation of their estates. Through gaining administrative positions in the provincial bureaucracy and controlling the local commercial network centered in the village of Paucartambo, these local elites cornered scarce financial resources and marketing opportunities. In Colquepata, on the other hand, weaker clientage ties fostered a more open market than in the Mapacho Valley (Fonseca and Mayer 1988). Proximity to Cuzco, moreover, has provided Colquepata producers with the additional benefit of transportation costs that are 30 percent to 50 percent lower than those faced by their Mapacho neighbors.

The development of wetland ridged fields depended not only on favorable social and economic conditions but also on the availability of suitable ecological habitats. Colquepata is located in a geomorphologic region where phyllite (a slightly metamorphosed mudstone similar to shale) has weathered to form an undulating badland topography at intermediate elevations (see Map 5.1).[5] A weak rock, phyllite permits rapid surface weathering, and gravity and runoff carry its weathered products into the local depressions or swales described above (see also note 5). The badlands topography controls the size, distribution, and fertility characteristics of swale sites, thus determining their suitability for wetland ridged-field cultivation.

Although swales constitute a microrelief feature, their ecology differs considerably from the adjacent upland surfaces that hold sufficient moisture for agriculture only during the rainy season. Swale sites, on the other hand, collect sufficient water as both runoff from the surrounding upland and as an inflow of groundwater from bedrock layers so that perennial saturation is a characteristic feature. The bog soils of swales (histosols) are further distinguished by high organic matter content (10 percent to 15 percent), moderate acidity (pH 3.5-4.8), and low-medium availability of macro- and micronutrients. Because they are saturated with moisture, uncultivated bogs can be used for agriculture only if significantly modified.

Uncultivated bogs, known locally as "virgin bogs" (*purun wayllares*), support a temperate flora that resembles the "moor community" of nearby high-elevation grasslands (Weberbauer 1943), although the intermediate-elevation location of the Colquepata bogs creates ecological conditions that depart from the higher environments. Many useful plants found in uncultivated swales (Table 5.2), such as duckweed (*unutumera*), water cress (*okururu*), and alga (*llulluchu*), are absent or less common in high-elevation bogs. Colquepata inhabitants, of course, recognize the distinct vegetation of bog habitats. The local name of bogs (*wayllar*) is derived from the habitat's most common plant species (*waylla, Distichia muscoides*), a mat-forming perennial herb. Colquepata inhabitants also distinguish and name plants based on their association with these habitats (for example, *waylla totora*, "*totora* of the *wayllar*").

Colquepata inhabitants attribute sinister properties to the bogs that punctuate their barren, rolling landscape. Like streams and other water sources, bogs are central features in myths of Quechua origin. They also are feared as sources of sickness and evil. Some bog animals, such as the common toad (*hamp'atu*), are thought to be imbued with similar malevolences. In commenting on the first-time plowing of a bog site in the Colquepata community of Sonqo, anthropologist Catherine Allen writes, "Other Sonqo *Runakuna* (Quechua-speaking peasants) looked on somewhat aghast, for the marsh [*wayllar*] is known to be *saqra* (demonic) and liable to swallow people or to make them sick" (1988:39). Even so, Colquepata peasants gathered wild plants and grazed livestock in the uncultivated bogs before their agricultural conversion.[6]

The widespread incorporation of "bog" (*wayllar*) into the names of both local

TABLE 5.2 COMMON UNCULTIVATED PLANT SPECIES IN COLQUEPATA WETLAND HABITATS

	Quechua Plant Names	English Plant Names	Scientific Plant Names
From Uncultivated Bog:			
	unutumera	duckweed	*Limnanthes sp.*
	okururu	water cress	*Lepidium sp.*
	llullucha	alga	*?*
	waylla totora	rush	*Juncus sp.*
From Cultivated Bog:			
	yuyu/nabo	wild mustard	*Brassica campestris*
	qowimirachi	philaree	*Erodium sp.*
	michi-michi	shepherd's purse	*Capsella bursa-pastoris*
	waylla llantin	plantain	*Plantago sp.*

places and a wide array of settlement names represents the long-standing salience of these habitats to Andean inhabitants. In Colquepata, one encounters many sites referred to as *wayllarpata* ("the flat area with a *wayllar*"). Even the names of Colquepata communities refer at least indirectly to the *wayllar*. For example, *choqo* ("infiltration") serves as the root of Chocopía.[7] A cursory examination of Andean toponyms reveals that *wayllar* sites figured prominently not only in Colquepata but indeed throughout the sierra of present-day Peru. *Wayllar*, which the linguist Lira (1944:1,127) defined as "turf" (*cespedal*), was widely adopted in the naming of towns and villages in the Peruvian highlands. Examples include Huaylas (Ancash), Huayllabamba (Cuzco), Huayllate (Apurímac), and Huayllay (Junin), spellings that were rendered by Spanish-speaking administrators.

In transforming intermediate-elevation bogs into agricultural fields, Colquepata cultivators drew on the knowledge gained from collecting useful plants and cultivating

TABLE 5.3 PRODUCTION ZONES AND THE SPATIAL ORGANIZATION OF AGRICULTURE IN COLQUEPATA

Quechua Designation	Main Crop(s)	Plowing Technology	Planting Surface	Crop Cycle
Kheshuar	Maize	Oxen	Furrows	late planting of maize
Yuñglla	Barley	Oxen	Furrows	main planting of potatoes
	Ulluco Potatoes (mixed native and improved cultivars)			
Loma	Potatoes (only native cultivars)	Foot plow	ridged	main planting of potatoes
Maway	Potatoes	Foot plow	wetland ridged	early and middle plantings

similar high-elevation sites. The cultivators also combined this knowledge with expertise gained from agricultural experience in other quite different environments. Before the late 1960s, most Colquepata peasants used three main agricultural production systems (*kheshuar*, *yuñglla*, and *loma* [Table 5.3]), each covering a distinct array of areal habitats known as a "production zone" (Mayer 1985)[8]. Each production zone is characterized by its set of crops, agricultural calendar, technology, and agricultural techniques, and the techniques and technologies of potato production vary widely among the different zones.

Cultivators who developed wetland ridged-field agriculture in Colquepata during the 1970s drew on their experience producing potatoes in other field systems. The use of improved potato varieties and agrochemicals in wetland production particularly benefitted from prior experience with commercial potato agriculture in two zones known as the *kheshuar* and *yuñglla* (Table 5.3). At the same time, differences in wetland ridged-field development that could have emerged from contrasting agricultural "biographies"—that is, variation in experiential knowledge among households—was limited by the minor ecological and spatial division of agricultural production in Colquepata. As in many Andean peasant communities, the ecologically dispersed holdings of households guaranteed that practical knowledge about agriculture was broad based.

Nonetheless, two main social cleavages have accompanied the rise of wetland ridged-field agriculture. First, only some households have access to wetland fields. In the Peasant Community of Chocopía (Map 5.1), located within the core of wetland cultivation, only one-half of all households possess a wetland field. In Sipascancha, on the geographical margin of wetland production, that fraction drops to one-fifth. Cultivation of a wetland field has depended primarily on the de facto ownership of a suitable site, few of which currently remain uncultivated. Even before the development of wetland agriculture, intermediate-elevation bogs were controlled by nearby households, which used them to graze their livestock. Thus the part-time herders who lacked access to these favored pastures—especially the poorest households in communities—later were unable to obtain suitable sites for wetland farming.

A second social division in wetland agriculture has separated members of the household on the basis of gender. Before the expansion of early-planting production and other commercial work activities (principally off-farm migration and the commercialization of barley), tasks in Colquepata agriculture corresponded to clearly demarcated gender roles. Women primarily carried out the tasks involving seeds (storage, selection, planting) and part of the harvest, and men worked in earth moving (plowing, hilling) and completed the heavy work of harvest such as lifting. Although gender no longer divides agricultural tasks so neatly in Colquepata (some women, for instance, now plow with oxen), gender differences still persist in wetland ridged-field agriculture. The case-study description of agricultural activities given below ("The Annual Round") shows how the gender-related division of labor leads to a different constellation of agricultural knowledge among Colquepata men and women.

LANDSCAPES OF PRACTICE AND KNOWLEDGE: TRANSFORMING WETLAND HABITATS

Labor and Technology

Prior experience with potato production and the organization of labor among Colquepata inhabitants helped structure how wetland ridged-field agriculture was formed and disseminated. The experience of households in producing potatoes on upland sites has provided a widely shared basis of expertise that subsequently became incorporated into wetland agriculture. Formation of this shared base of knowledge was shaped by the ways in which labor has been organized in Colquepata agriculture. Although individual households have controlled the production of single fields, they must recruit additional labor for certain tasks.[9] Wetland producers especially require outside labor when social and ecological motives for completing work in as short a period as possible reinforce the high labor demands of a particular task. For reasons discussed below, they prefer that plowing, hilling, and harvesting be completed simultaneously by several workers.

Households recruit outside labor for wetland ridged-field agriculture through both reciprocal labor exchange (*ayni*) and daily wage payments (*jornal*). Among wetland producers in Colquepata, reciprocal labor exchange is particularly common. Because such exchanges are not restricted to identical tasks or production systems, producers of wetland ridged fields are able to swap labor frequently with persons accustomed to working in upland parcels. Through the practice of labor exchange as well as constraints on the owner's capacity to manage labor during agricultural work, the development of the wetland production system has been bound closely to the upland system of ridged-field cultivation. The description of an "Annual Round" in the following section shows how several elements of wetland agriculture closely resemble attributes of upland production, similarities due not to fine-tuned ecological adaptation but rather to limitations imposed by knowledge, experience, and the social organization of labor.

The major features distinguishing wetland ridged-field cultivation from its upland counterpart include the agricultural calendar, the larger requirement for labor, and the considerable reliance on commercial farm technologies. The two production systems are planted at different times of year (Table 5.4). The prevalence of a dry-season production calendar for wetland ridged fields, intended to reap the benefits of high off-season prices, is in sharp contrast to the rain-fed main planting in surrounding uplands. Because the wetland plantings mature more rapidly than the rain-fed upland fields, agricultural work in the former must be concentrated in a shorter period of time. Furthermore, although cultivators strive to modify hydrological conditions in both field systems, they arrange the actual timing of tasks quite differently. Cultivators of wetland ridged fields aim to minimize soil moisture, whereas ridged-field producers in upland sites seek to maximize this same condition. For this reason, plowing of bog sites takes place during the dry season and shortly before planting (about one month). Upland-type ridged fields, on the other hand, are plowed at the end of the rainy season, six months before planting.[10]

The great demand for labor in wetland ridged fields stems primarily from the need to construct and maintain networks of drainage canals. As a result, the wetland parcels require a number of tasks not found in other ridged fields: the elaboration of drainage canals, pulverization, grading, and scooping (Table 5.4). In addition, the completion of certain specific activities in wetland ridged fields demands more labor than the same tasks in upland-type ridged fields. Plowing and weeding in the waterlogged sites especially require extra labor, for the complicated field design slows the rate of tilling and humid conditions favor rampant weed growth. Overall, the cultivation of a wetland ridged field requires about 30 percent to 50 percent more labor than a ridged field of equal area in an upland area.

Technology employed in wetland ridged-field agriculture consists of two contrasting tool kits. The first set differs little from the implements used in ridged fields of the adjacent uplands and includes the foot-driven hoe or "foot plow" (*chakitaklla*), various hand hoes (*allachu*, *raukana*, *k'uti*, *lampa*), a sickle (*ichuna*), and a crusher (*q'asuna*), each of which appears to have been developed before the earliest arrival of Europeans (Gade 1975). The only European-style implement used is the common pick (*pico*), a secondary tool. Most households in Colquepata own a

TABLE 5.4 CULTIVATION TASKS AND CALENDARS OF WETLAND AND UPLAND-TYPE RIDGED FIELDS IN COLQUEPATA

Task	Quechua Term	Wetland-ridged Field	Approx. Date	Ridged Field	Approx. Date
Preliminary Canals	*kunkasqa*	+	April 1	-	
Plowing	*yapusqa*	+	July 1	+	April 15
First Pulverization	*q'asunasqa*	+	July 1	+	April 15
Planting	*tarpusqa*	+	July 20	+	October 1
Second Pulverization	*yuwisqa*	+	July 20	-	
Grading	*challkisqa*	+	August 5	-	
Scooping	*upasqa*	+	August 5	-	
First Hilling	*hallmasqa*	+	September 15	+	December 1
Second Hilling	*harasqa*	-		+	January 15
Weeding	*ichunasqa*	+	(as needed)	+	(as needed)
Harvest	*pallasqa*	+	January 15		June 1

+ = present
- = absent

complete set of these instruments that are manufactured from Andean hardwood trees and the sinew of livestock hides. Handles and shafts are ideally shaped from the native arboreal saxifrage (*chachacoma, Escallonia resinosa*), while suitable though less preferred woods for these tools are *quishuar* (*Buddleia incana*), *q'euña* (*Polylepis racemosa*), and *eucalipto* (*Eucalyptus globulus*). The major nonlocal parts of the wetland-agriculture tool kit are nationally manufactured shares, important in the various hoes necessary for cultivation.

Agrochemicals and improved potato varieties constitute the second technological package employed in wetland fields. Colquepata cultivators rely on this pair of relatively costly inputs to a noticeably greater degree in their wetland parcels than in other field systems. The cultivators procure agrochemicals (pesticides, fungicides, fertilizers) from privately run stores and the outlets of government agencies and nongovernmental development organizations in Colquepata and Cuzco. The most common fertilizer types are mixtures of ammonium nitrate, super-phosphate, and potassium chloride. Organophosphate and organochloride pesticides (*Paratión, Aldrex,* and *Aldrin*), as well as standard potato fungicides (*Antrocol, Poliron Combi*), also are applied regularly. Reliance on both granular and aerosol applications requires that wetland agriculturalists either own or obtain the use of a backpack sprayer. Finally, only the seed of so-called improved potato varieties are planted in wetland fields to maximize yield and resistance to fungal blights and to reduce the maturation period. Exclusive planting of improved varieties in wetland fields contrasts markedly with the upland parcels at high elevations (above 3,900 meters), where typically the potato crop is based on unimproved cultivar types (landraces).

The Annual Round

The following account describes the annual round of wetland cultivation by a household consisting of Juan Santos Quispe (age forty-two), his wife, Fortunata Wallpa Amao (age thirty), their two sons (ages fourteen and eleven) and two daughters (ages ten and five).[11] The family lives in the peasant community of Chocopía, which is one of the main areas of wetland production in Colquepata District. Their neighbors consider them to be "middle peasants" (*campesinos medianos*), an intermediate economic rank within the community. The family owns one pig, eight llamas, and twenty-two sheep, and it possesses eight fields with a total area of approximately two and one-half hectares. Six of these eight parcels are usually cultivated during a single growing season. Since the mid-1970s the family also has tended early-planting potatoes in a 200 square meters wetland field located in the community at 3,650 meters.

Work in the wetland field begins in May and June with the preliminary excavation of several secondary drainage canals (*kunkas*). The secondary canals are positioned laterally with respect to the direction of the slope, and their locations eventually guide farmers in plowing the field (Figure 5.1).[12] Juan and his fourteen year-old son use footplows to dig these secondary canals, which they orient parallel to the others directly above and below. Although relatively shallow (.2 meters), the secondary canals nonetheless drain the field sufficiently to permit plowing. This step is crucial because if left waterlogged, the field cannot be mounded into elevated beds through use of the footplow, which requires moderately drained soils to be effective as a levering device. An operational knowledge of microenvironmental conditions in the field, especially hydrology and topography, is necessary in forming the secondary canals. To assess the ecology of field sites, Juan probes the soil with the footplow to note soil moisture and observes the distribution of indicator plants such as the mat-forming *waylla* plant (*Distichia muscoides*) which grows only in the wettest portions of the Colquepata bogs.

Subsequent work in the wetland field entails excavation of the main canals, known as *orqones*, or "mother canals" (*canales madres*), which capture the runoff that drains from the secondary canals. Mother canals run parallel to one another in the topographic low sections of the field, and they act as its primary drainage conduits (Figure 5.1).[13] To assure adequate drainage, secondary canals consistently strike an acute angle with respect to mother canals (Figure 5.1). Although the wetland field of Juan and Fortunata contains only one mother canal, sites that are more topographically complex require several of them. In fields of exceptionally diverse microrelief, mother canals are separated by only a few meters. Although the spacing of secondary drainage canals and mother canals varies considerably, their geometric relation to one another remains constant and creates a typical herringbone pattern (Figure 5.1).

The herringbone configuration is used by Juan and other Colquepata cultivators to manage drainage in topographically and hydrologically variable sites. Both mother canals and secondary canals are located according to soil humidity, which varies complexly as a function of several environmental factors, including slope, topography, soil type, and catchment area. The herringbone pattern also is well suited to the constraints of local farm tools. Most notably, the beds of fields mounded in the herringbone pattern run partly along the slope so that the footplow can be used efficiently. Moreover, the herringbone pattern of drainage canals evident in wetland parcels is also found in other types of agricultural fields and in various textiles, which suggests that the pattern might constitute a fundamental organizational concept in the spatial imaginations of local inhabitants. The herringbone pattern, therefore, takes form through functions related to ecology, farm technology, and spatial symbols.

After preliminary drainage, several weeks or even months typically pass before plowing. Cultivators plow, or more precisely mound, by constructing "lazy beds" or ridges that are identical in form to the basic elements of upland ridged fields. One or more footplow teams of three persons, each known as a *masa,* complete this task.[14] Juan contracts labor for plowing through reciprocal labor exchange and wage payments. Each team includes a pair of men (the *chakitakllakuna*) wielding footplows and one woman (the *rapachaq*), who forms the planting beds by hand. Like most field owners, Juan and Fortunata work alongside the others. Juan coordinates the work primarily by laying out the field design—in the form of secondary drainage canals and mother canals—before plowing. At the time of mounding, he assigns quotas to the plow teams based on equal allotments of ridges and gives them few other instructions or exortations. The workers use techniques that are customary in upland ridged fields, adjusting these practices slightly to build higher beds in modifying the bog habitat. The design of the field by Juan–rather than measures taken during plowing–assures the incorporation of the distinctive drainage system into the wetland ridged field.

A single plow team frequently tills as much as 500 square meters (including canals) in one day. As the team moves uphill orienting itself to the previously constructed secondary canals, the plowers dislodge and lift the sod, which is then mounded into beds. The areas from which sod is removed become tertiary drainage canals (*wachuwaykus*), which, like ridges, are made to follow the slope (Figure 5.1). Runoff from the tertiary canals drains into the secondary canals and subsequently into the mother canals. Although the plowing of a wetland field resembles that of an upland ridged parcel, the former requires closer attention to detail and forces cultivators to construct shorter segments of beds for drainage. Furthermore, secondary drainage canals and mother canals usually must be deepened following their preliminary excavation. These extra tasks slow the tilling of wetland ridged fields considerably, so that it takes more than twice as long as the plowing of upland sites (67-84 person-days per hectare versus 20-54 person-days per hectare [the latter figure is from Knapp 1984:197]).

Distances between constructional features of the herringbone shape wetland field range widely both among and within sites. The most highly variable parameter is the number and spacing of secondary drainage canals, which are spaced as near to one another as one meter and as far apart as twenty or more. The variable spacing of secondary canals is commonly a response to differences in both soil humidity and the capacity of the farm household to supply labor. Some households are small and unable to recruit the nonhousehold labor necessary for the construction of intensive field designs. Such cultivators sometimes must plow wetland sites without elaborating the secondary and mother canals. The resultant design (referred to as "pass-through ridge" or *pasaq wachu*) in effect does not differ from an upland-type ridged field and thus invariably jeopardizes yield in the wetland parcel because of poor drainage.

The basic drainage features—tertiary, secondary, and mother canals—are combined to form scores of design variants, especially in large wetland fields and those parcels that contain a complex topography. Sets of short secondary drainage canals, for instance, can be contained within larger secondary canals. One visually striking design variant is found in fields that consist of a bog in one portion and an upland environment in the other section. Although these fields are cultivated as one unit—beds and alleys are constructed continuously through both sections—wetland agricultural techniques are applied only where necessary. Disparities in the skills of plowmen along with topographic complexity also add minor variations to the design of wetland ridged fields. In some fields, for instance, certain areas unintentionally remain without ridges, and plowmen then mound short planting beds, known as "infant" beds (*uñas*), to use the full area of the field. "Infant" beds especially characterize uneven terrain where cultivators cannot easily gauge the parallel orientation of ridges.

Planting of potato seed in Juan's wetland ridged field closely resembles sowing in his upland parcels. The planting work is carried out by teams of two to three persons, and it begins as Juan, astride the planting beds, opens small holes with either the footplow or a pick.[15] Fortunata controls planting as she drops two or three small tubers in each hole. She—and sometimes her ten-year-old daughter—then spreads fertilizer on the seeds and covers the hole. The social composition evident in the planting team evidences the centrality of the household, and perhaps especially the woman, in organizing farm production. Like several other field activities, planting involves little if any labor from outside the household. The social makeup of planting teams also indicates attitudes toward work and gender: men till and women plant. These social roles in agriculture are altered by Colquepata households only if forced to do so by a critical shortage of labor, as has recently occurred more frequently because of the increase in commercial farming and the expanded importance of off-farm economic activities.

Substantial reliance on improved, high-yielding seed varieties and agrochemical inputs distinguishes wetland agriculture from the other farm production systems in Colquepata and situates cultivators there more fully in the market economy. Wetland agriculturalists currently rely on the improved *mariva* variety and secondarily on the varieties *yungay* and *mi Perú*, all seed lines developed by national and international crop programs. During planting, cultivators apply chemical fertilizers and pesticides, particularly those appropriate to the humid soils that characterize the field environments of the modified bogs. One precaution taken by Juan and most other wetland producers is to dust a concentrated pesticide over seed tubers, the planting hole, and the nearby surface of the bed. This agrochemical compound targets various larval pests (generically "potato worms") that abound in the humid soils. Because elevated soil humidity leaches granular fertilizers and thus diminishes their effectiveness, many wetland cultivators instead spray a foliar fertilizer.

Potato seedlings are especially vulnerable to excess soil moisture in the rooting zone. To lessen the amount of soil humidity and its potentially damaging effects, wetland cultivators customarily undertake a pair of tasks during the planting period: pulverization, which many complete twice, and the scraping and grading of drainage canals. In the field belonging to Juan and Fortunata, either she or one of the older children pulverize planting beds within a few days after plowing by breaking up clods of hardened soil and sod on the surface with a wooden mallet (*q'asuna*) or hand hoe (*kuti*). The texture of the surface soil is frequently made finer and more even, and thus better for seedling emergence, through pulverizing a second time. Although many cultivators do not pulverize the planting surfaces in upland-type ridged fields, they can ill afford to forgo this task in wetland fields because of the pronounced thickness of the sod and its propensity to harden during drying.

The grading of drainage canals in wetland fields is intended to precede the rainy season when the threat of waterlogging looms large. Grading requires considerably more labor than pulverization and is a more critical step in determining the success of the wetland crop. Like other cultivators of wetland fields, Juan does not rely on labor teams for grading, but instead, he completes it either alone or with the help of his fourteen-year-old son. Although they grade their moderate-size field in one or two days, a large wetland ridged field can demand several days of this work. Juan schedules grading according to humidity conditions in his field—a relatively humid one—where water starts to puddle in certain below-grade canals as soon as one week after planting. If grading is delayed too long during this crucial period, a large portion of the wetland crop can rot (*tapurasqa, selesqa*). The danger of rotting is significant, and households short on labor occasionally lose an entire crop when they are unable to grade drainage canals in a timely fashion.

During the first step of grading (*challkisqa*), Juan maneuvers his footplow or pick at the base of all drainage surfaces (tertiary, secondary, and mother canals) to loosen the soil. His attention focuses on above-grade portions that would otherwise impede drainage and raise the water table into the root zone of potato seedlings. The second step (*upasqa*) is carried out by his fourteen-year-old son, and it entails scooping soil from the drainage surface onto planting beds. Because 1 to 3 centimeters of fresh soil cover the beds after scooping, this step must occur before the seedlings emerge. In addition to ensuring adequate drainage, grading delays seedling emergence and thus reduces the risks of frost and hail. Following emergence, the young seedling becomes susceptible not only to these meteorological threats but also to pest attack. To reduce damage that might be done by the potato flea beetle (*Epitrex sp.*, *piki-piki*), Juan sprays the organochloride *Aldrin* on young seedlings within two weeks of emergence. Depending on field conditions and seedling status, he also applies a foliar fertilizer.

Two major tasks—weeding and hilling—are scheduled between the grading of the drainage canals and the harvest. Although both tasks require considerable labor, wetland cultivators in Colquepata organize the work force quite differently for each activity. Nonhousehold labor is rarely recruited for weeding, despite a typically large labor requirement because of rampant growth in wetland fields. Much weeding is undertaken by women, who work alone or in small groups, pulling weeds from the roots or cutting them using sickles at periodic intervals. Soil humidity and ecological disturbance in wetland fields favor invasion by dozens of weedy species, including many annual grasses and herbs of Old World origin (Table 5.2). (In contrast, only a few native species of the wetland flora have survived the conversion of their habitat to agricultural land.) Wetland cultivators reduce weed cover periodically not only because of outright competition with crop plants but also because weeds enhance the environment for potato pests and diseases. Greatest damage is often incurred by Potato Late Blight (*Phytophtera infestans*), a major fungal pathogen that thrives in humid conditions, such as those accentuated by weed cover. For this reason, wetland cultivators maintain a ready arsenal of fungicides.

Soil is mounded around the base of potato plants only once after planting, a task referred to as hilling, or *halmasqa*. Because soil is mounded initially during grading, wetland production does not necessitate the otherwise standard second hilling, or *harasqa*. Hilling of the planting bed depends on more nonhousehold labor than any other task. Additional workers are obtained primarily through reciprocal labor exchange and only secondarily through wage payment. Juan, like other field owners, customarily provides *coca* leaves three times a day and a midday meal with cigarettes. Although the various workers in the parcel coordinate their efforts, they are not regularly assembled as a group. Nevertheless, Juan exercises a minimum of verbal direction and authority, for the work is allocated according to a system of areal quotas, usually delimited by such field features as rows. Workers use the broad-blade hoe (*lampa*) to scrape soil from the canals and mound it onto the ridges. After completing a designated section of the field, a worker waits for the others before beginning the next section.

The timing of harvest is crucial in the production of wetland fields. Although most upland fields accumulate moderate pest loads by the end of the growing season, the problem plagues wetland sites most severely. Outbreaks of the Andean potato weevil (*Premnotrypes sp.*, Q. *papa quru*, Sp. *gorgoho de los Andes*) are common, and they frequently pressure farmers to harvest their parcels as soon as possible and even earlier than would otherwise be optimal. To obtain sufficient labor for timely harvest, most cultivators recruit workers from outside the household, managing the labor in a manner similar to that used during mounding. As with mounding, before the harvest work is allocated based on an area or number of beds. Environmental differences in fields are accommodated in part by a variety of short-handle hoes. Three major hoe types (*allachu*, *k'uti*, and *raukana*) assist the harvesting under wetter and drier and flatter and steeper conditions. The slightly longer handle of the *allachu* is notable because it furnishes greater leverage in the heavy wetland soils.

During the harvest, soils unloosed from the elevated planting beds fill most canals and create a slightly undulating surface that must be reengineered before the upcoming planting. Another common modification of wetland parcels following harvest involves the enlargement of the planting area. Motivated by the relatively high commercial return of early-potato production in his wetland field, Juan has expanded the size of his field by leveling its surface along the outermost edges, thereby subjecting an increased area to subsurface moisture. For many other cultivators, however, the rock outcrops around sites constrain such field expansion. Juan further alters his wetland field between growing seasons to prepare for subsequent plowing by digging shallow drainage ditches to discourage the invasion of a rhizomatous mat-forming plant, known as *q'emillu*.[16] Invasion by *q'emillu* poses two disadvantages. Its edible rhizome attracts swine that invariably root through such fields and destroy other more valuable forage, and *q'emillu* forms dense mats that impede plowing.[17]

Crop rotation and field fallow play minor roles in wetland agriculture insofar as cultivators typically produce potatoes for two to three consecutive years. Moreover, those who possess swales below 3,600 meters usually plant a second crop after the potato harvest, especially barley that is intended for fodder. Juan demonstrates a slightly different, but not uncommon, strategy. After potato production, he sows his swale site with oats and thus postpones fallow until the fourth or fifth year.[18] Fallow rarely exceeds one year, at which time a potato crop is again planted. The practices of rotation and fallow vary widely among cultivators. Many growers, for instance, have reduced the allotment of secondary crops in the rotation cycle, but few have eliminated both minor crops and fallow. Juan and other cultivators share the opinion that periodic fallow reduces the otherwise damaging—and costly—pest and disease loads of wetland crops.

CONCLUSION: THE TRANSFORMED LANDSCAPE

The present-day agricultural landscape of Colquepata District has been transformed dramatically by the cultivation of wetland ridged fields. Although wetland farming accounts for less than 10 percent of the cultivated area in the region, it has become a cornerstone of commerce. Wetland cultivation has required the modification of the agricultural calendar and labor schedule of farm households such that during the height of the dry season in July and August wetland fields now constitute the major agricultural activity. Overall, their production receives a greater level of capital and labor endowments than the other local production systems. Perhaps most notably, the spatial and ecological concentration of wetland fields in the intermediate-elevation zone between 3,500 and 3,900 meters distinguishes the rural landscape of Colquepata. This contemporary cultivation system offers several revealing features when compared to other production types in the Andes, past and present.

Previous studies of Andean wetland agricultural systems have focused on the landform attributes of prehistoric remains (for example, Denevan, et al. 1987; Turner and Denevan 1985). Their pioneering work on the classification of wetland agriculture can be used to identify taxonomically the contemporary cultivation of bog sites in Colquepata. Mathewson (1985), for instance, has proposed a tripartite nomenclature based on the form, function, and habitat of field landforms. According to Mathewson's proposal, the wetland fields in Colquepata can be identified as ridge/drainage/swale-bog wetland.[19] This identification clearly distinguishes the system in Colquepata from most prehistoric remains of wetland agriculture found in the Andes. The large abandoned fields along Lake Titicaca's north shore, for example, are described as platform/drainage/montane lake basin (Smith et al. 1968).

When compared to the management of small-scale irrigation in many Andean communities, as described for example by Mitchell (1976), the wetland agriculture of Colquepata is distinct, for it involves limited social coordination of physical resources and land use. Even where large swales are covered by several contiguous fields belonging to separate households, the control of runoff rarely poses the need for direct social coordination, for mother canals and secondary canals are oriented to divert excess water into gullies and rivulets that form along edges of the swale. Noncompliance in these practices, however, has caused at least one case where a downslope cultivator claimed recompense for damages. Community authorities supported the claimant based on customary practice in similar cases involving irrigation abuses and animal damage. Nonetheless, few other cases of social conflict or coordination mark wetland ridged fields. Yet, notwithstanding the absence of direct social regulation, there is a clear social basis of practical knowledge that is crucial for understanding the contemporary cultivation system.

Social shaping of the practical knowledge employed in wetland agriculture is evidenced by the degree of variation found in certain field features. The spacing and location of canals within the herringbone pattern, for instance, differ greatly. Various contrasting combinations of tertiary, secondary, and mother canals also distinguish wetland fields, although this variation occurs within limits defined by the similarity of field elements. Wetland cultivators share working knowledge of a basic array of landforms that in effect act as building blocks for the many different field designs, whereas the individual landforms themselves differ remarkably little. The form of lateral canals, for instance, scarcely varies among fields, and even ridges show few differences except height. This lack of variation within individual features occurs in single fields as well as in the broader agricultural landscape. To understand why some features vary and others do not requires consideration of how and why cultivators developed wetland agricultural practices.[20]

Many of the cultivators who had initiated the agricultural use of the intermediate-elevation wetlands in Colquepata during the 1960s and 1970s had already worked in the high-elevation counterparts of such sites. These instances of transferring techniques from one environment to another are probably commonplace. As geographer William Denevan notes, "I suspect that most techniques of environmental manipulation are long present within a culture before wholesale adoption occurs. They are known to and used by individuals or subgroups as secondary techniques or for unique situations" (1983:402). Indeed, the tillage and drainage techniques that inhabitants transferred from the high-elevation wetland fields were found to function well in the lower environments. Yet, in addition, the Colquepata farmers altered their techniques as they became familiar with production in the intermediate-elevation wetlands. The new forms of environmental modification in wetland cultivation thus blended existing techniques and recently adopted alterations.

The transfer of field pattern and cultivation techniques to intermediate-elevation bogs at first drew on the prior experience of Colquepata farmers in high-elevation wetland production, as well as on their nonagricultural experience with bog habitats. The farmers used their familiarity with bog plants to assess hydrological conditions, and they used this information to drain the topographically complex bogs by varying the spacing and location of canals within fields. Farmers also developed expertise in the optimal timing of cultivation tasks, which has depended greatly on their capacity to allocate labor on short notice. Together, ecological variability and differences in the labor-recruiting capacity of households led to significant differences in both the labor calendars of wetland cultivation and the layout of drainage canals. These features have remained distinctively variable aspects of the Colquepata agricultural landscape.

Certain techniques in wetland production, however, are marked by a uniformity that derives from common forms of labor organization. Reliance on many laborers, contracted either through reciprocal labor exchange or wage payment, encourages standardized techniques. The effectiveness of labor in these situations depends on shared practical knowledge, so that the field owner does not direct the task but instead works alongside the contracted labor. This organization of labor precludes all but minor departures from the possible designs inherent in the basic array of field features. Labor-intensive production also promotes the rapid communication of successful innovations, for laborers carefully observe the practices carried out in a field, frequently inquiring about cultivation techniques and production status.

Colquepata cultivators also needed new knowledge to adapt the production strategy for high-elevation wetlands to the bog sites at intermediate elevations. Although the Ministry of Agriculture (Ministerio de Agricultura) assigned an extension worker to Colquepata in the early 1970s, cultivators received little technical assistance. Instead, most extension workers attempted to exploit the peasant cultivators through gaining control of their land, marketable products, and labor. Colquepata cultivators, therefore, acquired knowledge of agrochemical and improved seed use not from government agents, but through experimentation and communication, much of it fostered by observation and work experience in other fields.[21] One demonstration of the widespread sharing of such knowledge was that most wetland producers had adopted the same potato variety (*mariva*) by the late 1970s.

The ecological transformation of bogs to agricultural fields has not eliminated their nonagricultural use. Colquepata inhabitants still pasture livestock there, although planting now shortens the grazing period. Residents also continue to collect a wide array of plant species in the modified wetland, where weedy species of Old World origin especially thrive (Table 5.4). The inhabitants use many of these introduced taxa as potherbs, medicinals, and forage for domesticated guinea pigs. Invasive potherb species include wild mustard (*Brassica campestris*), referred to as both *yuyu* and *nabo*, and curly dock (*acelgas*). The principal guinea pig foods of Old World origin gathered in wetland fields are philaree (*Erodium sp.*), known as *qowimirachi* ("that which is fed to *qowis*"), and sheperd's purse (*michi-michi*, *Capsella bursa-pastoris*).

Recent emergence of wetland agriculture in Colquepata demonstrates how the variation in agricultural knowledge and practices is shaped through a range of work experience, which in turn is guided by the social organization of farm production. Although the agricultural strategies of peasant and indigenous cultivators are frequently innovative, such innovation should not be seen as unrestrained or independent. Instead, the generation and dissemination of knowledge about land use follows channels embedded in the organization of social life and particularly production. A distinctive fashion thus characterizes wetland cultivation in Colquepata, a fashion that is reflected in the transformation of the wetland landscape.

NOTES

Acknowledgments. The fieldwork for this study was completed between March 1986 and August 1987 with the support of the Fulbright Foundation, the National Science Foundation, and the Joint Committee on Latin American Studies of the Social Science Research Council and American Council of Learned Societies, with funds provided by the Andrew W. Mellon Foundation. The initial research (March-June 1986) was undertaken while I was employed as Field Supervisor for the "Changes in Andean Agriculture" project of Drs. Stephen B. Brush and Enrique Mayer. Discussion with Enrique Mayer stimulated many of the ideas in this chapter. I would like to acknowledge the cooperation of project members in the field, especially Leonidas Concha and the preceding Field Supervisor, the late Dr. César Fonseca Martel. César Fonseca, an anthropologist at the Universidad Nacional Mayor de San Marcos, died tragically while leaving Paucartambo after field research. This article is dedicated to his memory.

1. An apposite Quechua metaphor is *iskay uya* (literally "two face"). Meaning "deceptive" or "misleading," it might describe how many Andean cultivators have viewed seemingly attractive wetland environments. "Wetlands," following Turner and Denevan, are "areas transitional between terrestrial and aquatic systems where the water table is usually at or near the surface or land is covered by shallow water" (1985:12).

2. Sites of concentrated raised-field remains include the Lake Titicaca Basin (Kolata 1986; Smith et al. 1968), Peru's Mantaro Valley (Earle 1980; Hastorf and Earle 1985), northern Ecuador (Knapp and Denevan 1985), and the Sabana de Bogotá in Colombia (Broadbent 1968).

3. The elevated, narrow-ridge planting surface in Andean fields is usually described as a "lazy bed" in recognition of its resemblance to the field form employed in the unmechanized potato fields of Ireland (Evans 1973; West 1959).

4. Swales in Colquepata are minor topographic depressions that occur on moderate slopes at intermediate elevations (3,500-3,900 meters). They are small (less than 1 hectare) and irregularly shaped.

5. The phyllite of Colquepata shows cleavages in two directions, both along the thinly spaced bedding planes and at an oblique angle to the former. This evidence, in addition to a disoriented arrangement of hills and valleys (resembling the displacement of detached "rock ships"), suggests that the intermediate-elevation phyllite layer might have been displaced and metamorphosed as an overthrust belt in the distant geologic past. The phyllite belt of the Cordillera Oriental in southern Peru is found in an area of roughly 100 kilometers by 20 kilometers (see Map 5.1).

6. Cultivators continue to collect diverse native species from the remaining undisturbed wetlands for use as potherbs, medicinals, and fodder, especially for the domesticated guinea pig (*qowi*). Wetlands also supply much-needed forage for livestock during the dry season. The environment's namesake, *waylla*, is collected for use as fuel.

7. *Chocopía* probably represents a Spanish alteration of the Quechua *choqopi*, a "place of infiltration."

8. Mayer (1985) has defined production zones both as areal units of land use and as production systems represented by characteristic combinations of technical and social factors.

9. The need for outside labor varies according to several factors, including the size of the household and its field and the nature of the agricultural task. The capacity to recruit outside labor depends on the household's socioeconomic power.

10. Early tillage intended to maximize soil moisture is a dry-farming strategy similar to "dust tillage" in the Great Plains of the United States.

11. These are fictitious names of real people.

12. Colquepata inhabitants, like other speakers of Cusco Quechua, commonly use the term *kunka* to signify "throat" (Cusihuaman 1976:68; Lira 1944:331). The use of *kunka* with reference to the secondary drainage canals of wetland fields contrasts with the similar, yet distinct, meaning indicated by Beyersdorf: "a system of rows which cuts a field obliquely in half" (1984:48, translation mine). In the same compendium of Quechua agricultural terms, Beyersdorf notes that *kunkuna*, a word similar to *kunka*, refers to an indicator species of *wayllas* (1984:69). It is identified as *Distichia muscoides,* the species referred to as *waylla* in Colquepata. Thus, both *kunka* and the closely related *kunkuna* describe attributes of wetlands and their cultivation.

13. The use of the term *orqon* by Colquepata cultivators probably derives from the verb form *orqoy*, which means "to take." *Orqon*, the noun form, signifies "that which takes [water]."

14. In Colquepata agriculture the term *masa* applies to a three-person footplow team made up of any three people. The term can also be used to mean wife-taking male, a kin relation generally implying considerable work obligations in Quechua society (Mitchell, personal communication, November 13, 1989).

15. Colquepata inhabitants refer to this task as *toqosqa*.

16. Colquepata inhabitants do not currently modify wetlands to increase the area of dry-season pasture. The use of such systems in the high-altitude wetlands of the Puno province of Chucuito has been described by Palacios Rios (1977).

17. Inhabitants of other Cuzco regions refer to this plant as *qatay waqachi*, a term meaning "that which makes the son-in-law cry." This alludes to the customary practice of assigning footplow duties

to young adult males.

18. Oats, commonly planted in bog environments of northern Europe and the British Isles, tolerate the humid conditions of swale sites better than other cereal crops.

19. The case of Colquepata, where the field ridges ("lazy beds") and canals not only drain wetland fields of moisture but also increase frost avoidance, points out that such nomenclatural systems should allow a notation for secondary functions.

20. The study of specific differences in the variation of agricultural techniques, as well as their historical development and relation to social forms of labor, challenges broad generalizations associated with both modernization (see, e.g., Schultz 1964) and indigenist (see, e.g., Johnson 1972) perspectives on economic practices in nonindustrial societies. The modernization perspective emphasizes communal uniformity and "backwardness," whereas the indigenist perspective stresses individual decision making removed from social contexts.

21. Although Colquepata cultivators have become knowledgeable users of agrochemicals for agronomic purposes, they remain largely unaware of the safety precautions that should accompany their use. Agrochemical misuse represents a growing health hazard not only to cultivators but also to other community members, especially because of the contamination of water supplies.

REFERENCES CITED

Alberti, Giorgio, and Enrique Mayer
1974 *Reciprocidad e intercambio en los Andes peruanos*. Lima: Instituto de Estudios Peruanos.

Allen, Catherine J.
1988 *The Hold Life Has: Coca and Cultural Identity in an Andean Community*. Washington, DC: Smithsonian Institution Press.

Beyersdorf, Margot
1984 *Léxico agropecuario quechua*. Cuzco: Centro Bartolomé de Las Casas.

Broadbent, Sylvia
1968 "A Prehistoric Field System in Chibcha Territory, Colombia. *Ñawpa Pacha* 6:135-147.

Brush, Stephen B.
1977 *Mountain, Field, and Family: The Economy and Human Ecology of an Andean Valley*. Philadelphia: University of Pennsylvania Press.

Cusihuamán G., Antonio
1976 *Diccionario Quechua: Cuzco-Collao*. Lima: Ministerio de Educación and Instituto de Estudios Peruanos.

Denevan, William M.
1970 "Aboriginal Drained-field Cultivation in the Americas." *Science* 169:647-654.
1980 "Latin America." In *World Systems of Traditional Resource Management*, Gary A. Klee, editor, pp. 217-244. New York: John Wiley.
1983 "Adaptation, Variation, and Cultural Geography." *The Professional Geographer* 35:399-407.

Denevan, William M., Kent Mathewson, and Gregory Knapp
1987 *Pre-Hispanic Agricultural Fields in the Andean Region*. Oxford: British Archaeological Reports International Series 359 (i) and (ii).

Denevan, William M., and B. L. Turner II
1974 "Forms, Functions, and Associations of Raised Fields in the Old World Tropics." *Journal of Tropical Geography* 39:24-33.

Earle, Timothy K.
1980 "The Upper Mantaro Archaeological Research Project: An introduction." *Journal of New World Archaeology* 4 (1):1-49.

Evans, Esther E.
1973 *The Personality of Ireland*. Cambridge: Cambridge University Press.

Farrington, Ian S.
1985 "The Wet, the Dry, and the Steep: Archaeological Imperatives and the Study of Agricultural Intensification." In *Prehistoric Intensive Agriculture in the Tropics*, Ian S. Farrington, editor, pp. 1-9. Oxford: British Archaeological Reports, International Series 232.

Fonseca Martel, Cesár
1972 "La economía 'vertical' y la economía de mercado en las comunidades alteñas del Perú." In *Visita de la provincia de León de Huánuco*, Tomo 2, John V. Murra, editor, pp. 315-337. Huánuco: Universidad Nacional Hermilio Valdizan.

Fonseca Martel, Cesár, and Enrique Mayer
1988 "De hacienda a comunidad: El impacto de la reforma agraria en la provincia de Paucartambo, Cuzco." In *Sociedad andina, pasado y presente: Contribuciones en homenaje a la memoria de Cesár Fonseca Martel*, Ramiro Matos Mendieta, editor, pp. 59-100. Lima: FOMCIENCIAS.

Gade, Daniel W.
1975 *Plants, Man, and the Land in the Vilcanota Valley of Peru*. The Hague: W. Junk.

Hastorf, Christine A., and Timothy K. Earle
1985 "Intensive Agriculture and Geography of Political Change in the Upper Mantaro Region of Central Peru." In *Prehistoric Intensive Agriculture in the Tropics*, Ian S. Farrington, editor, pp. 569-595. Oxford: British Archaeological Reports International Series 232.

Johnson, Allen W.
1972 "Individuality and Experimentation in Traditional Agriculture." *Human Ecology* 1 (2):149-159.

Knapp, Gregory
1984 "Soil, Slope, and Water in the Equatorial Andes: A Study of Prehistoric Agricultural Adaptation." Ph.D dissertation, University of Wisconsin, Madison.

Knapp, Gregory, and William M. Denevan
1985 "The Use of Wetlands in the Prehistoric Economy of the Northern Ecuadorian Highlands." In *Prehistoric Intensive Agriculture in the Tropics*, Ian S. Farrington, editor, pp. 185-207. Oxford: British Archaeological Reports International Series 232.

Kolata, Alan L.
1986 "The Agricultural Foundations of the Tiwanaku State: A View from the Heartland." *American Antiquity* 51(4):748-762.

Lira, Jorge A.
1944 *Diccionario Kkechuwa-Español*. Tucuman, Argentina: Universidad Nacional de Tucuman, Departmento de Investigaciones Regionales, Instituto de Historia, Linguistica, y Folklore XII.

Mathewson, Kent
1984 *Irrigation Horticulture in Highland Guatemala: The Tablón System of Panajachel*. Boulder, CO: Westview Press.
1985 "Taxonomy of Raised and Drained Fields: A Morphogenetic Approach." In *Prehistoric Intensive Agriculture in the Tropics*, Ian S. Farrington, editor, pp. 835-845. Oxford: British Archaeological Reports International Series 232.

Mayer, Enrique
1979 *Land Use in the Andes: Ecology and Agriculture in the Mantaro Valley of Peru, With Special Reference to Potatoes*. Lima: International Potato Center, Social Science Unit Publication.
1985 "Production Zones." In *Andean Ecology and Civilization*, Shozo Masuda, Izumi Shimada, and Craig Morris, editors, pp. 45-84. Tokyo: University of Tokyo Press.

Mitchell, William P.
1976 "Irrigation and Community in the Central Peruvian Highlands." *American Anthropologist* 78: 25-44.

Orlove, Benjamin S.
1977 *Alpacas, Sheep, and Men: The Wool Export Economy and Regional Society in Southern Peru*. New York: Academic Press.

Palacios Rios, Félix
1977 "Pastizales de regadío para alpacas." In *Pastores de Puna: Uywamichiq punarunakuna*, Jorge Flores Ochoa, editor, pp. 155-170. Lima: Instituto de Estudios Peruanos.

Parsons, James J., and William M. Denevan
1967 "Pre-Colombian Ridged Fields." *Scientific American* 217: 92-100.

Sánchez Farfan, J.
1983 *Evolución y Technología de la Agricultura andina*. Cuzco: Proyecto Investigación de los Sistemas Agrícolas Andinos (IICA/CIID).

Schultz, Theodore W.
1964 *Transforming Traditional Agriculture*. New Haven: Yale University Press.

Smith, C. T., William M. Denevan, and P. Hamilton
1968 "Ancient Ridged Fields in the Region of Lake Titicaca." *The Geographical Journal* 134: 353-367.

Thrift, Nigel
1985 "Germs and Flies: A Geography of Knowledge." In *Social Relations and Spatial Structures*, D. Gregory and J. Urry, editors. New York: St. Martin's Press.

Turner, B. L. II, and Stephen B. Brush
1987 *Comparative Farming Systems*. New York: Guilford Press.

Turner, B. L. II, and William M. Denevan
1985 "Prehistoric Manipulation of Wetlands in the Americas: A Raised Field Perspective." In *Prehistoric Intensive Agriculture in the Tropics*, Ian S. Farrington, editor, pp. 11-30. Oxford: British Archaeological Reports, International Series 232.

Turner, B. L. II, and Peter D. Harrison
1983 *Pulltrouser Swamp: Ancient Maya Habitat, Agriculture, and Settlement in Northern Belize*. Austin: University of Texas Press.

Waddell, Eric
1972 *The Mound Builders: Agricultural Practices, Environment, and Society in the Central Highlands of New Guinea*. Seattle: University of Washington Press.

Weberbauer, Augusto
1943 "Phytogeography of the Peruvian Andes." In *Flora of Peru*, vol. 13, John MacBride, editor. Chicago, IL: Field Museum of Natural History.

West, Robert
1959 "Ridge or 'Era' Agriculture in the Colombia Andes." *Actas del XXXIII Congreso Internacional de Americanistas* 1:279-282.

Wilken, Gene C.
1987 *Good Farmers: Traditional Agricultural Resource Management in Mexico and Central America*. Berkeley: University of California Press.

Zimmerer, Karl S.
1988 *Seeds of Peasant Subsistence: Agrarian Structure, Crop Ecology, and Quechua Agriculture in Reference to the Loss of Biological Diversity in the Southern Peruvian Andes*. Ph.d dissertation, University of California, Berkeley.
1991 "Wetland Production and Smallholder Persistence: Agricultural Change in a Highland Peruvian Region." *Annals of the Association of American Geographers* 81:443-463.

CHAPTER SIX

Levels of Autonomy in the Organization of Irrigation in the Highlands of Peru

Inge Bolin
Malaspina University College

INTRODUCTION

In Peru, as in many other parts of the world, irrigation is fundamental for ensuring and increasing agricultural production.[1] In recent years, however, scholars and development agencies have recognized that successful irrigation management is not based on technology alone. To a large degree, it depends on the organizational skills and cooperative strategies of the people using the technology.

The most discussed aspect of irrigation today is management (Fairchild and Nobe 1986; Steinberg et al. 1983:66). Who manages? The state, the village, or an irrigation group? Or is there, perhaps, "an interplay of state and locality initiatives and actions" as Walter Coward (1986:491) suggests.

The impact of national and international decision making on locally organized irrigation is much disputed. Investigators have stressed the beneficial and disastrous effects of such interventions. Closely identifying aid with paternalism, Goodell (1985) believes that this type of interference undermines local autonomy. Austin argues that "people remain in a situation of domination not through lack of initiative to form corporate groups but through lack of resources to maintain those groups as viable organisations" (1985:258). Coward recognizes the good and bad sides of state intervention, implying that some outside bureaucratic arrangements facilitate local mobilization, whereas others retard it (1980:16).

Scholars generally agree, however, that development projects imposed on local populations negatively affect local initiative and cooperation, thereby seriously harming or even destroying group autonomy (Coward 1977; Gibson 1985; Lees 1974; Mazrui 1975). If, however, indigenous irrigators request a development project, and if, furthermore, they manage the project or participate actively in decision making, the chance for successful project implementation is high. Group autonomy may be weakened temporarily, but in the long run it will usually be strengthened. This view is supported by Kleinig, who believes that government measures taken in accordance with a group's initiative are "an expression of their autonomy and not a threat to it" (as cited in Goodell 1985:248).

Whether considered advantageous or not, we must recognize that state involvement in irrigation development is the rule and no longer the exception. Efforts must therefore be directed toward finding ways in which decision making best serves local irrigators. Unfortunately, we know little about the ways in which decisions regarding irrigation management are made. Fox deplores this minimal attention given to decision-making strategies because "such an evaluation would appear essential if efforts to strengthen institutional design are to be attacked on a systematic basis" (1978:24).

This research examines the effects of local autonomy and extralocal decision making on the management of irrigation in three regions of the Vilcanota Valley of Peru. In this study, autonomy is defined as a group's ability to make decisions without the involvement of other local and extralocal institutions (see Clark 1974 and Hoggart 1981:3 for similar definitions). Although before my fieldwork a variety of models relating to corporate group organization (Appell 1976; Lewis 1991; Olson 1965; Weber 1947) were considered, the precise theoretical focus evolved in the field. The great emphasis the water users placed on autonomy was quite unexpected. "We are autonomous" ("*somos autónomos*") was an expression that was regularly reiterated during interviews with irrigation authorities. It was not only the frequency with which this statement was made that was striking, but also the sense of pride and self-confidence that accompanied it. To the local population, group autonomy was a most important aspect of irrigation management.

THE QUESTION OF GROUP AUTONOMY

Although significant research has been done on group autonomy, investigators do not agree on the definition, locus, and limits of the concept. These theoretical problems have often led social scientists to reach contradictory conclusions about the same data. Geertz and Geertz, for example, maintain that the Balinese *subak* (irrigation association) is autonomous (1975:19-20), whereas Grader, in his study of the same locale, argues that it is under centralized control (1960:270, 287, 288).

The question of whether local autonomy exists in pure form with no extralocal influence has been raised in a variety of contexts. Leeds (1973) argues that autonomy does not express itself exclusively at the local level, since local and extralocal decision makers enter into a variety of relations ranging from cooperative to oppositional. March suggests that "social groups arrange themselves roughly on a continuum marked by complete autonomy at one extreme and complete integration into a larger aggregate at the other" (1980:3). Whether autonomy is more strongly expressed by having total control over a few important functions or weak control over a large number of functions is a point that further complicates the issue (Hoggart 1981:23).

Tiffany has found that autonomy is the most important characteristic of a group, since "many social modifications are initiated with changes in this property" (1979:72). Smith similarly argues that in order for groups to obtain full corporate status they must, among other things, exhibit autonomy within a given sphere (1974:94). Brown suggests that whenever a social unit exercises autonomous control, "organization and relatively stabilized procedures are required" (1976:21). He agrees with Smith that group autonomy need not be absolute but "must be sufficient to regulate at least some affairs of the unit" (Brown 1976:21).

This research considers group autonomy of singular importance in understanding irrigation organization. The degree of autonomy, its spheres of expression, its articulation with higher levels of decision making, and its tendencies to shift in the course of a development project to a considerable degree determine the extent to which outside intervention contributes to successful irrigation management.[2]

METHODS

Fieldwork was undertaken from August 1984 to May 1985. A survey of several communities in the Vilcanota Valley, department of Cuzco, was followed by the selection of the Rio Tigre canal system in the district of Cusipata, province of Quispicanchis, and the Rio Chicon and Rio Yanahuara canal systems in the district of Urubamba, province of Urubamba (see Maps 6.1, 6.2 and 6.3, where respective canal systems are discussed).

The communities along these three canal systems are ideal for this study because they share (a) similar geomorphological, topographical, ecological, climatic, and demographic conditions; (b) similar land tenure systems and a common history; and (c) similar relationships between upstream and downstream irrigators. Comparable social and environmental conditions allow for a more precise focus on group autonomy and its impact on the management of irrigation (see Eggan 1954; Hoggart 1981).

The collection of data included participant observation and interviews. Meetings of local and regional irrigation groups and of communal and municipal councils were attended as well as corvée work groups (*faena* in Spanish) and reciprocal work arrangements (*ayni* in Quechua). Informants included the executive members of all Irrigators Committees (*comités de regantes*) and the Irrigators Commission (*comisión de regantes*), directors of the Ministry of Agriculture, and development personnel.

I also interviewed other communal authorities (mayors, community presidents, village elders), as well as market vendors, storekeepers, and subsistence farmers. The people interviewed locally all live along the canal systems and irrigate their own parcels of land. The irrigation authorities are peasant farmers who, on the whole, do not own more land or animals than the average villagers.

In the evaluation of the data, the degree of autonomy is considered in relative terms; that is, the groups are compared to one another rather than to some "ideal" construct. The autonomy of each local irrigation group is defined in relation to other irrigation groups, the community, the district, and national levels of authority. In agreement with Hoggart (1981:32), quantitative data are not used, since neither the precise dimensions nor the determinants of variation in local autonomy are known.

THE ORGANIZATION OF IRRIGATION

Contemporary irrigation organization in the Vilcanota Valley is far from homogeneous. Although the "Regulations for Water Users" (Reglamento de Organización de Usuarios de Agua) were issued by the Ministry of Agriculture and approved through Supreme Decree (Decreto Supremo No. 005-79-AA [Ministerio de Agricultura 1980:101]) in June of 1979, the degree to which these laws have been accepted varies between regions and even between neighboring villages (Bolin 1987). We find a range of responses to these regulations, from acceptance to rejection, a situation well reflected along the canal systems in the three regions under study.

The Rio Tigre Canal System

The Rio Tigre canal system is located 90 kilometers south of the city of Cuzco (see Map 6.1). The Rio Tigre, which originates in the glacier of Chilec at an altitude of 5,000 meters, provides most of the irrigation water for this region.

In 1981 the municipal authorities of the district capital of Cusipata expressed their need for more irrigation water to the international development agency Plan Meris II, which was working in the Vilcanota Valley. The developers agreed to construct new irrigation canals and improve the old ones for Cusipata (3,314 meters) and its administratively dependent communities of Paucarpata (3,412 meters) and Tintinco (3,491 meters). The community of Colcca (3,300 meters), situated 6 kilometers north of Cusipata, requested help for its irrigation after the construction of the canals to Cusipata had already been started. Colcca belongs administratively to the district capital Quiquijana, 4 kilometers to the north. Both Cusipata and Colcca, which are located downstream at the lower part of the irrigation system, were in urgent need of water and welcomed the help. The inhabitants of the higher lying communities Paucarpata and Tintinco, many of whom speak only Quechua, were tied only superficially into the negotiations regarding the construction of the canals. They were not in favor of canal construction since they did not suffer from water shortage.

MAP 6.1 RIO TIGRE CANAL SYSTEM

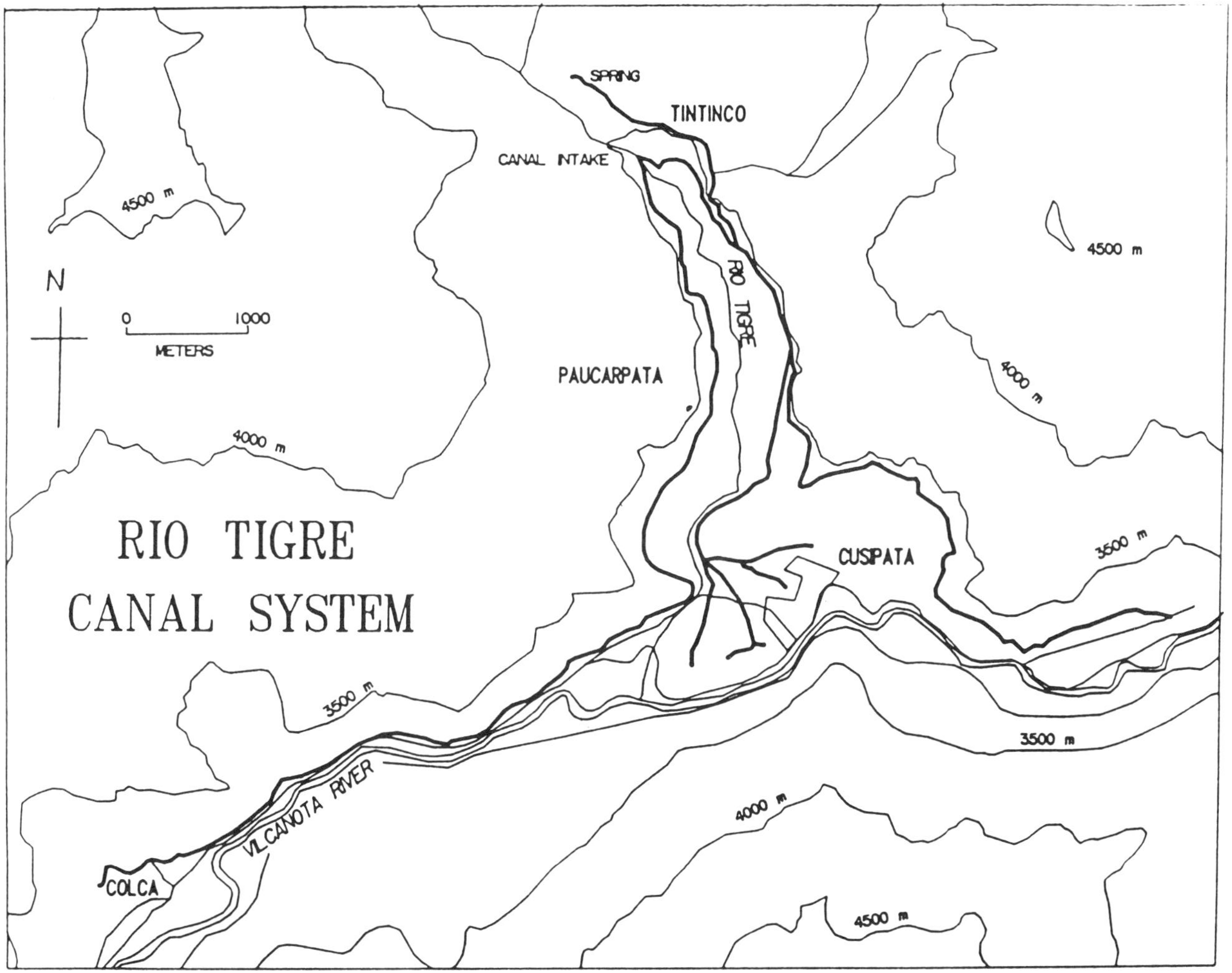
RIO TIGRE
CANAL SYSTEM
SPRING
TINTINCO
CANAL INTAKE
RIO TIGRE
PAUCARPATA
CUSIPATA
COLCA
VILCANOTA RIVER
N
0
1000
METERS
4500 m
4000 m
3500 m
3500 m
3500 m
4000 m
4000 m
4500 m
4500 m

Adapted from : Ministerio de Agricultura- Proyecto Cusipata

Irrigation development consisted of the construction of two principal canals (4.6 and 5.4 kilometers long respectively) along the right and left shores of the Rio Tigre. These canals are considered property of the state to which the irrigators have use rights. Part of this new canal system existed previously but required improvement, and part of it was newly constructed. The principal canals, as well as several lateral canals, one of which extends over a distance of 6 kilometers to Colcca, conduct water from the Rio Tigre to four villages–Tintinco, Paucarpata, Cusipata, and Colcca–irrigating an area of 476 hectares (see Table 6.1 for statistics).

The objectives of the development efforts agreed to by developers and municipal authorities were to improve the first crop, to encourage the planting of a second one, and to improve pastures by providing more plentiful water. The development agency also promised to introduce improved methods of agricultural production and livestock raising. These measures were intended to provide the 679 families (3,563 people) along the canal system, most of whom are small landowners, with a better and more varied diet and cash income through market activities. By creating more work within the agricultural domain, the standard of living within the communities was to be raised, thereby reducing seasonal and permanent migration to the overcrowded cities.

Instruction programs focused mainly on agriculture, livestock, health, hygiene, and use of the credit system. In addition, schools, kindergartens, community halls, and a first aid station were built in the communities and were very much appreciated by the recipients who provided most of the labor required for their construction. The development agency also introduced precise regulations for the distribution of water, which required peasants to arrange for a turn one day before irrigating. This improved method of distribution combined with the new plentiful supply of water eliminated the severe conflicts that had formerly arisen over access to water within and, even more markedly, between communities.

According to the "Regulations for Water Users," irrigators must participate at three organizational levels: through irrigators committees (*comités de regantes*) at the level of each community, through an irrigators commission (*comisión de regantes*) at the level of the canal system, and through an irrigators board (*junta de regantes*) at the level of the irrigation district (see Figure 6.1). This system is meant to permit local irrigators to participate at higher levels of organization and eventually in the formulation of water policies. Actual water user participation in policy-making, however, is not found in the Rio Tigre system, nor has it been observed by Solanes (1983:49) in his studies of irrigation policies elsewhere in Peru. The irrigators along the Rio Tigre have thus far (1985) only participated in the committees and the commission.

Each irrigation group consists of an executive board (*junta directiva*) and the general assembly (*asamblea general*). The executive board consists of a president, in charge of all major decisions, who acts as the organization's legal attorney; a vice-president who takes over the president's functions in his absence; a secretary who records the affairs of the organization by taking minutes during all meetings; a treasurer who is responsible for recording all financial matters and for collecting water fees; one or more water distributors (*tomeros*) who distribute the water during

TABLE 6.1 RIO TIGRE CANAL SYSTEM, DISTRICT OF CUSIPATA

Data on Irrigation Canals, Water Users, and Irrigated Land Area

Name of Village	Name of Canal	Maximum Capacity liter/second	Number of Water Users	Area irrigated in hectares	Average size of Holding
Tintinco			209	81	0.39
	Right Principal Canal	200			
	Left Principal Canal	350			
Paucarpata			59	94	1.59
	Right Principal Canal	200			
	Left Principal Canal	350			
Cusipata			290	207	0.71
	Right Principal Canal	200			
	Left Principal Canal	350			
Colcca			121	94	0.78
	Lateral Canal LD1	200			
TOTALS			**679**	**476**	**0.70**

Data from: *Ministerio de Agricultura*
Distrito Riego Cusco
Subsector Riego Cusipata, 1985

FIGURE 6.1 ORGANIZATION OF IRRIGATION ALONG THREE CANAL SYSTEMS IN THE DEPARTMENT OF CUZCO

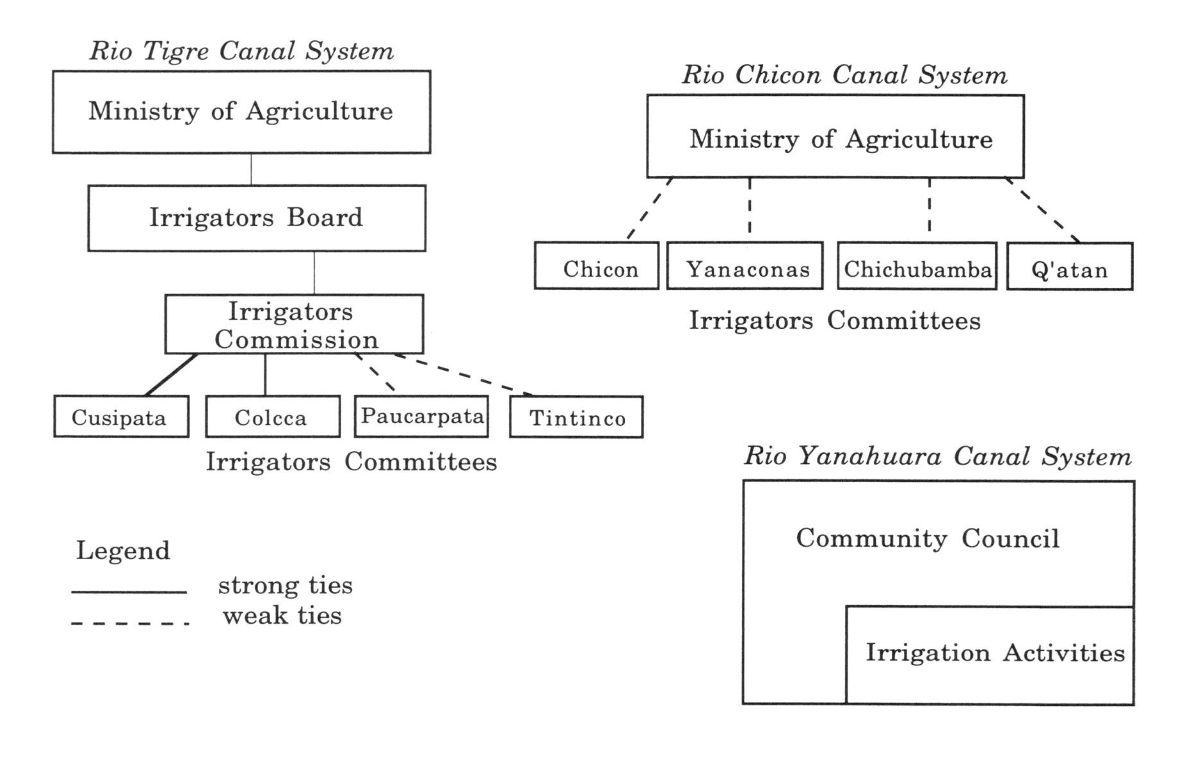

the dry season; one or more delegates (*vocales*) who inform the water users about all relevant matters and who replace board members in case of absences. Depending on the policies of the respective irrigation committee, a water distributor may or may not receive pay for his efforts. All other board members work without remuneration.

The functions of the executive board are to call meetings of all water users on fixed dates at least twice a year; to collect fees; to keep up-to-date books and accounts; to present a budget and work plan to the general assembly; and to implement decisions made by the general assembly. At the time of this research the members of the executive boards of the irrigators committees and the commission had started to familiarize themselves with their new tasks through discussions with representatives of the Ministry of Agriculture and development personnel.

The general assembly, which consists of all irrigators, approves the actions of the executive board. The assembly also determines irrigation fees and has the power to remove the members of the executive board.

The general assembly elects the executive board from its members, an act that irrigators consider one of their most important tasks. They do so at an election that must be announced ten days in advance and that requires an attendance of more than 50 percent of the irrigators. If attendance is lower, the election must be repeated and will then be held regardless of how many members of the general assembly are present. A representative of the Ministry of Agriculture presides over the election committee, but he cannot oppose any resolution taken by the general assembly. Any active irrigator, however, can object to the nomination of a candidate he believes does not fulfill the necessary requirements. In consequence, the members of the executive board along the Rio Tigre canal system are Indian peasants, people who are economically no better off than most villagers. As one of the peasants stated, "We elect people to the board who are like us because only they can understand our situation and therefore can represent our views."

The three levels of irrigation groups are legal corporate bodies under public law. Their presidents have the power of a legal attorney (*personero*). If conflict cannot be resolved by the president of a committee, the case is brought before the president of the commission. If conflict resolution is not achieved at this level either, the representatives of the Ministry of Agriculture for the particular district are supposed to intervene.

The main functions of the irrigators committees are to elect their representatives; to clean and repair the irrigation canals; to distribute water; and to settle disputes over irrigation. Even though the "Regulations for Water Users" specify that the irrigators commission is in charge of water fees, along the Rio Tigre, each irrigators committee collects and banks its own fees.

The irrigators commission, however, calculates the water fees and tries to resolve conflicts that cannot be resolved by an irrigators committee. The commission is also concerned with the study and improvement of irrigation and cultivation methods and the coordination of activities with the Technical Administration of the Irrigation District (Administración Tecnica del Distrito de Riego).

Membership in Irrigators Associations An irrigator or water user is legally defined as a public or private person who uses water for irrigation and is registered in the Book of Water Users (Padrón de Usuarios de Agua) of the Technical Administration of the Irrigation District. Newly accepted water users are normally the sons or daughters of irrigators who have inherited land and have set up their own households. If a peasant holds irrigated land in more than one community, he must be a member of each irrigators committee that distributes water to his fields. He cannot refuse to belong to a committee or to resign from it as long as he owns irrigated land within the boundaries of a respective community. In the case of a landowner-tenant relationship, it is the owner of the land who must belong to the committee, regardless of whether or not he works his own land.

Once accepted into an irrigators committee, irrigators have a number of rights and obligations. They have rights of access to irrigation water, canals, bridges across canals, and footpaths alongside canals. They are entitled to participate in meetings, elect representatives, and hold office. To perform effectively they have the right to receive information periodically on the organization, including access to the books and inventories. At the same time, a water user assumes the obligations to use water economically and efficiently; to contribute proportionally to the construction, improvement, and maintenance of the irrigation system (this rule is currently not enforced along the Rio Tigre canal system because developers take over the costs for canal construction and repair); to attend meetings; to pay water fees regularly; to advise the local water authority if he uses no water or only part of his allocation; and to vote and stand for election.

To be elected a member of the executive board of an irrigators association, a person must abide by the rights and obligations of a water user as stated above and, furthermore, be responsible, esteemed, literate, and have resided for at least two years in the community.

Institutional Funds Funds to operate the various irrigators associations are obtained primarily through the imposition of water fees. In the district of Cusipata, fees were calculated according to the benefits that water users derive and the construction costs of the new irrigation canals. Irrigators in Tintinco paid 200 *soles* (100 *soles* were equivalent to 1 U.S. cent) in 1984 and 1985 for each water application to one *topo* of land (0.33 hectares). In Paucarpata and Cusipata they paid 300 *soles*. In Colcca they paid 1,000 *soles* per *topo*, because the water that they desperately needed required the construction of a lateral canal 6 kilometers long. Because they were desperate, they were willing to pay a much higher amount for irrigation water.

Water fees are paid in cash to the treasurer once a year. According to the "Regulations for Water Users," the irrigators commission is responsible for the collection of fees, but in reality each committee handles its own funds. Ten percent of the fee goes to the Ministry of Agriculture, and the rest is deposited in the committee's bank account at the Cuzco Agrarian Bank (Banco Agrario de Cuzco). Fines in the amount of 5,000 *soles*, the equivalent of one day's wage labor paid by irrigators who do not participate in collective labor tasks, as well as loans, donations, and interest are added to the bank account. Although the development agency still pays for repairs and the salaries of water distributors, committee funds will have to cover these expenses once the local population assumes the total operation of the project.

Relationships Between Communities and Irrigators Committees Irrigators Committees have full autonomy over all irrigation tasks other than the construction of canals and associated bridges and footpaths. Such construction must be discussed with the respective municipal or community councils. The autonomy of irrigation authorities vis-à-vis municipal authorities has, in all other respects, been successful in preventing powerful local groups from interfering with irrigation as they had in the past (see also Gelles *intra*). Village councils give support and informal advice, but decision making about irrigation matters rests entirely with irrigation authorities, people who do not otherwise occupy important municipal or local government posts. This situation differs from the observation of Hunt and Hunt that "higher roles in the local stratification system are linked with instrumental decisions in irrigation" (1974:397).

MAP 6.2 RIO CHICON CANAL SYSTEM

Scale = 1:25000

CHICON

YANACONAS

Canal Yanaconas

Canal San Isidro Chicon

Canal Q'atan

Q'ATAN

Canal ChanChilley

CHICHUBAMBA

URUBAMBA

Canal Tullumayo

Canal Ccantuyoc

Canal Jatan Yarko

RIO VILCANOTA

N

3200

3100

3000

2900

RIO CHICON CANAL SYSTEM

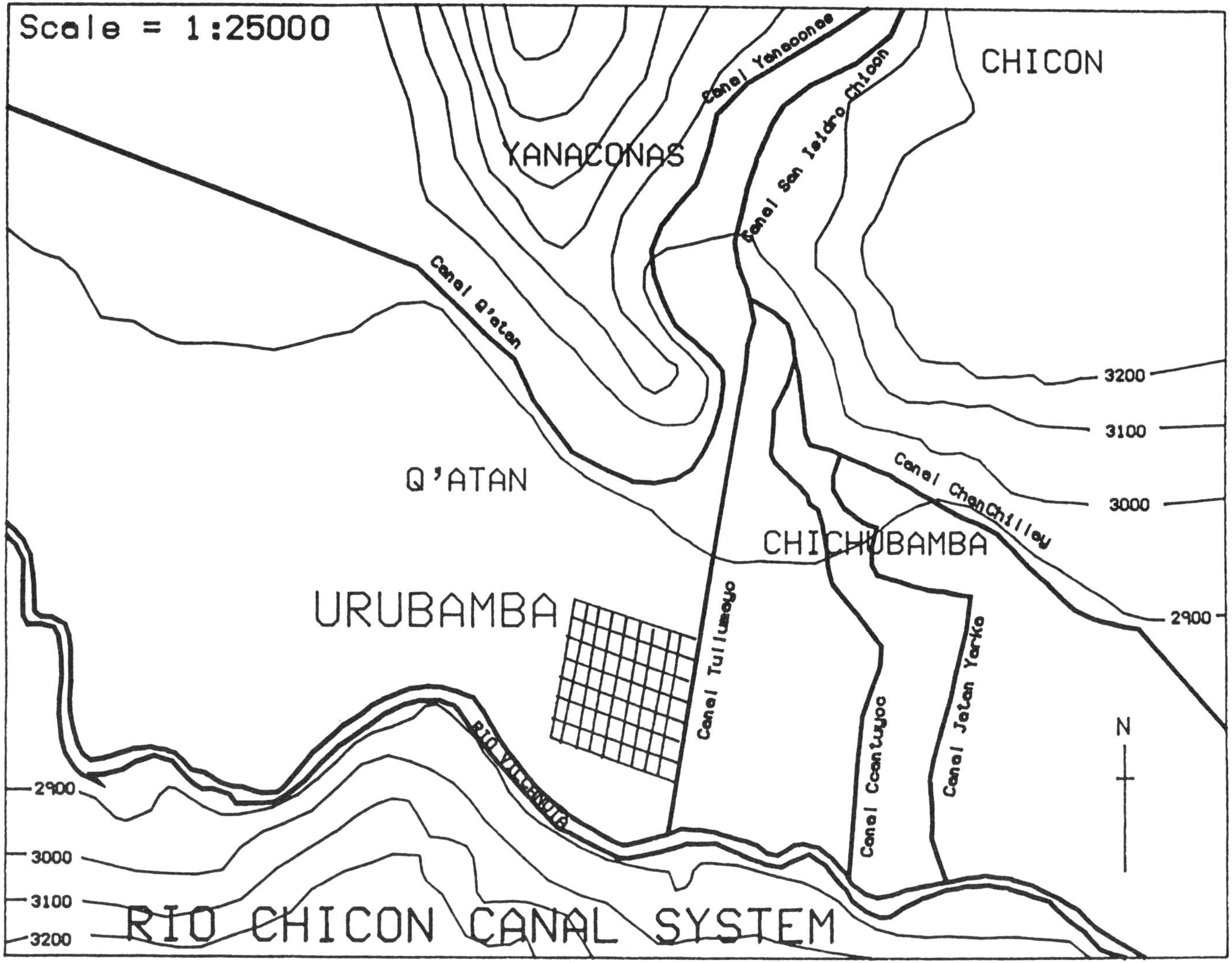

The Rio Chicon Canal System

The Rio Chicon canal system is located 70 kilometers north of the city of Cuzco (see Map 6.2). The villages of Chicon, Yanaconas, Chichubamba, and Q'atan, all of which belong administratively to the town of Urubamba, receive water from the Rio Chicon, a river that obtains its water from the glaciers of Mount Chicon at an altitude of 5,200 meters. Above the village of Chicon the river water is captured by an Inca canal system[3] that consists of two principal canals and several lateral and sublateral canals. The upstream villages of Chicon and Yanaconas receive water from the canals San Isidro Chicon and Yanaconas respectively. The downstream villages of Q'atan and Chichubamba receive irrigation water from five major canals–Q'atan, Tullumayo, Ccantuyoc, Jatan Yarka, and Chanchillay (see Table 6.2 for statistics).

The villages along the Rio Chicon have accepted the "Regulations for Water Users" to the extent that they have established an irrigators committee in each village. The structure and function of the committees, the prerequisites to qualify as a water user, water rights, and the procedures by which a new member is accepted and a new executive board is elected are in accordance with national regulations as described above. The president of each committee calls meetings twice a year to discuss the cleaning and maintenance of canals, the distribution of water, and water fees.

Since the Rio Chicon villages have not created an irrigators commission, they have not limited the autonomy of each irrigators committee. The committees of the four villages, however, do meet jointly once a year to discuss common issues, and they also meet every two years to vote for a new executive board (see Figure 6.1).

The irrigators committees also act independently of the community. Irrigation authorities make decisions about all routine irrigation tasks without soliciting the consent of the community council. They only involve the council when extensive repairs have to be made on the irrigation canals. (The last such occasion was in 1942 to repair landslide damage.)

The irrigators committees are fairly independent of the Ministry of Agriculture (see also Guillet 1987:414). The irrigation sector of this ministry is represented in Urubamba, but these officials are primarily concerned with technical matters. They are supposed to intervene in cases of conflict or organizational problems, but they rarely have to do so. Their presence is felt primarily when it is time to collect water fees.

Along this canal system each water user paid 200 *soles* (2 U.S. cents) in 1984 and 1985 for irrigating one *topo* of land, regardless of whether the land was located at the upper or lower part of the irrigation system. As in the Rio Tigre area, 10 percent of the water fees goes to the Ministry of Agriculture. The rest of the fees, along with fines and interest, is allotted to the committee's bank account. The funds are used for repair of the canal system and for the water distributor who is needed from August to October every year. In 1984 the water distributor received 15,000 *soles* (equivalent to U.S. $ 3.50) per month.

TABLE 6.2 RIO CHICON CANAL SYSTEM, DISTRICT OF URUBAMBA

Data on Irrigation Canals, Water Users, and Irrigated Land Area

Name of Village	Name of Canal	Maximum Capacity liter/second	Number of Water Users	Area irrigated in hectares	Average size of Holding
Chicon	San Isidro Chicon	80	68	73.02	1.07
Yanaconas	Yanaconas Chicon	30	131	43.37	0.33
Chichubamba	Tullumayo	50	72	43.44	0.60
	Ccantuyoc	50	56	20.58	0.37
	Jatan Yarka	50	156	53.82	0.35
	Chanchillay	50	78	24.26	0.31
Q'atan	Q'atan	60	132	46.47	0.35
TOTALS			**693**	**304.96**	**0.44**

Data from: *Ministerio de Agricultura*
Distrito Riego Cusco
Subsector Riego Cusipata, 1985

MAP 6.3 RIO YANAHUARA CANAL SYSTEM

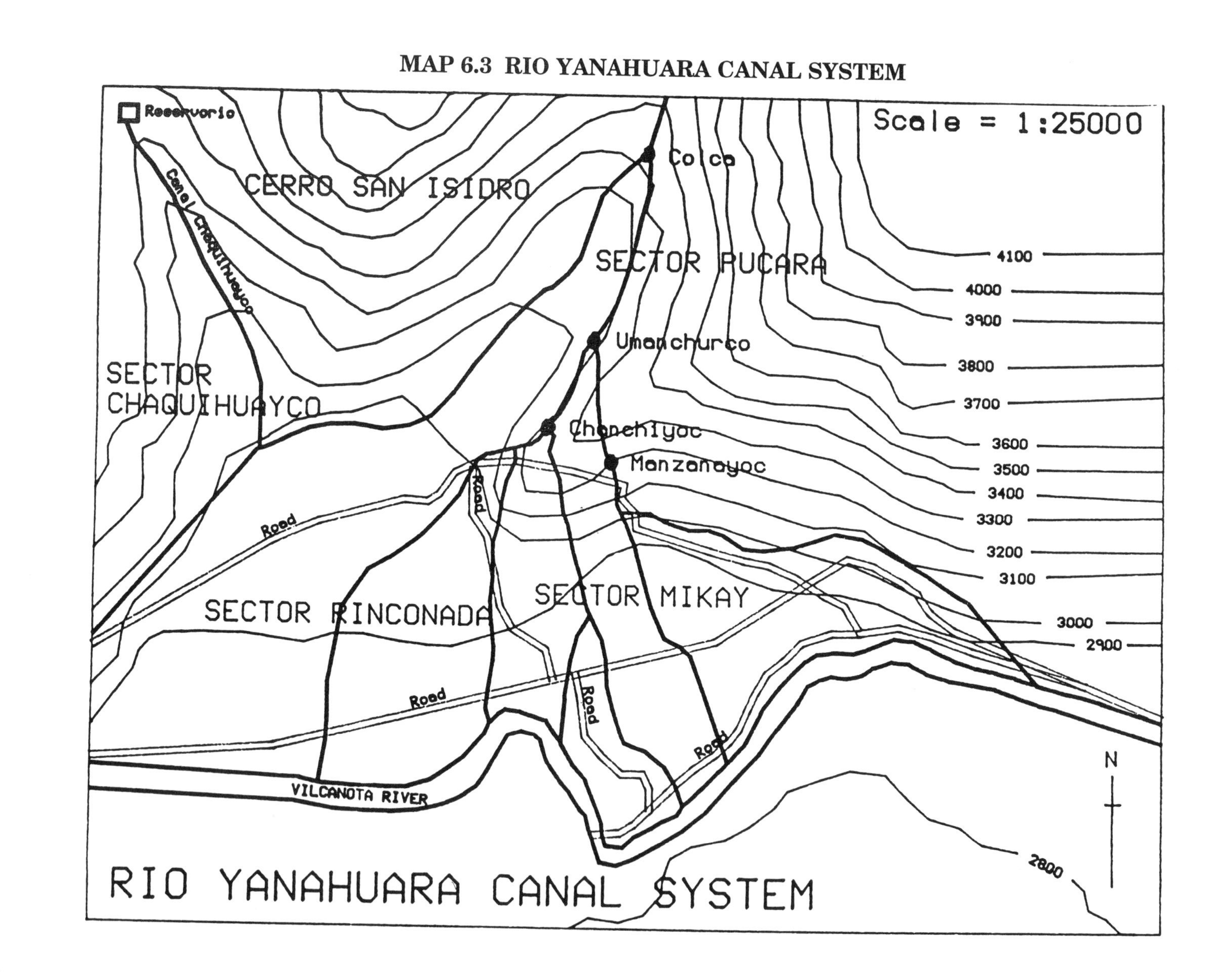

The Rio Yanahuara Canal System

The Rio Yanahuara canal system is located eighty kilometers north of the city of Cuzco (see Map 6.3). The lagoon Yurajcocha, twelve kilometers above the village of Yanahuara, collects water from the high lagoons Ouroray and Champacocha and from the snow fields of the surrounding mountains. The water from Yurajcocha drains into the ravine Pucara. Above the village of Yanahuara it is diverted into the Inca canals Colca, Umanchurco, Chanchiyoc, and Manzanayoc. A network of lateral and sublateral canals brings water to three of the four sectors of Yanahuara: Pucara, Rinconada, and Mikay. The fourth sector Chaquihuayco receives water from a small mountain stream of the same name. The 800 families of Yanahuara irrigate 1,000 hectares of land (see Table 6.3 for statistics).

Yanahuara has not sought official recognition as a peasant community (*comunidad campesina*), since most of the residents believe that this official status would interfere with their autonomy. Nor have the people accepted the "Regulations for Water Users" or formed an irrigators association. Irrigation matters are under the control of the community president and other authorities (see Figure 6.1). They organize canal cleaning, canal repairs, and the distribution of water according to customary law. Villagers meet once a year, or more frequently if necessary, to discuss irrigation in conjunction with other communal affairs. Local autonomy, therefore, remains intact in this canal system.

SUMMARY OF ORGANIZATIONAL ISSUES ALONG THE THREE CANAL SYSTEMS

International developers have tried to organize the irrigation groups along the Rio Tigre to conform to the "Regulations for Water Users" (see Figure 6.1), but have not been completely successful (Bolin 1987). The committees from the communities of Colcca and Cusipata that had initiated their own requests for irrigation development work most efficiently. Their members cooperate enthusiastically at local and regional levels. The committee of Paucarpata and especially that of Tintinco, communities that had not themselves requested aid because they have always had enough water and have benefitted little from development, show much less enthusiasm and sometimes even resist cooperation, especially at the level of the entire canal system. Although an irrigators commission has been established at the level of the Rio Tigre canal system, water users prefer to keep tasks such as the administration of funds at the level of each irrigators committee (Bolin 1990). Furthermore, most irrigators along the Rio Tigre canal system would rather deal with the water judge (*juez de agua*) to resolve conflict than with the presidents of the committees or the commission. Finally, the local water users are not yet represented on the Irrigators Board.

In the Rio Chicon canal system, the "Regulations for Water Users" have been implemented with even less success. This system has established irrigators

TABLE 6.3 RIO YANAHUARA CANAL SYSTEM, DISTRICT OF URUBAMBA

Data on Irrigation Canals, Water Users, and Irrigated Land Area

Name of Village	Name of Canal	Maximum Capacity liter/second	Number of Water Users	Area irrigated in hectares	Average size of Holding
Yanahuara			800	1000	1.25
	Colca	not known			
	Umanchurco	not known			
	Chanchiyoc	not known			
	Manzanayoc	not known			
	Chaquihuayco	not known			

Data from: villagers of Yanahuara, 1985

committees, but not an irrigators commission or an Irrigators Board. As is true for the Rio Tigre system, irrigators here prefer to use a water judge to resolve conflicts rather than the presidents of the irrigators committees. Connections with the Ministry of Agriculture are weak.

In the Rio Yanahuara system, the "Regulations for Water Users" find no application at all, and there is no connection to the Ministry of Agriculture. Irrigators here organize themselves according to customary law administered by local authorities.

Some aspects of the "Regulations for Water Users" have served the irrigators in the Rio Tigre and Rio Chicon systems well. Foremost among these is the high degree of autonomy assigned to the irrigators' associations, especially with regard to election procedures, canal maintenance, and the distribution of water. Concerning the complex task of water distribution, irrigation authorities of the lower parts of the Rio Tigre and Rio Chicon canal systems agree that, especially during disputes over water, an official document is more effective in demonstrating prevailing rules to offenders than are oral arguments. That the rights and obligations of a water user are clearly spelled out in the "Regulations" has facilitated the tasks of the irrigation authorities. In the absence of written rules concerning irrigation management in Yanahuara, on the other hand, farmers must communicate only by word of mouth about irrigation-related issues. This method has worked well for all activities other than the distribution of water. Because of great water scarcity during the dry season, water distribution in Yanahuara is fraught with tremendous conflict (Bolin 1985).

TRENDS IN AUTONOMY ALONG THE THREE CANAL SYSTEMS

To convey a clear picture of autonomy along the canal systems, three major tasks–election procedures, the maintenance of canals, and the distribution of water–are summarized below. The high group autonomy in election procedures and canal maintenance have resulted in efficient task performance along all canal systems. The success of water distribution, on the other hand, varies considerably among the three regions, even though it is also carried out with a high degree of autonomy.

Election Procedures

Elections are handled effectively and without problems by local irrigators along the three canal systems. Full group autonomy is vested in the biennial events during which the General Assembly of irrigators elects representatives to the executive board of the committees along the Rio Chicon and to the executive board of the committees and commission along the Rio Tigre. Along the Rio Yanahuara, full autonomy is vested in the community for the election of the communal representatives who deal with both community and irrigation matters.

Maintenance of Irrigation Canals

Maintenance of the canals, which includes both cleaning and repair, is done by means of corvée work groups (*faena*) in all three areas. The rules for participating in these work groups are very similar in the three study regions. Canal cleaning takes place each year at a specific date agreed on by the water users. One member of each household must participate in the activities, send a replacement, or pay a fine. All water users carry the same workload regardless of the amount of land owned or water used. People in the upper parts of the canal system participate in maintenance activities only to the point where the canals leave their territory, whereas irrigators living downstream work along the entire canal system. Disputes over these arrangements are rare.

Several minor variations are found within these general trends. To encourage participation in canal maintenance, the development personnel in Cusipata (along the Rio Tigre system) provide food and drink. In Yanahuara a sponsor (*pendonero*) provides refreshments to the workers. In an attempt to strengthen group solidarity along the Rio Chicon, the trusted municipal council of Urubamba encourages each water user to participate in the labor corvée rather than send a replacement or pay the fine.

Along the Rio Tigre the call for collective labor by the local authorities on the irrigators commission and committees is reinforced by the demands of development personnel. Along the Rio Chicon, the presidents of the four irrigators committees alone make the call, whereas in Yanahuara, communal authorities organize the collective work party. However organized, participation in canal maintenance is close to 100 percent along the three systems, primarily because the irrigation authorities insist on the strict imposition of fines for those who don't participate. The development agency along the Rio Tigre canal system still pays for materials needed to repair the canals. Along the Rio Chicon the irrigators committees pay for materials, and in Yanahuara, each water user makes a contribution toward the cost of needed supplies.

Distribution of Water

The distribution of water is much more complex in organization and execution than is canal maintenance.[4] Along the Rio Tigre canal system water distribution proceeds according to the "Regulations for Water Users." Development personnel enforce these rules in cooperation with the executive board of the irrigators committees and the commission. Since the water supply is plentiful throughout the year, scheduling turns presents few problems. The only significant step an irrigator must take to get water is to notify the water distributor about his or her intentions one day before irrigating.

Along the Rio Chicon the executive boards of the irrigators committees attempt to enforce the rule assigning equal rights to each water user, but they are often unable to do so during the dry season and at the beginning of planting time because of water scarcity. During these periods, people must assemble frequently at the intake, spending long hours waiting for their turn to irrigate (see Mitchell 1976, 1991:207-209). Disputes about water are common, but the conflict is normally resolved within each committee rather than by outsiders.

Extreme water scarcity in Yanahuara during the dry season, and especially at planting time, makes adequate distribution impossible. Consequently, water distribution does not occur in an organized fashion and often results in severe conflict, even though canal maintenance activities arranged through communal authorities show good results. Water scarcity is also a major reason why an irrigators association has not yet been established. Nobody could or would want to be responsible for the task of distribution given the virtual absence of water during the height of the dry season. Irrigators must normally spend considerable time waiting at the intake. Irrigation continues day and night, requiring several family members to guard the canal to discourage water theft. In Yanahuara, as in the other regions studied, conflict about irrigation matters is normally resolved locally.

DISCUSSION

Given the definition of group autonomy as a group's ability to make decisions without the involvement of other local and extralocal institutions, the preceding discussion has demonstrated that full local autonomy is not always present in irrigation management, even when local irrigators consider themselves to be autonomous. An analysis is therefore necessary to determine approximate degrees of autonomy that may affect irrigation management.

The fact that, along the Rio Tigre and Rio Chicon canal systems, the "Regulations for Water Users" have been accepted to various degrees by local irrigators and that extralocal decision makers (development authorities along the Rio Tigre and representatives of the Ministry of Agriculture along the Rio Chicon) intervene, provides for a range of situations in which local and national interests mingle, affecting group autonomy to differing degrees.

Although much of the literature describes autonomy as a simple dichotomy (decentralized local control versus centralized elite control), other investigators consider the relationship between local autonomy and extralocal decision making to be more complex. William Kelly, for example, uses the concept "articulation/autonomy" "to characterize the degree to which irrigation is linked to, or is independent of the state" (1983:883). Although Eva and Robert Hunt recognize that a decentralized system has local autonomy, they apply the term *centralization* to express a local system's articulation into a hierarchy of state supported and legitimized water-control institutions (1974:133).

Max Weber's (1947:148) model of autonomy provides most useful distinctions that help us analyze the autonomous aspects of irrigation organization in the three regions. This model distinguishes between the origin of the rules governing a group (autonomy versus heteronomy) and the locus of authority of the leaders (autocephaly versus heterocephaly). In Weber's terms, "autonomy" occurs when the order governing the group has been established by its own members on their own authority, whereas "heteronomy" characterizes the situation in which the order governing the group has been imposed by an outside group. In "autocephaly," leaders act by the authority of the autonomous order of the corporate group itself, whereas in "heterocephaly," they are under the authority of outsiders.

Weber has accounted for a range of possible situations in which, for example, a group may be heterocephalous in one sphere and autonomous in another (1947:148). He has further postulated that all four characteristics may be present in the same situation to some degree.

The following analysis defines approximate expressions of autonomy, heteronomy, autocephaly, and heterocephaly along the three canal systems. The criteria used in making the rankings–weak, medium, strong, and full (see Figure 6.2)–reflect the extent to which decision making is exerted from within or from outside the irrigation groups. A medium ranking, for example, means that decision making is approximately equal between local irrigators and government authorities.

Analysis of Irrigation Organization in Light of Max Weber's Model

In the Rio Tigre canal system, the irrigators committees are strongly heteronomous according to Weber's model, since "the order governing the group" (the "Regulations for Water Users") has been imposed by an outside agency, the Ministry of Agriculture. Only along this canal system, where development efforts are currently taking place, are the national regulations closely followed. Here two of the three decision-making levels (commission and committee) specified in the regulations have been established. Water distribution in these communities also takes place in accordance with national regulations (see Figure 6.2 for an approximate representation of autonomy and heteronomy in local irrigators committees). The Rio Chicon system is less heteronomous, since an irrigators commission has not been established, and overall the "Regulations for Water Users" are not enforced to the same degree but are more directly adapted to local conditions.

The irrigators groups along the Rio Tigre and Rio Chicon canal systems are partially autocephalous. The executive board in both regions acts by the autonomous order of the corporate group itself in all routine activities (election of representatives, canal maintenance, and distribution of water) but is under the authority of extralocal decision makers (the Irrigation Sector of the Ministry of Agriculture) in executing such nonroutine activities as canal construction and conflict resolution. In the Rio Chicon area, however, autocephaly is more strongly expressed because the executive board is less dominated by higher-level authorities than along the Rio Tigre, where local groups make most decisions in cooperation with development personnel. It is expected, however, that autonomy and autocephaly will increase along the Rio Tigre when the irrigators take full project responsibility into their own hands.

The situation differs considerably along the Rio Yanahuara canal system. Here, customary laws are dominant and village authorities alone are in charge of irrigation. Despite considerable conflict over water during the dry season, the community relies only on local leaders (the justice of the peace and the community

FIGURE 6.2 EXPRESSIONS OF AUTONOMY, AUTOCEPHALY, HETERONOMY AND HETEROCEPHALY OF THREE IRRIGATION COMMITTEES, DEPARTMENT OF CUZCO

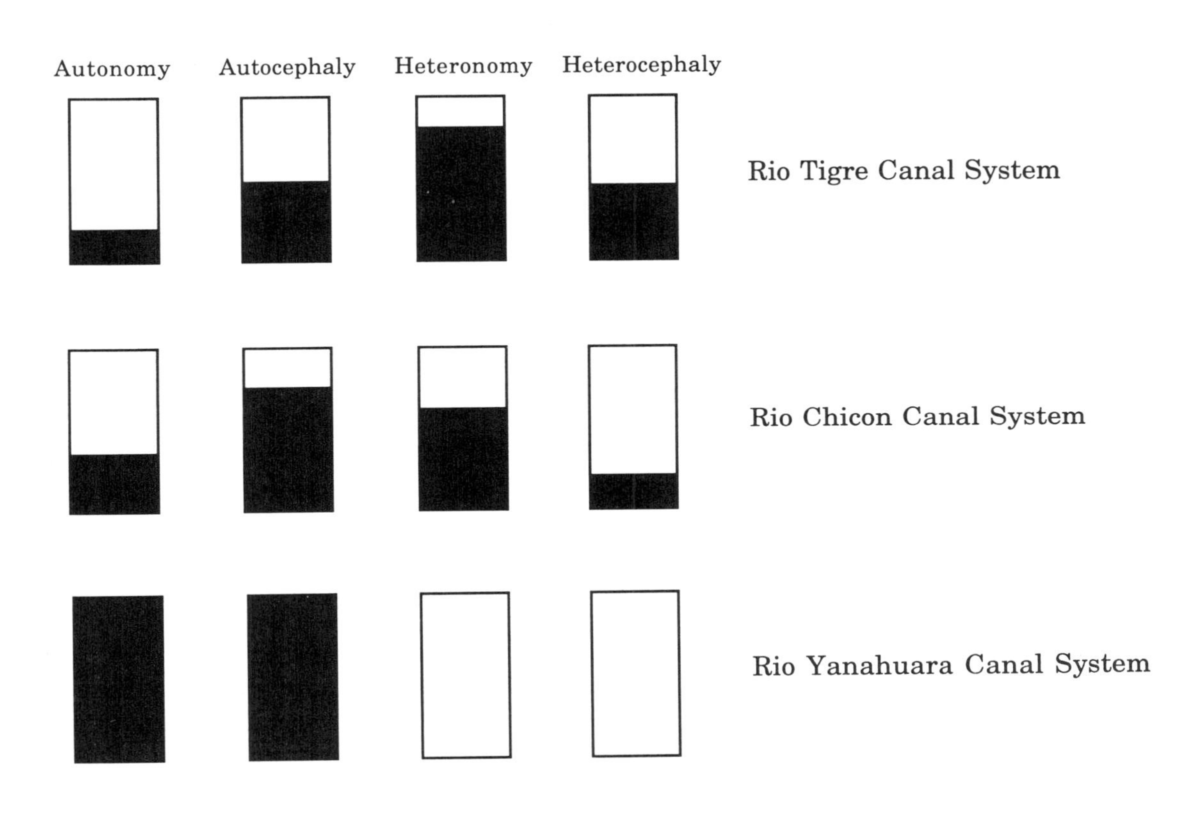

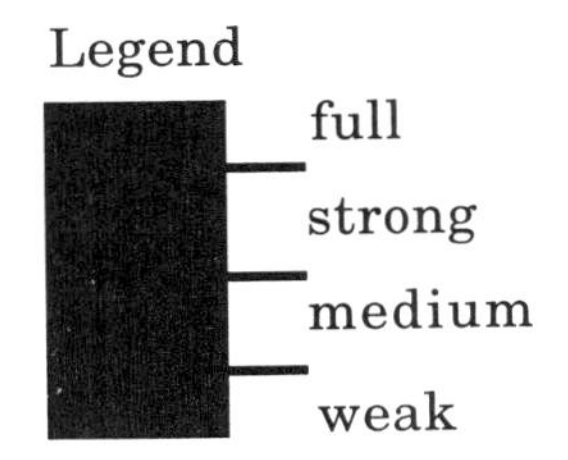

president) to resolve it. The situation in Yanahuara, therefore, is autonomous in that "the order governing the group" was established by its own members on their own authority, and autocephalous in that the community authorities act "by the autonomous order of the corporate group" (Weber 1947:148).

My analysis of the three regions suggests that Weber's model and many case studies ignore an important aspect of heteronomy. To better represent the effects of extralocal interventions–especially in development situations–another dimension must be added to Max Weber's model. The concept of heteronomy must distinguish between nonlocal intervention requested by the local group and that imposed by an outside agency against the will of the local people.

Outside development efforts requested by the local irrigators (as was the case in Colcca and Cusipata) are likely to be successful because they conform to local needs. When local people either direct the project or strongly participate in decision making, the chance for success is high. Although outside intervention may temporarily weaken local autonomy and autocephaly, as has happened along the Rio Tigre, these characteristics are likely to be strengthened in local groups over the long run. Where state regulations for water users reflect local desires for managing irrigation, they are expressive of the autonomy of local groups.

CONCLUSIONS

This study has examined the extent to which autonomy affects the successful management of irrigation. An analysis of irrigators associations relative to Max Weber's classification of corporate groups has indicated that we are dealing with a broad spectrum of situations in the three canal systems studied. Along the Rio Yanahuara, a highly autonomous and autocephalous community regulates irrigation activities. In contrast to Yanahuara, we find lower degrees of autonomy and autocephaly along the Rio Chicon, since irrigators committees were established using some of the decision-making procedures from the "Regulations for Water Users." The Rio Tigre committees are even less autonomous and autocephalous because local decisions are made in conjunction with the irrigators commission and development personnel.

The data have shown that autonomy and autocephaly are important to the successful organization of irrigation in so far as they allow farmers to manage their affairs with pride and self-confidence and in accordance with prevailing social and environmental conditions. Yet, a high degree of local group autonomy and autocephaly in itself does not necessarily determine successful irrigation management. Instead, these characteristics must be viewed in relation to social factors, such as the terms under which cooperation between a group and other local or extralocal institutions takes place, and economic factors, mainly available resources.

The study has further indicated that autonomy and autocephaly need not be concentrated at the local level (irrigators committees) to assure proper functioning of irrigation. It is important, however, that local interests are represented at the higher level through local representatives elected by the farmers themselves or through other mechanisms that respect local interests. Strong irrigators associations (*subaks* in Bali and *zanjeras* in the Philippines), for example, have developed where responsibility is aggregated through leaders from small units based on canal layouts to progressively larger units, which eventually cover the entire irrigation system (Bagadion and Korten 1985:73).

The study has also indicated that, although irrigators committees are supposed to be autonomous, cooperation between irrigation and communal authorities can be beneficial to irrigation management when based on good relations.

Different irrigation tasks executed with the same degree of autonomy (as is the case along the Chicon and Yanahuara canal systems) do not always produce the same results. Thus, activities such as the election of representatives to committees and commission and the maintenance of canals have been successful along all canal systems studied. Nonetheless, water distribution is problematic along the Rio Chicon and Rio Yanahuara.

State intervention has frequently been portrayed as detrimental to local autonomy. This study, however, has indicated that a lessening of local autonomy through outside intervention need not be debilitating. Without the economic or technical resources to improve their canal systems, irrigators must often depend on outside assistance. Such assistance can increase local autonomy and autocephaly when local irrigators request the help and when they are given the right to assume the decision-making tasks formerly carried out jointly with development personnel. It was apparent throughout this research that in those cases where local irrigators were given the opportunity to make decisions regarding development, their committees functioned better (Colcca and Cusipata) than in those cases where irrigators were only marginally involved in the planning of the project (Tintinco and Paucarpata). My analysis also suggests that official regulations (such as the "Regulations for Water Users") can be of benefit to irrigation management if they are flexible enough to respond to local needs and adapt to the local physical and social environment.

More will be learned about the above issues when developers leave the Rio Tigre project and go on to assist farmers along the Rio Chicon and Rio Yanahuara canal systems (an event scheduled for 1990). These new projects have arisen solely from initiatives of local irrigators. They are expected to proceed under local direction using local organization and labor for the construction and improvement of the canal systems. The developers have agreed to supply building materials and to give technical advice. It will be especially interesting to document this approach that combines farmer's initiative and state assistance in such a way that local interests are preserved.

NOTES

Acknowledgments. This essay is based on research in Peru that was completed in 1985 and made possible by grants from the Wenner Gren Foundation for Anthropological Research, the GTZ-Deutsche Gesellschaft für Technische Zusammenarbeit in Germany, and the Fund for Support of

International Development Activities at the University of Alberta in Canada. For constructive advice on earlier versions of this paper the author would like to thank the editors of this volume. Bob Slobodian and Glen Langford were very helpful with the drawing of the maps. For support during my fieldwork, my special thanks go to H. T. Lewis from the University of Alberta, Plan Meris II, and the Misión Tecnica Alemana in Cuzco, and above all to the irrigators of the Vilcanota Valley.

1. Fairchild and Nobe state that within the last fifty years irrigated land in developing countries has increased threefold. In the last ten years "roughly 40 percent of all increases in food production in developing countries have come from expanded irrigation" (1986:356).

2. Irrigation management is successful when water users cooperate according to accepted rules and conflict does not disrupt irrigation tasks.

3. A canal system can be defined as Inca when its construction matches that of surrounding terraces and house foundations that have been dated as Inca.

4. In all three areas, as throughout the Andes, rights to land and water are interrelated in that water is allocated in proportion to the amount of land owned and worked by each peasant family, a type of water allocation known as the "Syrian model" by Glick (1970).

REFERENCES CITED

Appell, George N.
1976 "The Rungus: Social Structure in a Cognatic Society and its Ritual Symbolization." In *The Societies of Borneo: Explanations in the Theory of Cognatic Social Structure*. Special Publication, 6. George N. Appell, editor, pp. 66-86. Washington, DC: American Anthropological Association.

Austin, Diane
1985 "Response to Grace Goodell, Paternalism, Patronage and Potlatch: The Dynamics of Giving and Being Given to." *Current Anthropology* 26:257-258.

Bagadion, Benjamin U., and Frances F. Korten
1985 "Developing Irrigators' Organizations: A Learning Process Approach." In *Putting People First: Sociological Variables in Rural Development*. Michael M. Cernea, editor, pp. 52-90. New York: Oxford University Press.

Bolin, Inge
1985 "Die Organisation der Bewässerungswirtschaft im Andenhochland von Peru." Unpublished report to the *Deutsche Gesellschaft für Technische Zusammenarbeit (GTZ) GmbH*, Eschborn, Deutschland.
1987 *The Organization of Irrigation in the Vilcanota Valley of Peru–Local Autonomy, Development and Corporate Group Dynamics*. Ph.D. dissertation, Department of Anthropology, University of Alberta, Edmonton, Canada.
1990 "Upsetting the Power Balance–Cooperation, Competition and Conflict along an Andean Irrigation System." *Human Organization* 49(2):140-148.

Brown, D. E.
1976 *Principles of Social Structure–Southeast Asia*. London: Gerald Duckworth & Co. Ltd.

Clark, Terry N.
1974 "Community Autonomy in the National System: Federalism, Localism, and Decentralization." In *Comparative Community Politics*. Terry N. Clark, editor, pp. 21-45. New York: Halstead Press Division, John Wiley & Sons.

Coward, Walter E.
1977 "Irrigation Management Alternatives: Themes from Indigenous Irrigation Systems." *Agricultural Administration* 4:223-237.
1980 "Irrigation Development. Institutional and Organizational Issues." In *Irrigation and Agricultural Development in Asia*. Walter E. Coward, editor, pp. 15-27. Ithaca: London: Cornell University Press.

1986 "State and Locality in Asian Irrigation Development: The Property Factor." In *Irrigation Water Management in Developing Countries: Current Issues and Approaches*. R. K. Sampath and Kennth C. Nobe, editors, pp. 491-508. Boulder, CO: Westview Press.

Eggan, Fred
1954 "Social Anthropology and the Method of Controlled Comparisons." *American Anthropologist* 56:743-763.

Fairchild, Warren, and Kenneth C. Nobe
1986 "Improving Management of Irrigation Projects in Developing Countries: Translating Theory into Practice." In *Irrigation Management in Developing Countries: Current Issues and Approaches*. R. K. Sampath and Kenneth C. Nobe, editors, pp. 353-412. Boulder, CO: Westview Press.

Fox, Irving
1978 "Institutions for Water Management in a Changing World." In *Water in a Developing World*. Albert E. Utton and Ludwik Teclaff, editors, pp. 9-24. Boulder, CO: Westview Press.

Geertz, Hildred, and Clifford Geertz
1975 *Kinship in Bali*. Chicago: University of Chicago Press.

Gibson, Nancy
1985 *Aid, Trade and the Extension of Capitalism to the Third World*. M.A. thesis, Department of Anthropology, University of Alberta, Canada.

Glick, Thomas
1970 *Irrigation and Society in Medieval Valencia*. Cambridge, MA: Harvard University Press.

Goodell, Grace
1985 "Paternalism, Patronage, and Potlatch. The Dynamics of Giving and Being Given To." *Current Anthropology* 26:247-266.

Grader, C. J.
1960 "The Irrigation System in the Region of Jembrana." In *Bali: Studies in Life, Thought and Ritual*. J. L. Swellengrebel, editor, pp. 267-288. The Hague: W. van Hoeve.

Guillet, David
1987 "Terracing and Irrigation in the Peruvian Highlands." *Current Anthropology* 28:409-430.

Hoggart, Keith
1981 *Local Decision-Making Autonomy. A Review of Conceptual and Methodological Issues*. University of London King's College, Department of Geography. Occasional Paper No. 13. London:

Hunt, Eva, and Robert C. Hunt
1974 "Irrigation, Conflict and Politics–A Mexican Case." In *Irrigation's Impact on Society*. Theodore E. Downing and McGuire Gibson, editors, pp. 129-157. Tucson, AZ: University of Arizona Press.

Kelly, William W.
1983 "Concepts in the Anthropological Study of Irrigation." *American Anthropologist* 85:880-886.

Leeds, Anthony
1973 "Locality Power in Relation to Supralocal Power Institutions." In *Urban Anthropology*. Aidan Southall, editor, pp. 15-41. New York: Oxford University Press.

Lees, Susan H.
1974 "The State's Use of Irrigation in Changing Peasant Society." In *Irrigation's Impact on Society*. Theodore E. Downing and McGuire Gibson, editors, pp. 123-128. Tucson, AZ: University of Arizona Press.

Lewis, Henry T.
1991 *Ilocano Irrigation: The Corporate Resolution*. Asian Studies at Hawaii, 37. Honolulu: University of Hawaii Press.

March, James
1980 *Autonomy as a Factor in Group Organization*. New York: Arno Press.

Mazrui, Ali
1975 "The African University as a Multinational Corporation: Problems of Penetration and Dependency." In *Education and Colonialism*. Philip G. Altback and Gail P. Kelly, editors, pp. 331-354. New York: Longman.

Mitchell, William P.
1976 "Irrigation and Community in the Central Peruvian Highlands." *American Anthropologist* 78(1):25-44.
1991 *Peasants on the Edge: Crop, Cult, and Crisis in the Andes*. Austin: University of Texas Press.

Olson, Mancur
1965 *The Logic of Collective Action*. Cambridge, MA: Harvard University Press.

Ministerio de Agricultura, Peru
1979 *Reglamento de Organización de Usuarios de Agua*. Annual No. 4. Oficina de Difusión Tecnologica.
1980 *Proyecto Cusipata*. Dirección General Ejecutivo del Programa Nacional de Pequeñas y Medianas Irrigaciones.

Smith, Michael C.
1974 "A Structural Approach to Comparative Politics." In *Corporations and Society*. Michael G. Smith, editor, pp. 91-105. London: Duckworth & Co. Ltd.

Solanes, Miguel
1983 *Irrigation Users' Organizations in the Legislation and Administration of Certain Latin American Countries*. Rome: Food and Agriculture Organization of the United Nations.

Steinberg, David I., Cynthia Clapp-Wincek, and Allen G. Turner
1983 *Irrigation and Aid's Experience: A Consideration Based on Evaluations*. A.I.D. Program Evaluation Report No. 8. Washington, DC: U.S. Agency for International Development.

Tiffany, Warren
1979 "New Directions in Political Anthropology: The Use of Corporate Models for the Analysis of Political Organization." In *Political Anthropology–The State of the Art*. S. Lee Seaton and Henry J. Claessen, editors, pp. 63-78. New York: Mouton.

Weber, Max
1947 *The Theory of Social and Economic Organization*. London: The Free Press of Glencoe, Collier-Macmillan Ltd.

CHAPTER SEVEN

Canal Irrigation and the State: The 1969 Water Law and Irrigation Systems of the Colca Valley of Southwestern Peru

David Guillet
Catholic University

The irrigation systems of highland Peru present interesting examples of resource management that originated in noncapitalist Inca and pre-Inca societies, and that were subsequently incorporated into the Spanish colonial empire and independent nation-state. In 1969, a new water law, the Ley General de Aguas, enacted by decree under the regime of President Juan Velasco Alvarado, represented a radical attempt to change the relationships between local irrigation systems and the state. It sought to improve irrigation administration by strengthening local-level, special-purpose irrigation associations, linking these associations to the state and providing a corpus of regulations to manage water resources more efficiently.

The subject of this paper is the implementation of the 1969 General Water Law in the Colca valley of southwestern Peru. Highland communities have responded variably to this law (see Bolin *intra*; Seligmann and Bunker *intra*; and Gelles *intra*). Some have adopted it wholeheartedly, others partly, and still others not at all. I focus on three aspects of the implementation of the law in the Colca Valley: (1) continuance of traditional forms of water management, (2) provisions establishing a priority in the uses of water, and (3) provisions for emergency water regimes. I use this case to explore the nature of highland Andean irrigation systems and their articulation with the state.

HIGHLAND PERUVIAN IRRIGATION SYSTEMS AND THE STATE

The history of highland irrigation systems suggests the importance of two interacting spheres of control.[1] During Incan hegemony, the first and smaller unit was a cluster of fields sharing a common source of water and often a water judge (*cilquiua*) responsible for distribution, named during November at the onset of the irrigation season (Guaman Poma de Ayala 1980[1584-1614]:1,058). When water was scarce, irrigation within the cluster was by turns, "according to the order of the plots of land, one after another," and water was allocated by "...the number of hours supply he needed for the amount of land he had..." (Garcilaso de la Vega 1966[1609] I:248).

The second level organized groups of clusters. During Inca hegemony, this level included the corporate descent group (*ayllu*) associated with land, ceremonial sites (*huacas*), and water sources. *Ayllus,* in turn, were incorporated into radial (*ceque*), quadripartite (*suyu*), and dual (*saya*) divisions.

The city of Cuzco and its immediate hinterland exemplifies the application of the Incan model of hydraulic organization above the cluster level. The Huatanay River divided land between mountain and valley and left and right, with Hanan, the right side, higher in sociopolitical rank than Hurin, the left side. Hanan and Hurin had separate irrigation systems and water officials who monitored boundaries. Hanan Cuzco and Hurin Cuzco were divided in half into Chinchaysuyo/Antisuyu and Collasuyu/Cuntisuyu, respectively, producing a quadripartition. A set of lines (*ceques*) radiating from a center in the city further divided each quarter into wedge-shaped divisions, creating irrigation subdivisions each with its own source of water. These subdivisions, in turn, were associated with social groupings ranked according to tripartite principles and may have served as a basis for the recruitment of labor. Although seemingly complex, rigid, and reflecting on-the-ground divisions informed by a cognitive model of spatial organization, the resulting system was quite flexible, able to adjust to topographic and other environmental constraints.

Efforts were made to protect the irrigation cluster from abuses by *ayllu* political authorities (*curacas*) (Guaman Poma de Ayala 1980[1584-1614]:1,058). Water judges, rotated annually, were charged with ensuring an equitable distribution of water. By their ability to withhold water, water judges held check over the hereditary *curacas* who otherwise would have been able to monopolize the limited water, taking it from the poor to do so.

In Cuzco, Spanish colonial administration dismembered the highest levels of Incan organization–the dual moiety and quadripartite divisions–but in other respects it retained the lower levels of the Inca system, the irrigation cluster, *ayllu* and wedge-shaped *ceque* divisions (Sherbondy *intra*). The autonomy of the irrigation cluster was challenged, however, in the reforms instituted by Viceroy Toledo in the 1570s, which placed water management under the authority of local government.

Following independence, irrigation management continued to be subordinated to local government structure in the 1892 Municipal Law and, in the 1933 constitution, through links to a prefectural hierarchy. The autonomy of water management of precolonial Peru was effectively curtailed until it was reestablished by the state through the 1969 General Water Law (Guillet 1989). I will take up this development below. I turn now to a précis of the irrigation systems of the Colca Valley.

THE COLCA VALLEY

The Colca Valley lies on the semiarid, western side of the western branch of the Andean Cordillera (see Map 7.1). With a mean rainfall of 414 millimeters, it falls within the generally accepted range (150-250 millimeters to 250-500 millimeters) of semiarid ecosystems (Moran 1979). A series of stepped geomorphic surfaces separated by steep scarps or hill slopes characterize the valley; the surfaces are alluvial terraces of the Río Colca and its tributaries. Denevan and his colleagues (1986:49) have identified eight separate surfaces: the narrow channel and floodplain of the Río Colca, a low alluvial terrace subject to flooding, and the remainder, consisting of either gravel fills, sand, and fine-grained sediments or erosional surfaces underlain by thin alluvium over bedrock. Villages tend to be located on these alluvial terraces overlooking the inner valley; the majority are concentrated in the middle section of the valley, from the villages of Chivay to Cabanaconde. Farmers depend directly or indirectly on cultivation of the terraced slopes and flats surrounding each village. Aside from a handful of mine workers, craft producers, and a few other occupations, the local economy is oriented primarily toward the production of foodstuffs for consumption and only secondarily toward the marketing of agricultural and pastoral surpluses.

The contemporary Quechua-speaking settlements of the middle Colca originated in a prehistoric Aymara-speaking Collaguas polity, an outlier of the circum-Lake Titicaca Colla kingdoms, conquered by the Inca under Mayta Capac (Málaga Medina 1977).[2] Incan reorganization of the Collaguas polity has been heralded as an unusually clear and faithful application of Incan cultural models found in Cuzco (Wachtel 1977:122-127; Zuidema 1964). Following the Spanish Conquest, the Colca Valley fortunately never came to be dominated by the large estates (*haciendas*) found elsewhere in the Peruvian highlands.[3] During the eighteenth century, attention focused on the newly discovered mines in the *puna* above the valley, rather than on the agricultural potential of the valley per se. In the late nineteenth and early twentieth centuries, an expanding export wool economy led to the formation of some large-scale wool haciendas in the *puna* and an increase in the numbers of alpacas in the herds of small-scale pastoralists but largely bypassed farmers in the valley below. In the present century, an improved road system has stimulated some commodity production for Arequipa markets, such as the external trade in maize in Cabanaconde, but distance and transportation costs have effectively constrained the spread of cash cropping (Gelles *intra*). Distance also prevents fresh milk from being produced in the Colca Valley for delivery to the Leche Gloria plant in

MAP 7.1 LARI AND THE COLCA VALLEY

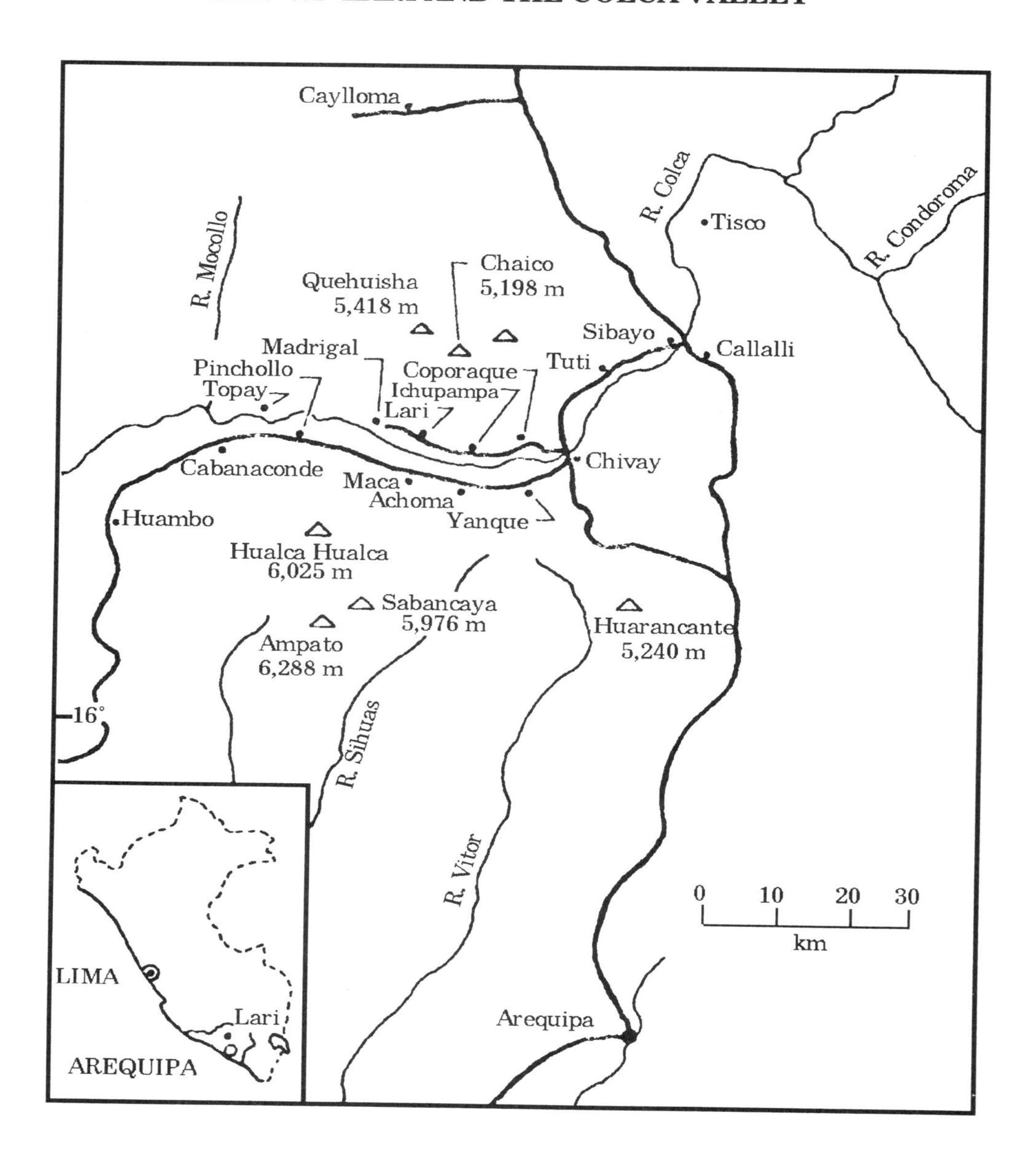

Arequipa. With the exception of limited cheese production, concentrated in Achoma and oriented to a local market, fresh milk production is quite low and almost entirely consumed in the home (Webber 1988).

Thus, the forces that have integrated Arequipa small-holders into capitalist production–the urban markets of the city of Arequipa, the world fiber trade, and capital-export capitalism (Love 1989)–have largely bypassed Colca Valley villagers. This is not to say, however, that Colca villages have noncapitalist economies: from at least the early colonial period, villagers participated in active land markets in which land was bought and sold, and they also traveled beyond the valley in search of wages to pay tribute assessments (Guillet 1992a).

VILLAGE IRRIGATION SYSTEMS IN THE COLCA VALLEY

The middle Colca Valley is a catchment basin for precipitation that falls on the peaks surrounding it. Villages capture the runoff and tap springs that arise in the lower slopes. The Colca River itself is too far below the villages to offer a cost-effective source of water for irrigation. The irrigation systems that have evolved in this setting are some of the most complex in the literature. The five Colca villages discussed in this chapter are, from east to west and by declining altitude: Yanque, Coporaque, Lari, Cabanaconde, and Tapay.[4] Table 7.1 presents data on selected socioeconomic characteristics of the villages.

Yanque, on the south side of the valley on the outskirts of Chivay, is unique among the five villages in possessing land on both sides of the Colca River. This land is irrigated with melt from the snowcaps of Huarancante and Mismi and water from small springs that is stored in reservoirs. On the north side of the river a short distance (7 kilometers) from Chivay lies Coporaque. Its water comes primarily from streams flowing from Huillcaya, a peak above the village, from a seep (*bofedal*) on the east side, and from a small stream on the northwest corner of the village. Fields are watered directly during the day from canals fed by their sources and from reservoirs that have stored water during the night. Farther down the north side of the valley, past Ichupampa, stands the village of Lari. The majority of its fields are irrigated by water originating in snowmelts and high altitude seeps that feed three separate canal networks. Small clusters of fields are also irrigated by streams emerging on the lower slopes. Fields are irrigated during the day, both directly and from water stored in reservoirs at night.

Cabanaconde, farther down the south side of the river, is the most distant of the villages from Chivay and the largest in population and amount of land in cultivation. Water for the irrigation of the maize fields in which it specializes comes from the peak of Hualca Hualca, the Majes canal, and a small spring; lower altitude fruit trees are irrigated from nearby springs. Unlike the other villages, reservoirs are not used and water moves directly to fields both day and night.

Last, across the Colca River from Cabanaconde lies Tapay. It is a remote village, even by local standards, with access via mule path from the neighboring village of Madrigal or by a strenuous descent from Cabanaconde to the river and a climb up

TABLE 7.1 LAND, WATER AND POPULATION IN FIVE COLCA VALLEY VILLAGES

Village	1981 Population	Cultivated Area (hectares)	Water Source Name	Water Source Liters/Second
Yanque	2313	510	Huarancante	60
			Misme	65
			Jatunyacu	60
			Umajaca	50
			Vizcaychani	25
			Curiña	20
Coporaque	1163	229	Sawara	88-630
			Chilliwitira	32-132
			Qachulle	24
Lari	1179	339	Surimana	15
			Callumayo	111
			Chaico	101
Cabanaconde	3421	1231	Hualca Hualca	120
Tapay	997	246	Ciprijina	50
			Tocayo	30

Note: For Cabanaconde, the figure for population and cultivated area includes the community of Pinchollo.

Source: Population and cultivated areas are derived from Denevan 1986, Table 2, p. 30. Water flow for Cabanaconde, Tapay, Yanque, and the Surimana canal in Lari is from C.I.P.A. 1981. Water flow for the Callumayo and Chayco canals is from Herman Swen, personal communication (averages for measured flows, November and December, 1988). Water flow for Coporaque is from Treacy 1989

the other side. The village is the only one of the five to attain the Andean ideal of the control of all major vertical production zones: aboriculture, for the fruits in which it specializes, from 2,250 meters to about 3,000 meters; maize to 3400 meters; potatoes, barley, and broad beans to 3,800 meters; and pastoralism in the *puna* above 4,000 meters. As elsewhere, water is obtained from the snowmelt of peaks above the village and from subsurface springs. These sources are much more numerous and dispersed in Tapay than in the other villages; there are 21 groundwater-fed springs and 32 off-takes from 6 runoff-fed streams.

In sum, all the village irrigation systems of the region exploit water from run-off of snowmelt from the upper peaks and from springs on the lower slopes. The villages vary considerably, however, in the relative contribution each type of water source makes to the overall system of irrigation and in the degree of dispersion and independence of these water sources. They also differ in the importance placed on nighttime storage in reservoirs.

THE 1969 GENERAL WATER LAW

On July 24, 1969, the reformist military government of General Juan Velasco Alvarado enacted by decree the General Water Law, Ley General de Aguas, replacing the Water Code, Cogido de Aguas, in effect since 1902. The law was one of several radical reform initiatives by a government heavily influenced by norms of social justice. Modified slightly in 1981, the law remains in effect today.

The law laid the groundwork for a far-reaching reorganization of water management, strengthening local-level, special-purpose irrigation management bodies and creating a hierarchy linking these bodies to the state. The guiding premise of the law is contained in Article One in which water is declared inalienable and the property of the state. All uses of water, other than domestic, require permission from the appropriate state agency (Article 4f). Private ownership of water and water rights is expressly prohibited.

Permission to use water is regulated, in turn, by a system of prioritization. Permissible uses of water include, in order of priority, from high to low: domestic needs, animal husbandry, agriculture, hydroelectric power, industry and mining, and other miscellaneous uses (Article 27). Within the range of agricultural uses, priorities are also established, including, again from high to low: irrigation of cultivable land with existing irrigation systems; irrigation of specific crops with surplus water; improvement of soils; and unspecified all-purpose irrigation (Article 42).[5] New land made available by the expansion through natural or artificial means of water sources can be transferred to private ownership by the state for purposes of housing and agrarian reform.

Permission to use water for agriculture is granted for a particular crop. To receive permission, one must be inscribed in a register (*padrón*), have maintained the irrigation infrastructure of one's fields in good repair, and have paid the tariffs and other fees levied by the water authority (Article 49). Once granted, a specified use may not become part of a title of ownership (Article 37).

For administrative purposes, the law divides regions into irrigation districts (*distritos de riego*) with an elected administrative body (*junta de usuarios*). Administrative bodies must submit an annual or biennial cropping and irrigation plan (*plan de cultivo y riego*) (Article 43, 136). Irrigation districts are administered by the Water and Irrigation Agency (*dirección general de aguas e irrigación*) of the Ministry of Agriculture.

Irrigation districts are subdivided, in turn, into sectors (*sectores*), each with a corresponding administrative body or irrigation commission (Comisión de Regantes) (Article 136). Irrigators pay tariffs, based on the volume of water used, which provide funds for the administrative costs of water distribution, irrigation infrastructure and the financing of hydraulic development studies (Article 12).

The 1969 General Water Law empowers the state, through its delegated authorities, to enter directly into local water management. The scope of their powers is wide. Under normal conditions, authorities can dictate the use to which water is put, giving them indirect control over crop production and urban land use. Their power increases during periods of water scarcity through the declaration of an emergency water regime, allowing them to radically modify existing systems of water distribution. This emergency water regime is one of the most important provisions of the law. It can be triggered by water scarcity, excessive water, or pollution. Water authorities are empowered under these provisions to dictate strict procedures to protect, control, and distribute water for the benefit of the collectivity, giving preference to domestic needs (Article 17). Emergency water regimes for reasons of scarcity (*Estado de emergencia por escasez*) are treated specifically in Articles 45 through 48. Once it is determined that the water supply is insufficient to meet the demands of registered landowners, cultivation and irrigation plans must be in accord with the most efficient structuring of the irrigation system and must give preference to the crops and fields that provide the greatest collective benefit (Article 45). Article 48 authorizes water authorities to establish special systems of distribution.[6]

The ability of the state to implement the provisions of the 1969 General Water Law is the issue to which I now turn.

IMPLEMENTATION OF THE 1969 WATER LAW

In 1970 the Ministry of Agriculture established an office in Chivay as part of the Velasco government's focus on the countryside. Since there were no *haciendas* in the Colca Valley to be expropriated or agrarian syndicate activity to monitor, the office could direct its attention, undistracted, to improving water management and implementing the new water law. The office moved quickly, naming an administrator and assembling a staff, including a water engineer (*ingeniero de aguas*), to manage the newly created irrigation district of the Colca Valley. By the middle 1970s, 18 sectors had been organized, each with an irrigation commission. Each commission had and still has a governing board with a president, secretary, treasurer, fiscal, and two aldermen (*vocales*). Water judges are also members of each commission. With the exception of water judges, members are elected, serve for two years, and can be removed by a vote of the majority of users. Irrigation commissions send representatives to the provincial administrative body, which meets at the Ministry of Agriculture office in Chivay. Most villages have a resident extension agent (*sectorista*).

Landowners are required to declare their holdings, pay a tax, and register with the irrigation commission to receive water. In 1986, the tax was approximately U.S. $1.00 per *topo*, a local unit of land measure, approximately 3,500 square meters (Treacy 1989: 316). An official of the irrigation commission collects the taxes and brings the proceeds to the Ministry of Agriculture office in Chivay. The office is obligated to return 90 percent of the tax to villages in the form of improvements to irrigation infrastructure, such as the lining of canals and reservoirs with cement and the installation of sluice gates. The irrigation section of the Ministry of Agriculture office in Chivay now helps resolve disputes over water and makes villagers aware of the provisions of the water law.

Persistence of Traditional Forms of Irrigation Organization

Despite the creation of irrigation commissions, linked to the Ministry of Agriculture, traditional forms of irrigation organization persist to a greater or lesser degree in each community (see Table 7.2). Yanque represents the best example of this, approximating closely the irrigation organization of Incaic Cuzco. The Colca River divides land into halves with Hanansaya on the right side looking upstream, toward the rising sun, and Hurinsaya on the left, as in Ulloa Mogollon's 1581 description of the province (1965 [1581]).[7] Each division has a separate irrigation system; Hanansaya draws its water from Chucura Mountain, and Hurinsaya, from Misme Mountain. This Hanansaya-Hurinsaya division is duplicated within the confines of the contemporary village.[8] In this settlement, an imaginary line runs north-south through the plaza, the church, and beyond. It divides village land into Hanansaya on the east, or upriver, and Hurinsaya on the west, downriver. The church is located across this line so that the altar and apse lie in Hanansaya and the main entrance is in Hurinsaya. In important ceremonies, the faithful separate spatially according to moiety affiliation. The church is further divided lengthwise by a line running east-west, parallel with the river. The intersection of these lines creates four quarters called *barrios* making a quadripartite division (Benavides 1983:28, 190;Valderrama and Escalante 1988).

Each of Yanque's two irrigation systems has its own set of irrigation clusters (*tomas*) that share a common water source and set of water judges (*regidores*). The state respected the integrity of each system by giving each an irrigation commission. The resulting organization is presented in Table 7.3.

In the traditional irrigation system of Coporaque, called the *saya* system, land is not divided into Hanansaya and Hurinsaya sections (see Table 7.2). Individuals, however, identify themselves as members of one or the other moiety, and it is this personal designation of *sayas* that enters into irrigation organization. Water from the main canal is shifted between two field areas for four days at a time and, within each field area, shunted from east to west. During each four-day period, *regidores* distribute water by moiety, with water "accessible to Anansaya members for two days, then rotated to Urinsaya members for two days . . . producing two separate, identical systems separated by a time dimension" (Treacy 1989:322).[9] Moiety affiliation also comes into play during the annual cleaning of the canals, when villagers ascend to the source of the headwaters. Splitting into halves, Hanansaya members clean one leg, Hurinsaya the other. They descend, cleaning as they go, to meet and celebrate at the point where the two canals join. Then, united, they proceed to clean the major (Sawara) canal.

TABLE 7.2 IRRIGATION ORGANIZATION IN FIVE COLCA VALLEY VILLAGES

	Yanque	Coporaque	Lari	Cabana	Tapay
Moiety Principles in:					
Water Distribution	+	+		+	
Contiguous Spatial Divisions	+		+		+
Non-Contiguous Field Divisions				+	
Social Organization	+		+	+	
Dominant Water Source:					
Surface Water	+	+	+	+	
Spring Water					+
Irrigation Organization:					
Water Judges	+	+	+	+	+
Irrigation Commission	+	+	+	+	

Note: In Cabanaconde, water judges are the only individuals identified as belonging to Hanansaya or Hurinsaya and then only during their service as water judge.

Sources: See Footnote 4.

In Lari, land is divided into Hanansaya and Hurinsaya. The owners of land are also considered members of Hanansaya or Hurinsaya. However, *saya* divisions play no role in water transport, allocation, or distribution, with one exception–the recruitment of labor during communal work parties (*faenas*) called to repair irrigation infrastructure. On these occasions, labor is recruited by teams (*cuadrilla*) drawn from lists of members of each moiety. There are no other traditional irrigation officials above the level of the *regidor*. The irrigation commission, established by the state, functions as an active village-based irrigation association.

In Cabanaconde, one finds areas of contiguous fields associated with Hanansaya and Urinsaya, but within each division, there are fields of the other moiety.[10] Thus, along one canal some fields are Hanansaya, and others are Urinsaya. Notwithstanding the breakdown of spatial division by *saya*, moiety principles do function in terms of water distribution. Each *saya* has water judges who perform as their counterparts elsewhere in Colca villages, overseeing claims to water, monitoring the established order of distribution, ensuring that offtakes are opened and closed according to schedule, and mediating conflicts. Water judges are named from among the ranks of large landowners.

Interestingly, water judges are the only individuals identified as belonging to Hanansaya or Hurinsaya, and then only during their service. Each farmer possesses land in both Hanansaya and Hurinsaya and receives water for each field from the corresponding water judge. One water judge works a shift of four days and

TABLE 7.3 IRRIGATION ORGANIZATION IN YANQUE

Irrigation Commission	Hurinsaya	Hanansaya
President	+	+
Vicepresident	+	+
Secretary	+	+
Treasurer Fiscal	+	+
First Vocal	+	+
Second Vocal	+	+
Traditional Organization		
Regidor de Aguas/Yaku Alcalde	+	+
Kamachikusqa yana (Water ritual specialist)	+	+
Rondador (Caretaker of the Misme Canal)	+	
Rikuk (Water ritual assistant)	+	+
Misme Regidor	+	
Qochapata Regidor	+	
Huaranqante Regidor		+
Hatun Yaku Regidor		+
Pampaqocha Regidor		+
Churkina Regidor		+
Vizcachani Cabecilla		+
Piyuto Cabecilla		+
Pukyo Cabecilla		+
Pirichu Cabecilla		+

four nights, after which he is replaced by the other.

Although a district (*distrito*) like its Colca neighbors, Tapay lacks their large, nucleated, central settlements. Instead, its administrative center (known as Tapay) is small, and most of the sparse population of 983 is scattered across the landscape in nine hamlets (*anexos*). Most of the hamlets belong to one of the two moieties. The hamlets of Paclla, Llatica, Pure, and Tocallo, however, are not affiliated with the *saya* system.

Most hamlets control their own sources of water. *Regidores* are the only water officials, and irrigation clusters are not incorporated into any moiety or wider levels of organization. Many independent water sources minimize the need for coordination and allow the few conflicts over water distribution to be resolved within the irrigation cluster. In only two cases, Tapay and Cosñihua, do hamlets share water from the same source, in this case from the Sepregina River. Tapay, upstream of Cosñihua, takes water first. Although this order could conceivably lead to upstream-downstream conflicts, a spring adds considerable water to the amount Cosñihua gets from the river, mitigating the potential for conflict.

Although Tapay's irrigation system is decentralized with respect to water transport, allocation, and distribution, it is unified through a shared belief system. Ritual sacrifices (*pagos*) are made to spirits that inhabit each water source in concordance with a complex spatial and temporal calendar culminating on All Saints Day. Through these rituals the decentralized collection of irrigation clusters is integrated into a spatial and temporal network.

Irrigation organization in the five Colca Valley villages is summarized in Table 7.2. In Yanque, each moiety has its own irrigation system with a set of irrigation clusters and water judges. The state has respected this system by according each moiety system its own irrigation commission. The nucleated settlement of Yanque also has a quadripartite division of *barrios*. In Coporaque, although there are no *saya* land divisions, *saya* affiliation governs water distribution and labor recruitment. In Lari, nothing imposes itself between irrigation clusters and a village-level irrigation association in terms of water allocation and distribution. Moiety affiliation is used only for the recruitment of labor for the maintenance of irrigation infrastructure and water ritual. In Cabanaconde, large contiguous blocks of land are divided along moiety lines, but some of the opposite moiety's fields are interspersed among them. Water sources are not divided by moiety. Despite the weakness of moiety spatial divisions, each moiety names its own set of *regidores*. In the decentralized system of Tapay there are no levels of organization above the irrigation cluster other than in ritual.

Prioritized Uses of Water

The implementation of the provisions of the law concerning the priority of permissible uses of water is an important issue in the Colca Valley, to which I turn next. Officials have assigned low priority in water allocation to several types of abandoned and fallowed land. One category of land, *terreno eriazo*, has no water rights. It includes uncultivable peaks, outcroppings, and barren slopes, cultivable land that has never been cultivated, and cultivable land once cultivated but abandoned in a distant past. Cultivable *terreno eriazo* may be extended water if new sources of water are located or existing sources tapped more efficiently by a decree of the Ministry of Agriculture. Fallow land (*terrenos en descanso*) consists of two types. Owners of fields in fallow up to three years plus one day can receive water by paying a fine for each field and reinscribing in the respective register (*padrón*). Fields in fallow more than three years and one day lose all rights to water. To recover rights, one must go to the Ministry of Agriculture office in Chivay, show title to the property, give the reasons for abandonment, and pay a fine to the local irrigation commission. These provisions present legal obstacles to long-term fallowing, a regenerative practice of traditional agriculture.

These provisions of the 1969 General Water Law, however, do help communities defend control of their uncultivated land. Highland communities traditionally manage *terreno eriazo* for grazing and as a source of firewood, herbs, and medicinal plants. It is not uncommon, however, for local officials to illicitly sell parcels of *terreno eriazo* to individuals, or for powerful individuals to simply seize such parcels (see Gelles *intra*). Unless a community has the corporate status of a legally recognized peasant community, it can do little to counteract these transfers. The 1969 General Water Law, by restricting the access of water to *terreno eriazo*, gave irrigation-dependent communities such as those in the Colca Valley an important resource to defend themselves.

Lari and Cabanaconde provide cases of legitimate and illegitimate forms of conversion of *terreno eriazo* to private control (Guillet 1989). In Lari, a Ministry of Agriculture decree in the 1970s freed *terreno eriazo* in two locations for sale by lottery and granted the land limited water rights. The subsequent use of these fields by their owners has been carefully monitored by the Lari irrigation commission and the Ministry of Agriculture. In other instances where *terreno eriazo* has been obtained illicitly, the irrigation commission has been able to deny the illegally seized lands irrigation water, so that they are used only as unirrigated natural pastures.

A similar strategy appears to have been employed in Cabanaconde. Five powerful families had controlled the offices of mayor, governor, and president of the village irrigation association until the 1960s. These families were literate and maintained contacts with powerful officials in Chivay and Arequipa. During the period 1930-55, members of the families began to sell, mostly among themselves, tracts of *terreno eriazo* in a fertile but unirrigated area of the village. To capitalize on their investment, the new owners needed water. This presented a problem, since to extend water to the new lands meant reducing the amount available for already cultivated fields. The community refused permission to do so. Even after a prolonged legal battle in which the families used their contacts with powerful individuals in Arequipa and Lima, they were able to obtain water for only less than a hectare of land (Gelles *intra*).

Emergency Water Regimes

The last aspect of the 1969 General Water Law we take up are its provisions for emergency water regimes. Increased population growth in this century has greatly intensified demands on water in Colca villages (Cook 1982). Heightened demand in conjunction with periodic droughts has created the conditions for invoking these provisions of the water law.[11] To deal with this increased demand, the Ministry of Agriculture attempts to "rationalize" village irrigation systems in two ways. The first is to replace irrigation systems in which social considerations determine distribution with one in which location of the field is the determining factor. This system of the continuous irrigation of one adjacent field after another is known by various names: *mita*, *mita global,* and *riego canto a canto* in Coporaque, *mita* in Lari and *de canto* in Cabanaconde.[12] The ministry attempts to introduce such continuous irrigation wherever and whenever it can, preferably as *the* system of irrigation, or failing that as an emergency water regime. Second, the Ministry of Agriculture also tries to reduce demand on water by setting limits on the amount of land a household can cultivate. Ministry officials have restricted, in some cases, the water available to households for irrigation and have tried to eliminate the irrigation of alfalfa in favor of subsistence crops.

This two-part strategy is fine-tuned to the situation in each village.[13] In Coporaque, when the drought regime is in effect, water shifts, as usual, between field areas and rotates across canals, but water distribution by moiety affiliation is dropped. Water judges are repositioned, and for two to five days, water goes to the west field area then rotates to the east. The number of irrigators that can be accommodated under continuous irrigation is usually considerably higher than under the moiety system, thus conserving water. Treacy, for example, found that in 1983-84 under the moiety system, 205 people from Urinsaya irrigated during the month of October (with a high of 22 people in one day), while during the same month in 1981-82 under continuous irrigation, 291 farmers irrigated (with a high of 59 in one day) (Treacy 1989: 328).

In the 1960s the Ministry of Agriculture attempted to institute continuous irrigation as the norm in Coporaque. It proved highly unpopular in practice because of the limitations it imposed on a farmer's access to water. Although increased numbers of people could irrigate daily, farmers with large acreage fields could not obtain sufficient water to irrigate all their fields. Large landowners complained that the system was unfair to them, so that the government abandoned continuous irrigation as the norm in 1972 after only ten years.

Nonetheless, continuous irrigation persists in Coporaque as an emergency regime because of its clear advantages as a water conservation method. It is one of two pillars of a drought strategy instituted successfully by the Ministry of Agriculture. The other imposes limits on the number of *topos* a household can cultivate during drought years. In a drought, such as that in 1985-86, a household was allowed only enough water for one *topo* each of broad beans and potatoes, and two *topos* each of maize and barley.[14] As a result, bottomland fields usually seeded in barley are often temporarily abandoned in dry periods.

Lari, plagued with low moisture-retentive soils and an inadequate supply of water, has the most difficult problem of the five villages in meeting its demands for water. Consequently, it had instituted a continuous irrigation system well before efforts by the Ministry of Agriculture to impose it elsewhere in the Colca Valley.[15] The Ministry of Agriculture, however, has introduced limits on water during droughts. Under normal conditions, water is distributed in a quantity sufficient to cover the requirements of a field, irrespective of size or crop. At the suggestion of the water engineer from the Ministry of Agriculture, limits have been imposed by water judges during droughts on the amount of water distributed to a user according to crop and field size. During the serious drought of 1983-84, for example, two days of continuous irrigation were set as the maximum for the irrigation of barley fields by one owner in an irrigation cluster. Although fought vigorously by large landholders each year, limits nonetheless have been imposed during several years of the 1970s and 1980s.

Treacy contends that continuous irrigation was probably Incaic in origin (Treacy 1989: 174-181). His argument is based on an east-to-west rotation pattern of irrigation and planting and the physical layout of the irrigation system, with canals emanating from central points and the close association of house sites with canals. He suggests that this corresponds to *ceque* organization:

> The arrangement of the alleged ceques in Coporaque may have involved sight lines from reservoirs (nodes) to settlements on *qhopos* (hills) near canals, allowing an observer to note which canal was in use on any day. Most importantly, the heart of the organizational principles of ceque lines is discernable in the Coporaque system, especially in terms of establishing a locus of responsibility (irrigation administration from reservoirs), and assigning irrigation use periods according to a semi-circular march through canals in space and time (1989:181).

In Cabanaconde, farmers have adopted continuous irrigation during the high moisture months of January through April. At this time of low water demand, they drop the saya field sequence and distribute water along a canal without regard to moiety status. From June to December, however, the most important period for the irrigation of maize, farmers use the traditional *saya* system and have opposed Ministry of Agriculture efforts to replace it with continuous irrigation. Even though Gelles reports (*intra*) that Cabanaconde had gained six full days when continuous irrigation had been tried during an irrigation cycle in 1947, farmers oppose its use from June to December because it is associated with attempts of powerful elites to wrest land from the community and because it is also viewed as a slower method of water distribution.

During continuous irrigation, water is distributed by paid water judges (*controladores*) instead of the unpaid ones (*yaku alcaldes*) found in the *saya* system. Since the two unpaid water judges in the *saya* system compete with each other to finish irrigating first, water moves more rapidly (according to informants), allowing them to meet very narrow schedules for irrigating crops. Delay can jeopardize a crop in this frost-ridden environment with its short growing season. This role of competition in "advancing" water brings into focus the agronomic function of moiety organization in the Colca Valley, a region with a well-documented (but now banned) tradition of ritual battles along moiety lines. In the continuous system of water distribution, on the other hand, water judges are paid by the hour, a system of payment that informants say causes them to prolong the distribution of water.[16]

In brief, implementation of the provisions for emergency water regimes has varied in the valley. The state has successfully introduced continuous irrigation as one pillar of an emergency water regime, perhaps unconsciously resurrecting an older, pre-Columbian form of water distribution. It has been unable, however, to expand it as the dominant system in those villages where moiety principles are firmly established. Two reasons come to mind. Individuals who benefit from moiety distribution coalesce to resist continuous irrigation. Moiety competition may motivate water judges to advance quickly through the irrigation cycle. The other pillar of the state's emergency water regimes, limiting water demand directly or indirectly through restrictions on land under cultivation, has also had significant effects.

DISCUSSION

In the twenty years in which the water law has been in effect, the state has been unable to fully implement its provisions in the Colca Valley. Traditional systems of water management persist, despite efforts of the Ministry of Agriculture dedicated to reorganizing and "rationalizing" them. This is not to say that the state has failed entirely in its efforts. The eighteen irrigation commissions that have been established encompass virtually all the agricultural villages of the province. In these instances, the state has successfully removed water management from municipal and civil administrations and turned it over to special-purpose bodies. Although the state has not been able to replace traditional forms of water management entirely, it has helped support the autonomy of water management against attacks made by powerful elites, thereby bucking a trend that had been established with the Spanish Conquest.

Why do traditional social and economic forms persist when confronted by a state policy designed to replace them? Some observers answer this question by pointing out links between these traditional forms and capitalist forces at the regional and national level and then showing how their preservation is consistent with the logic of capitalist development or the needs of capital.[17] Establishing the links necessary for such an argument in this case is difficult at best. As we have seen, small landholders in the Colca Valley are only weakly integrated into capitalist circuits of the Arequipa region, unlike those in the Arequipa countryside. Nor is the revenue obtained through the taxation of water a significant income stream for the state, since 90 percent returns to the villages to support infrastructure investments.

Recent critics of this position suggest that dynamism resides not in the dominant classes of core capitalist countries or peripheral states, but among local coalitions, classes, and actors (Roseberry 1989). This suggestion helps explain the persistence of traditional water management. Colca Valley communities pursue an economic logic based on small-scale subsistence cultivation in a fragile and demanding environment. This logic is not one of capitalist development or the needs of capital. Since at least the early colonial period villagers have been involved in an active land market in which land is bought, sold, rented, used as collateral for loans, and leased. Rather, their logic is one of appropriate water management–the principles of water transportation, allocation, and distribution–which have stood the systems in good stead through years of both adequate and inadequate rainfall. Competent water management and well-adjusted land use has produced the more than 1,500 years of sustainable agriculture in the Colca Valley, not an easy feat in this mountainous region plagued with erosion and loss of soil fertility.

The state's successes have been in areas that return autonomy to village water management or enhance group survival. Thus, the state has been successful in many villages in implementing continuous irrigation and limits on water during emergency water regimes, although it has failed to impose it as the normal water distribution system in Coporaque and Cabanaconde. It is only in Lari, among the villages studied here, that continuous irrigation and limits on water have been accepted as the norm.

Where the state has failed in implementing the law, it has usually been the result of coalitions of large landowners who resist egalitarian principles of water distribution.[18] These elites, however, have been unable to act together outside the village to oppose the law in general. Village elites are physically isolated from one another and are relatively weak in being able to act beyond the limits of their village. Given their economic base in the fiber trade rather than agricultural production, for example, the Chivay elite has little interest in the disposition of water disputes in other Colca Valley communities. Thus, no regionwide class of powerful landowners has been able to exert its influence disproportionately on the implementation of the water law.

Similarly, the state has been able to regulate land expansion and urban and rural use in the valley. Officials have been able to achieve this by specifying the conditions under which transfers of *terreno eriazo* to private use are legitimate and by controlling the access of these lands to water. The ability of the state to make these decisions has thereby empowered villages to withhold water from "owners" of land obtained illicitly.

Traditional forms of irrigation derived from noncapitalist pre-Colombian societies in the Colca Valley are best understood not as shelters from capitalist penetration, but as successful adaptations to agriculture in a difficult environment. Consequently, the state has been unable to eliminate them. The state, however, has been able to implement a more efficient system to deal with the periodic droughts and the demands within villages to expand cultivable land. Finally, the state has been successful in those cases where it has emphasized the autonomy of local irrigation systems in their efforts to combat the demands of local elites. In an odd twist, then, the 1969 General Water Law has helped communities return to a successful adaptation of their pre-Colombian past.

NOTES

Acknowledgments. I wish to thank William P. Mitchell and Paul Gelles for their insightful comments on an earlier draft of this essay. I remain responsible for all errors of fact or interpretation.

1. See Sherbondy 1982, 1987, and *intra* for irrigation organization in Incaic Cuzco. Changes in irrigation systems subsequent to the Spanish Conquest are only beginning to be understood (see Sherbondy's chapter in this volume). For an extended discussion of the concept of irrigation cluster, see Guillet 1992a.

2. The Cabana people, at the southwest end of the middle Colca Valley, were not ethnically related to the Aymara-speaking Collas (see Gelles *intra*). Although the people of Yanque-Collaguas and Lari-Collaguas spoke Aymara, the Cabaneños spoke Quechua, the language of Cuzco (Ulloa Mogollón 1965). Other traits linking Collagua with Colla culture (origin myths, hats, and mode of head deformation) were different among the Cabana. Today, the distinctions between Cabanaconde and the other higher villages of the middle Colca Valley remain. Villagers from Cabanaconde speak a Quechua with a different pronunciation and vocabulary, wear a distinct folk dress, and are separated by a natural topographical boundary. For positions that support the divergence of the Cabana from the Collagua people, see Zuidema (1964) and Manrique (1985: 28ff). Gelles cites phenotypical and cultural differences between contemporary Cabaneños and other villagers stemming from a larger number of Spaniards in the late colonial and early republican periods (1988).

3. Sources for the reconstruction of the post-Conquest history of the Colca Valley include: Flores

Galindo 1977; Gomez Rodriguez 1978; Malaga Medina 1977; Manrique 1985.

4. Sources include the chapters in this volume by Treacy, Gelles, Guillet, and Paerregaard as well as Gelles 1988; Guillet 1987, 1989, 1992a, 1992b; Paerregaard 1989; Treacy 1989; Valderrama and Escalante 1986, 1988.

5. Article 42 of the 1969 water law: "Podrán otorgarse usos de aguas para agricultura en el siguiente orden: a) el riego de tierras agrícolas con sistemas de regadio existente; b) el riego de determinados cultivos con aguas exedentes en tierras agrícolas con sistemas de regadio existente; c) mejorar suelos; y d) irrigación."

6. Specifically mentioned are *mitas*, which will be discussed later in the paper, and *quiebras*, the closure of an upstream offtake of a river or a canal to shunt water to lower offtakes. Codicil to the water law.

7. I am indebted to María Benavides for this observation.

8. In the Toledan reforms, Yanque's population was resettled into a nucleated village (*reduccion*) on the right side of the river, and the original settlement on the left side of the river (known today as Yanque Viejo) was abandoned.

9. The pattern described is for the Sawara Qantumayo system. According to Treacy (1989: 320), two smaller systems function independently in a similar manner.

10. See Gelles *intra*.

11. Rainfall patterns in the Colca Valley reflect year-to-year unpredictability with occasional stretches of years of extremely low rainfall. For background, see Guillet 1987 and Treacy 1989. According to Caviedes (1982), unusually wet years in arid northern and central Peru (El Niño events) are often coupled with below average precipitation in the southern highlands and altiplano of Peru and altiplano of Bolivia. The pattern was particularly evident in El Niño, 1982-84, and the contemporaneous drought in the southern highlands. For material on this most recent, and arguably most serious, drought to affect the region, see Claverías and Manrique 1983.

12. See Gelles *intra*; Guillet 1987; Treacy 1989:327ff.

13. Yanque does not have an emergency regime of continuous irrigation, although the irrigation of alfalfa is prohibited during droughts (Valderrama and Escalante 1988: 68-70).

14. The *topo* is an indigenous Andean land measure used in the Colca Valley at least through the mid-nineteenth century as an all-purpose measure. It is a relative rather than an absolute measure, referring to the amount of land a team of men wielding a footplow (*chakitaclla*) could prepare in one day. *Topo* continues today as an occasional measure for land and work assignments during corvée labor projects. See Guillet (1992a:33-36) for a discussion of land measures used today in the Colca Valley.

15. One older Lareño informant says that a system of rotation along moiety lines similar to that of Coporaque's antedated the current system. Another suggests the transition to the current system dates from the 1940s, which would coincide with experiments with continuous irrigation in Cabanaconde about the same time (Gelles *intra*). It isn't clear how they arrived at the idea of continuous irrigation. It may be the resurrection of an older system in response to growing demands on water triggered by the population growth that began in the 1940s. In many areas of the Andes where a moiety consists of a contiguous geographic territory associated with a single irrigation system, continuous irrigation during the main planting is the norm. See, for example, Mitchell's (1976) discussion of irrigation in Ayacucho and Treacy's (1989:174-181) discussion of Coporaque.

16. Gelles suggests that the rather weak fit between informant's statements that de *canto* distribution is ultimately unable to provide the motivation necessary for quick irrigation and evidence that, in fact, it is quicker is because of a reluctance to do away with the ritual complex and equitable distribution associated with the *saya* system (Gelles *intra*).

17. See Orlove (1986) for a discussion of current theoretical positions on the persistence of traditional social and economic forms.

18. Whether such groups constitute classes with interests reducible to those of capitalists or capital is debatable. I have argued that limits placed on water during droughts produce the weak forms of stratification encountered in Lari and elsewhere in the Colca Valley (Guillet 1989; INP 1983:14).

REFERENCES CITED

Benavides, Maria

1983 *Two Traditional Andean Peasant Communities under the Stress of Market Penetration: Yanque and Madrigal in the Colca Valley, Peru.* M.A. thesis in Latin American Studies, University of Texas at Austin.

Caviedes, César N.

1982 "On the Genetic Linkages of Precipitation in South America." In *Fortschritte Landschaftsokologischer und Klimatologischer Forschung in den Tropen: Festschrift zum 60 Geburtstag von Professor Wolfgang Weischett,* pp. 55-57. Freiburg: Freiburger Geographische Hefte, No. 18.

C.I.P.A. (Centro de Investigación y Promoción Agraria)

1981 *Problematica agropecuaria de la provincia de Caylloma.* Arequipa: Centro de Investigación y Promoción Agraria.

Claverías, Ricardo and Jorge Manrique, editors

1983 *La sequía en Puno: Alternativas institucionales y populares.* Puno: Instituto de Investigaciones para el Desarrollo Social del Altiplano.

Cook, Noble David

1982 *The People of the Colca Valley: A Population Study.* Boulder, CO: Westview Press.

Denevan, William M.

1986 "Introduction: The Río Colca Abandoned Terrace Project." In *The Cultural Ecology, Archaeology, and History of Terracing and Terrace Abandonment in the Colca Valley of Southern Peru.* William M. Denevan, editor, pp. 8-46. Technical Report to the National Science Foundation (Anthropology Program). Madison: Department of Geography, University of Wisconsin.

Denevan, William M., John Treacy, and Jon Sandor

1986 "The Physical Geography of the Coporaque Region." In *The Cultural Ecology, Archaeology, and History of Terracing and Terrace Abandonment in the Colca valley of Southern Peru.* William M. Denevan, editor, Technical Report to the National Science Foundation (Anthropology Program), pp. 47-59. Madison: Department of Geography, University of Wisconsin.

Flores Galindo, Alberto

1977 *Arequipa y el sur andino siglos XVII-XX.* Lima: Editorial Horizonte.

Garcilaso de la Vega

1966 [1609] *Royal Commentaries of the Incas and General History of Peru.* 2 Volumes. Austin: University of Texas Press.

Gelles, Paul

1988 *Los hijos de Hualca Hualca: historia de Cabanaconde.* Arequipa: Centro de Apoyo y Promoción al Desarrollo Agrario, Serie Aportes No. 1.

1989 "*Dual Organization and Irrigation in an Andean Peasant Community.*" Paper read at the Annual meeting of the American Anthropological Association, Washington DC.

Gomez Rodriguez, Juan

1978 La reforma agraria *en Caylloma.* Lima: Pontificia Universidad Catolica.

Guaman Poma de Ayala, Felipe

1980 [1584-1614] *El primer nueva crónica y buen gobierno.* 3 Vols. Mexico, DF: Siglo Veintiuno.

Guillet, David

1987 "Terracing and Irrigation in the Peruvian highlands." *Current Anthropology* 28(4):409-430.

1989 "The Struggle for Autonomy: Irrigation and Power in Highland Peru." In *Human Systems Ecology.* Sheldon Smith and Ed Reeves, editors, pp. 41-57. Boulder, CO: Westview Press.

1992a *Covering Ground: Communal Water Management and the State in the Peruvian Highlands.* Ann Arbor: University of Michigan Press.

1992b "The Impact of Alfalfa Introduction on Common Field Agro-pastoral Regimes: Quechua Agro-pastoralists in Southwestern Peru." In *Plants, Animals and People: Crop-Livestock Systems Research on the SR-CRSP Sociology Project*. Constance McCorkle, editor, pp. 111-124. Boulder, CO: Westview Press.

Instituto Nacional de Planificacion (INP)

1983 *Diagnostico microregional de las provincias altas de Arequipa-Caylloma*. Arequipa: Instituto Nacional de Planificacion.

Love, Thomas F.

1989 "Limits to the Articulation of Modes of Production Approach: The Southwestern Peru Region." In *State, Capital, and Rural Society: Anthropological Perspectives on Political Economy in Mexico and the Andes*, Benjamin S. Orlove, Michael W. Foley, and Thomas F. Love, editors, pp. 147-179. Boulder, CO: Westview Press.

Málaga Medina, Alejandro

1977 "Los collaguas en la historia de Arequipa en el siglo XVI." In *Collaguas I*. Franklin Pease, editor, pp. 93-129. Lima: Pontificia Universidad Catolica.

Manrique, Nelson

1985 *Colonialismo y pobreza campesina: Caylloma y el valle del Colca, siglos XVI-XX*. Lima: DESCO.

Mitchell, William P.

1976 "Irrigation and Community in the Central Peruvian Highlands." *American Anthropologist* 78:25-44.

Moran, Emilio

1979 *Human Adaptability*. North Scituate, MA: Duxbury Press.

Orlove, Benjamin

1986 "Barter and Cash Sale on Lake Titicaca: A Test of Competing Approaches." *Current Anthropology* 27:85-106.

Paerregaard, Karsten

1989 "Exchanging With Nature: T'inka in an Andean Village." *Folk* 31:53-75.

Roseberry, William

1989 "Anthropology, History, and Modes of Production." In *State, Capital, and Rural Society: Anthropological Perspectives on Political Economy in Mexico and the Andes*. Benjamin S. Orlove, Michael W. Foley, and Thomas F. Love, editors, pp. 9-37. Boulder, CO: Westview Press.

Sherbondy, Jeanette E.

1982 *The Canal Systems of Hanan Cuzco*. Ph.D. dissertation, University of Illinois.

1987 "Organización hidráulica y poder en el Cuzco de los Incas." *Revista Española de Antropologia Americana* 17:117-153.

Steward, Julian

1955 *Theory of Culture Change*. Urbana: University of Illinois Press.

Treacy, John

1989 *The Fields of Coporaque: Agricultural Terracing and Water Management in the Colca Valley, Arequipa, Peru*. Ph.D. dissertation, Department of Geography, University of Wisconsin.

Ulloa Mogollón, Juan de

1965[1581] "Relacion de la provincia de los Collaguas." In *Relaciones Geograficas de Indias*, Volume 1. Marcos Jiminez de la Espada, editor, pp. 326-333. Madrid: Biblioteca de Autores Españoles.

Valderrama Fernández, Ricardo, and Carmen Escalante

1986 "Sistema de riego y organización social en el valle del Colca–caso Yanque." *Allpanchis* 27:179-203.

1988 *De tata mallku a la mama pacha: Riego, sociedad y ritos en los andes Peruanos*. Lima: Centro de Estudios y Promocion del Desarrollo.

Wachtel, Nathan
1977 *The Vision of the Vanquished: The Spanish Conquest of Peru through Indian eyes, 1530-1570*. New York: Barnes & Noble.
Webber, Ellen Robinson
1988 "Alfalfa and Cattle in Achoma: A Study of Stability and Change." In The *Cultural Ecology, Archaeology, and History of Terracing and Terrace Abandonment in the Colca Valley of Southern Peru*. William M. Denevan, editor, pp. 91-111. Technical Report to the National Science Foundation (Anthropology Program). Madison: Department of Geography, University of Wisconsin.
Zuidema, R. T.
1964 *The Ceque System of Cuzco: The Social Organization of the Capital of the Inca*. Leiden: E.J. Brill.

CHAPTER EIGHT

Why Fight Over Water? Power, Conflict, and Irrigation in an Andean Village

Karsten Paerregaard
University of Copenhagen

Studies of the relationship between irrigation and sociopolitical organization have focused in recent years on issues of centralization and decentralization (Hunt and Hunt 1976; Kelley 1983). These concepts have been defined in terms of who controls water sources and water distribution, be it local groups of irrigators, local elites, or the state. Only recently have these discussions taken up the role of belief and ritual in irrigation organization (Lansing 1987). This development is salutary and apposite to the ethnology of highland Andean irrigation systems in which beliefs and ritual surrounding water are widespread.

In this chapter, I address the manner in which beliefs and ritual unify an irrigation system reliant on a large number of independent water sources and offtakes from different rivers that are managed in a decentralized manner. Villagers in Tapay, the subject of this study, believe that the flow of water is controlled by spiritual beings that inhabit each of their water sources. Maintaining good relations with these spirits through periodic ritual offerings is crucial to securing adequate water flow and, therefore, agricultural success. Although water distribution seldom provokes disputes, contested communication with water spirits sometimes leads to conflict. I argue in this chapter that conflicts concerning irrigation in Tapay are played out for symbolic reasons rather than for utilitarian reasons. This argument is substantiated by an analysis of efforts by a Protestant congregation to challenge the moral and religious precepts of the Catholic majority. Through an analysis of this conflict, I explore the nature of power manifested when Tapeños fight for control of water.

TAPAY AND THE COLCA VALLEY

Tapay is a district in Caylloma Province of the department of Arequipa in southwestern Peru. The total population of 983 is distributed among 287 households found in dispersed household clusters in the herding zone (the tundralike *puna* above 4,000 meters) and in nine nucleated hamlets located in the agricultural zone (below 4,000 meters). Tapay hamlet (149 inhabitants living in 47 households) is the capital and the historical center of the district of the same name (see Map 8.1). Tapay hamlet is being challenged for dominance, however, by the rapidly growing hamlet of Cosñihua (168 inhabitants living in 53 households).[1]

The population of Tapay is found along several altitudinal belts ranging from 2,150 to 5,000 meters in a vast area punctuated by several mountain ranges. The subtropical climate from 2,250 to 3,000 meters lends a junglelike ambience to the lowest belt. This zone specializes in the fruit cultivation that constitutes the backbone of the village economy. Tapay is the main producer of fruit in the regional barter system.[2] This subtropical area contains the hamlets of Paclla (2,250 meters at the valley floor), Chugcho (2,400 meters), Llatica (2,500 meters), Pure (2,650 meters), Cosñihua and Malata (2,600 meters), Puquio (2,700 meters), and Tapay (3,000 meters). The next higher belt, from 3,000 to 3,400 meters, is devoted almost entirely to maize agriculture. The hamlet of Tocallo (3,800 meters) in the highest agricultural zone specializes in potatoes, barley, and broad beans. Puna herders living above 4,000 meters produce fresh and dried meat and wool.

Like other Colca villages, Tapay is divided into two moieties: Hanansaya (Quechua "upper moiety"), including the central hamlet of Tapay, Puquio, and Chugcho, and Urinsaya (Quechua "lower moiety"), including Cosñihua and Malata (see Map 8.1).[3] These moieties are created by the Seprigina River, a natural boundary and one of the main sources of water in Tapay (see Map 8.2). Hanansaya is considered the more important of the two: its location upstream allows Tapay hamlet to take water before Cosñihua and Malata in Urinsaya. In addition, the Fiesta of the Virgin de Candelaria, an event of particular ritual importance to the village, is celebrated traditionally in Hanansaya.[4] These facts are interpreted by Tapeños as signs of Hanansaya's superiority.

Not all hamlets are included in the dual moiety structure. Paclla, Llatica, Pure, and Tocallo belong to neither Hanansaya nor Urinsaya. Despite their seemingly peripheral status with respect to the moieties, these hamlets are closely linked economically and politically to the rest of the district. Paclla, Llatica, and Pure residents have the same production orientations as their Hanansaya and Urinsaya neighbors and participate on an equal footing with other Tapay villagers in the regional barter system.[5]

MAP 8.1 TAPAY PRODUCTION ZONES

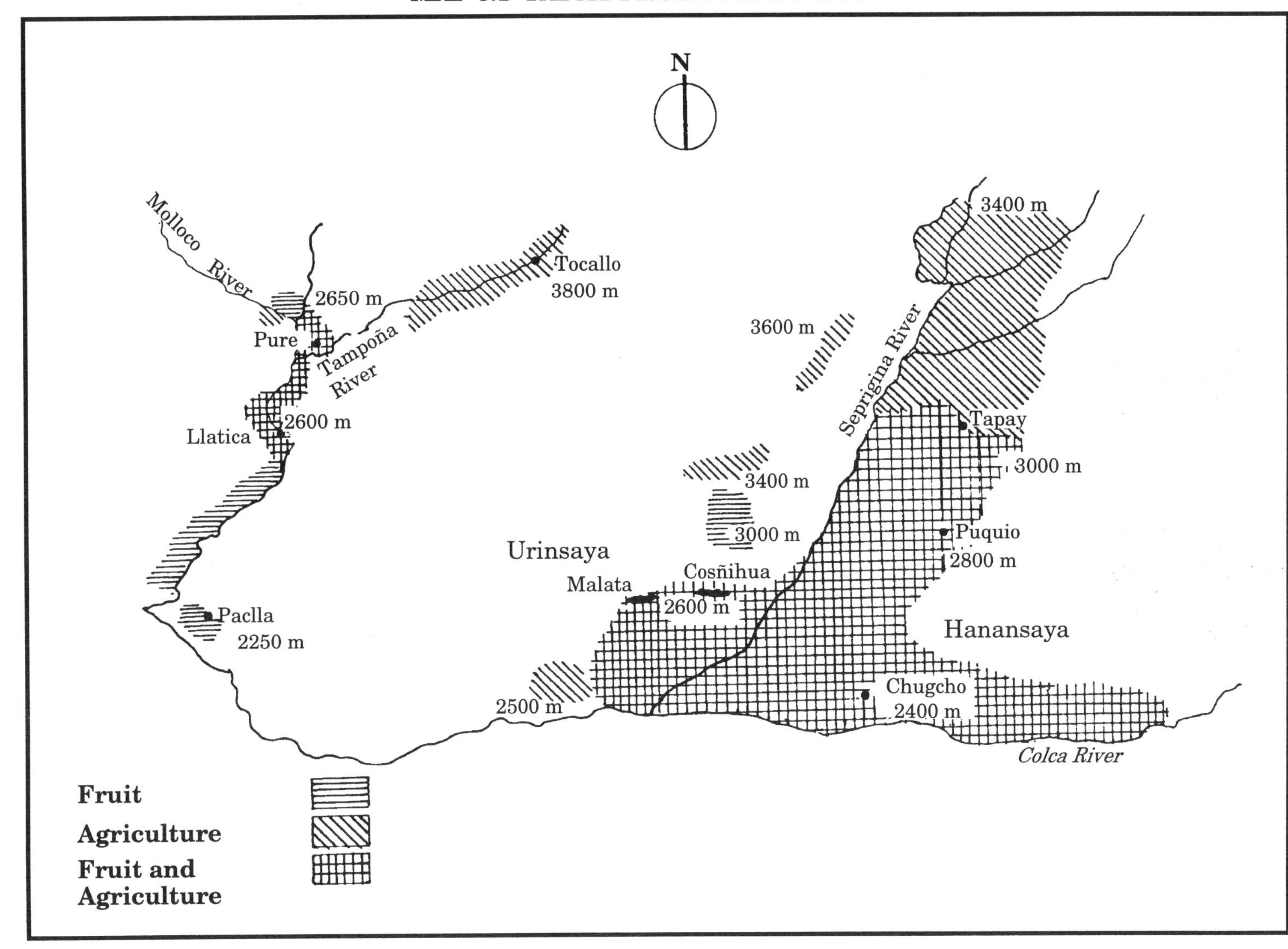

MAP 8.2 TAPAY IRRIGATION SYSTEM

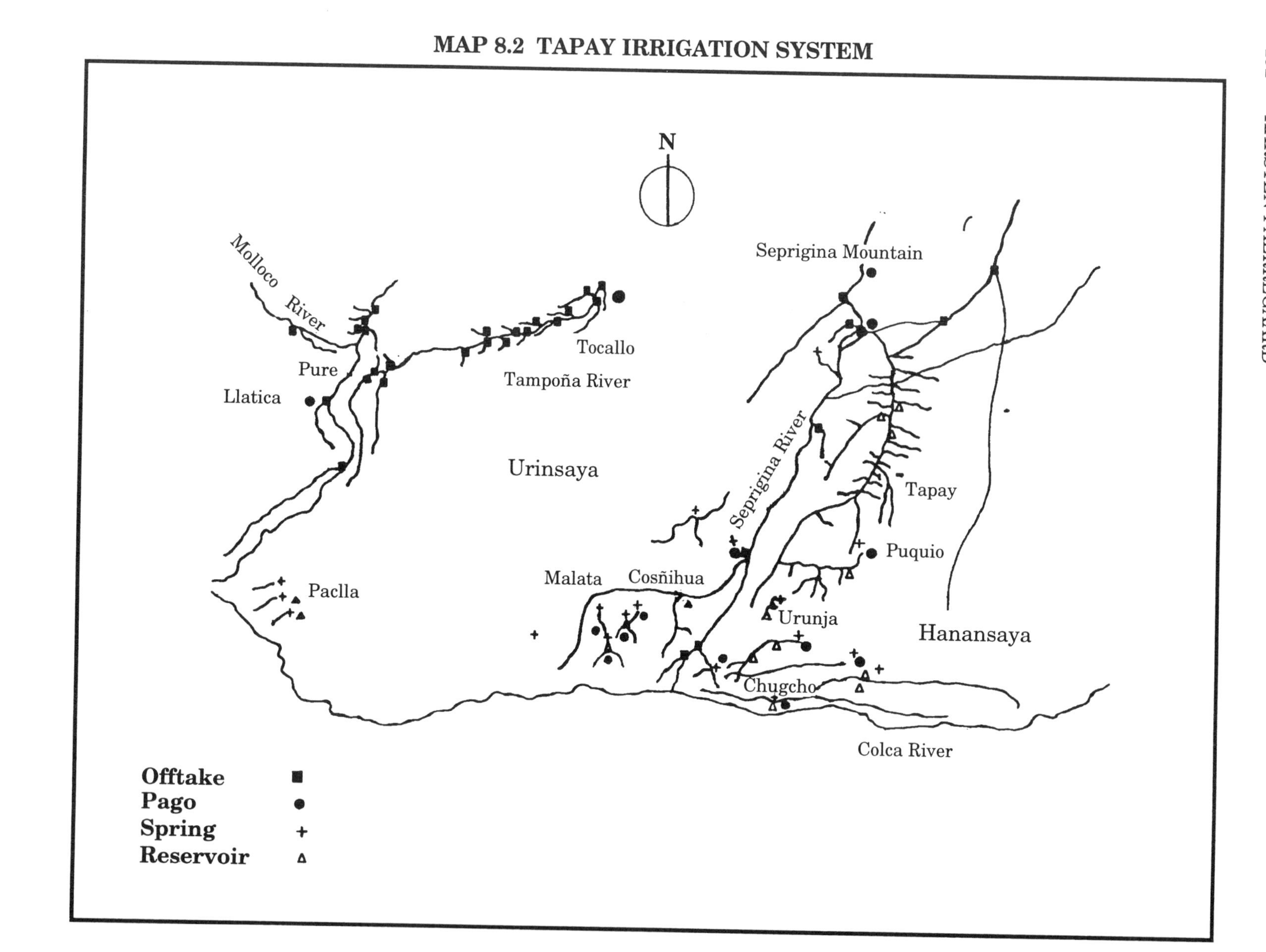

In sum, although Tapay has much in common with other Colca Valley villages, there are significant divergences. Its population is dispersed across the landscape in small hamlets rather than concentrated in one large nucleated settlement. Although it possesses a dual moiety structure, not all hamlets are included. The village produces fruit rather than the grains produced by other valley villages, which distinguishes it in the regional bartering system. Tapay is also isolated geographically from the rest of the Colca villages. They are all connected to one another by a regional road network, but Tapay is accessible only by foot or mule.[6] Finally, the village diverges significantly from others in the nature of its irrigation system, the subject to which I now turn.

WATER SOURCES AND IRRIGATION ORGANIZATION IN TAPAY

In most Colca Valley villages, irrigation systems are supplied by water from a few main surface or ground water sources (see Gelles *intra*; Guillet *intra*; and Treacy *intra*). In Lari, water is transported from the upper puna via three separate main canals (Guillet 1987:413). Yanque is supplied by two different high altitude sources, one on each side of the Colca River (Valderrama and Escalante 1988:61-77). Irrigation in Cabanaconde, across the Colca River from Tapay, is supplied by three main water sources and one spring, used mainly for potable water (Gelles 1988:13-16).

In the village of Tapay, there are 53 sources dispersed across the landscape: 21 independent springs and 32 offtakes from six different rivers and streams (see Map 8.2). Five hamlets control their own sources of water. In only two cases do different hamlets take water from the same river or mountain. These are Tocallo and Pure, which share the Tampoña River, and Tapay hamlet and Cosñihua–the two main settlements of Hanansaya and Urinsaya respectively, which share the Seprigina River. This arrangement between Hanansaya and Urinsaya follows a general pattern in Andean communities where dual organizations often are based on the irrigation system (Fock 1981b:405; Isbell 1978:139; Mitchell 1976:26; Sherbondy 1986:43; Skar 1982:150; Valderrama and Escalante 1986:181; Zuidema 1986:190-191,197).

Tapay is unique in the Andes for its lack of conflict over water, a common problem elsewhere (Gelles 1986:125, 1988:22-23; Guillet 1987:413; Mitchell *intra*; Skar 1982:151; Treacy 1987:425; Valderrama and Escalante 1986:183-187, 200). Even when different hamlets in Tapay use the same river as a water source, conflict over that water is minimized by the presence of supplementary water sources. For example, Tocallo's position 1,000 meters above Pure allows farmers there to take water first, but the reduced flow to Pure is mitigated by Pure's access to several independent sources of water. Similarly, since irrigation in Urinsaya does not depend entirely on the Seprigina River, conflicts with Hanansaya over water rarely arise. A spring located between the water offtake and Cosñihua adds a considerable amount of water to the supply after Hanansaya has taken its share from the river. This reduces competition over water to a minimum.

Minimal competition over water is true even in times of water scarcity, as happened with the late arrival of the rains in January of 1988. At that time, Hanansaya cultivators located on the upper agricultural levels suffered a water deficit, and they either sowed their fields in vain or did not risk sowing at all. In Urinsaya, however, people believed that more water than usual was flowing from the supplementary spring, reducing any possible tensions over Hanansaya's use of river water.

Other springs act similarly, reducing upstream-downstream tensions in years of drought. Local streams also meet all the needs for fruit growing, which is concentrated in the lower altitudes of Puquio, Chuqcho, Cosñihua, and Malata. When the rains failed to come in January 1988, for example, only a few people on these fruit-producing levels complained about the lack of water, even though farmers in the upper altitudes were experiencing a water shortage.

Tapay also differs from the rest of the Colca Valley in the organization of irrigation. In Tapay, like elsewhere in the valley, water rights are associated with land rights. A Tapay villager has access to water and participates in the local water-user group wherever he owns land. Neighboring villages, however, use a hierarchy of water authorities at the offtake (*tomo*), moiety, or village level to distribute and manage water (Gelles 1988:16-25; Guillet 1987:413-414; Valderrama and Escalante 1986:198). The people of Tapay differ in that they maintain their canals, allocate water, and resolve water conflicts without any superordinate authority roles. No central authority in or outside the village controls irrigation.

In sum, three features are important to an understanding of Tapay's irrigation system. First, the population is distributed in such a way that they are able to use a number of dispersed sources of water. Second, agrarian production is adapted to the altitudinal differences in water sources and irrigation systems. In the lower altitudes, where springs feed local irrigation systems, production is dedicated to fruit growing. Agriculture predominates on the upper levels where snowmelt supplies the needed water. Third, irrigation is associated with the division of the village into two opposing moieties. The Seprigina River divides the village into halves and provides irrigation water to both moieties.

CENTRALIZATION AND DECENTRALIZATION IN THE ANDES AND IN TAPAY

In discussing the concepts of centralization and decentralization in irrigation it is convenient to distinguish at least five tasks of major importance: new construction of canal systems, routine maintenance of infrastructure, water allocation, resolution of conflicts, and linkages to ritual organizations and religious institutions (Hunt and Hunt 1976).[7] Although new construction projects requiring the participation of several hamlets or an entire moiety are rare, the maintenance of the physical infrastructure, the allocation of water, and the resolution of conflicts are problems that Tapeños must deal with on a regular basis. These three tasks are all managed in Tapay by water users of the same offtake, reservoir, or spring. Two water distributors (*regidores*) are elected for each water source during annual cleaning and water ritual celebrations.[8] The *hatun regidor* (major *regidor*) is in charge of distribution from July until January; the *huch'uy regidor* (minor *regidor*) holds office during the rainy season. Canal and reservoir maintenance and conflict resolution are managed by users of the water source.

In addition to water distributors, one water judge (*juez de agua*) is elected from each moiety. His function is to resolve conflicts arising between a water distributor and users. No other village authorities, including municipal officials, enter into water conflict resolution. Each water-user group, irrespective of its size or location, operates independently. Water distributors and water judges are not integrated into an administrative hierarchy. Not even in Tapay hamlet–the central hamlet and administrative center of the village–is irrigation subject to administrative or political authorities. Thus, even though the people of Tapay hamlet drawing water from the Seprigina River outnumber users elsewhere in the village, this numerical dominance has no effect on the decentralized nature of irrigation. Water distributors and water judges are the only irrigation authorities in the village, and they are not integrated into a villagewide, centralized irrigation association. No central power among the hamlets or in the village manages the irrigation tasks in question.

THE SPIRITUAL POWERS OF WATER

No central authority organizes water ritual. Instead, irrigation rituals are arranged locally, at the hamlet, moiety, and village levels. These rituals are, nonetheless, important. Tapeños believe they are subject to the power of spirits residing in the sources of water. Propitiating these spirits is crucial to adequate water flow, and the failure of Protestant villagers to perform these rituals has led to serious conflict within the village.

In Tapay, every spring is thought to be protected by a spirit to whom users must make a ceremonial offering called *pago*.[9] This sacrifice, known throughout the Andes, must be made to the water spirit several times a year at a water source, be it offtake, spring, or reservoir (Barthel 1986:156-157; Isbell 1978:162; Ossio 1978:382-384; Skar 1982:225; Zuidema 1986:183-184). These ceremonies please the spiritual powers and ensure the continual flow of water, on which the harvest hinges.

Ceremonial offerings require several ritual elements to satisfy the powers that control the water: seeds and leaves of the coca plant (*Erythroxylum coca* var. *coca*), *conuja* (an unidentified herb from the puna), *qorilibro* and *qolqelibro* (small objects of gold and silver), *pichuwira* (fat from the breast of llama), maize of three different colors, *cochayuyo* (sea weed), and starfish. Some of the objects are burned at the spring, canal offtake, or at the peak of the mountain water source. Others, such as a special thick fermented beer (*aqa*) made from maize of three different colors, are consumed during the ceremonies. Wine and seawater are also taken during the ceremonies.

The seaweed, starfish, and seawater components of the ritual merit special comment. Their common origin in the sea is associated with a belief in the union of otherwise independent water sources. The idea of the Earth floating on a sea that unites all parts of the world is common in many parts of the Andes, (Bastien 1985:604; Cáceres 1986:114; Fock 1981a:315; Sherbondy 1982:3-4).

As can be seen in Table 8.1, moiety offerings are repeated twice, and in some cases three times.[10] They are given in different places and at different times the first time they are offered. The second offerings are held in the same locations as the first and partially overlap with them in time. A number of offerings in Hanansaya (Puquio, Chugcho, Pallajua, and Huilcasco) are repeated a second time on the same date, but at different locations. The second offering of Urinsaya coincides in time and place with the third offering of Hanansaya. All offerings move in a sequence toward the same geographical spot, Seprigina Mountain, and they all finish at the same time, November 1, All Saints' Day. This advance of offerings and their eventual culmination on the same day transforms a dimension in space into a dimension in time, a familiar phenomenon in the Andean world (Fock 1981a:313).

Ritual Sequence, Moiety Ceremonial Offerings

Although the people in Hanansaya make three offerings a year, those in Urinsaya make only two. Except for one, all four hamlets outside the moiety structure make a first offering at places and times that differ from the rest of the village. I have no knowledge of either a second or a third offering in these hamlets. Because of this, Paclla (which has no official offerings at all), Llatica, Pure, and Tocallo are not included in my model.

November 1, All Saints' Day (Todos Santos), is considered the most important of all rituals associated with irrigation. The *regidor* and a water ritual specialist visit the summit of Seprigina Mountain and offer a sacrifice to the spiritual powers controlling the flow of water. All Saints' Day is the celebration of the departed. The dead are divided into two classes: the recently deceased and the ancestors (*gentiles*) or grandparents (*abuelos*) who lived in Tapay before the Incas and the Spaniards. *Gentiles* also refer to the living spirits that emanate from the human skeletons and bones found in the numerous pre-Hispanic graves located in the village.[11] Because the skeletal materials are not perceived as really dead, they bring fear to villagers. A common belief among older informants is that water flows because the *gentiles* urinate, and villagers are virtually at the mercy of the ancestors who, at a whim, can move water from one hole to another or withhold it altogether. One can quell their anger and persuade them to release water only by performing *pagos* at the right time and place.

Tapeño perception of spiritual rather than human powers controlling water has roots in a pan-Andean cosmology (Bunker and Seligmann 1986:171; Sherbondy 1982:7-8, 1986:42; Zuidema 1986:192). Because the irrigation system is fed by many independent water sources, the local clusters of water users are not subject to central control and water allocation rarely causes conflict. Instead, people fight over their communication with water spirits, so that the political arena deals with belief and ritual rather than water distribution. To illustrate the nature of this spiritual conflict involving water, I turn to recent attempts by Protestant cliques to modify

TABLE 8.1 WATER RITUAL CALENDAR

Date	Saint's day	Activity
June 24	San Juan	Cleaning of and *first offering* to springs in Chugcho, Pallajua and Huilcasco (all of HANANSAYA)
July 29	Santa Maria Magdalena	Cleaning of and *first offering* to offtakes in central hamlet (HANANSAYA)
August 5	San Clemento	Cleaning of and *first offering* to spring in Puquio (HANANSAYA)
August 8	San Clemento	Cleaning of and *first offering* to spring in Urunja (HANANSAYA)
August 24	Santa Marta	Cleaning of and *first offering* to spring and offtakes in Cosñihua and Malata, (URINSAYA)
August 30	Santa Rosa de Lima	Cleaning of and *first offering* to offtakes in Pure
September 8	La Virgen de Natividad	*Second offering* to springs in Puquio and Chugcho, Pallajua, Urunja and Huilcasco (HANANSAYA)
September 10	La Virgen de Dolores	Cleaning of and *first offering* to offtakes in Llatica
September 20	San Fransisco	Cleaning of and *first offering* to offtakes in Tocallo
October 4	San Fransisco	*Second offering* to offtakes in the central hamlet (HANANSAYA)
November 1	Todos Santos	*Second offering* by URINSAYA and *third offering* by HANANSAYA to the mountain peak supplying the water to offtake in the central hamlet, the two hamlets in URINSAYA and to all springs in HANANSAYA.

traditional irrigation organization and abolish water ritual.

THE EVANGELICALS AND CONTROVERSY OVER WATER

Although they make up a scant 5 percent of the village population, Evangelical Protestants (*Evangelistas*) are an emerging political force in Tapay. They take every opportunity to oppose the dominant Catholic majority, focusing on two issues to mark their opposition: abstinence from alcohol and abolition of ritual and ceremony. By stressing work, discipline, and "progress" as moral values, and arguing that these are the "true" ways to obey God, they place themselves in opposition to a mystic and ritualized Catholic church, which they consider irrational and sinful.[12] The majority of the Evangelicals live in the tiny hamlet of Puquio in Hanansaya moiety, although several reside in Tapay hamlet and Chugcho.

In 1986, a small group of Protestants attempted to gain control over irrigation in the central hamlet. In that year, representatives from the Ministry of Agriculture visited the village to institute a tax on water. A list of all landowners using the irrigation system was required, and the state authorities gave the villagers the option of compiling it through irrigation committees to be formed in each moiety and the hamlets of Llatica and Pure. Although Tapeños had paid land taxes for many years, they resisted this effort to tax their water. In Hanansaya, Catholic water users did this passively by refusing to get involved with the irrigation committee. Consequently, the Evangelicals filled all the important positions on the committee.

What eventually led to clashes between Evangelicals and Catholics was not the issue of the taxation of water but the work of canal cleaning in the central hamlet at the end of July. The young, dynamic, and dedicated Evangelical president of the committee took charge of organizing the labor corvée (*faena*) and insisted on taking a roll call in the morning at a fixed time. Anyone absent from the labor corvée was fined. Fines had never been levied against absentees in the past, because no central power had ever organized a roll call until the Evangelicals initiated one and thereby provoked a problem.

Although intended to coerce people to work, the roll call did not succeed. At first, the young Evangelical leader did have some success. A few participants arrived on time and actually started to follow his orders, but discipline soon lapsed. As more people arrived at the work scene, attention focused–as it always had in the past–on the election of the water distributor for the coming irrigation cycle. Nonetheless, the young Evangelical leader continued to insist on keeping people working. When it became apparent that the Evangelicals intended to control the *faena*, Catholics became irate. In Tapay, as in other parts of the Andes, the irrigation *faena* is considered a fiesta as well as an obligation to work (Gelles 1986: 117; Isbell 1978: 139). Until the Evangelicals took control of the corvée, Tapeños had given the same importance to the election of the new water distributor and the celebration of the fiesta in honor of Santa Magdalena, as they had to the cleaning of the canals. Indeed, they were motivated to work by music, drinking, and dancing.

Such opposing worldviews–Protestant work and discipline versus Catholic symbols and rituals–had to clash. The Evangelicals used the new committee on irrigation as a resource, transforming the annual cleaning into an arena to contest the traditional moral system. Their premier goal was to convert the rest of the population. To us it matters little if they succeeded or not. What matters is that the battle for souls took place during the water cleaning and rituals.

A second instance of conflict between Evangelicals and Catholics occurred in January 1988 at the height of a drought. Widespread anxiety about the scarcity of water caused people to question the natural order. The Catholics blamed the drought on the Evangelicals–one of the *regidores* of the hamlet of Chugcho in Hanansaya was a member of the denomination. Catholics attributed the drought to the Protestant attempt to abolish the ritual and religious customs attached to the irrigation cleaning and election of *regidores*. Water rituals reflect the vital importance of irrigation. Catholics believed that the spirits were angered and withheld water because the Evangelical *regidor* had neglected to make offerings to one of the springs in Chugcho.

CONCLUSION: WHY USE WATER TO PROVOKE CONFLICTS?

The key to understanding the nature of irrigation in Tapay lies in the nature of the conflicts over water. In a system in which the sources of water are numerous and dispersed and water distribution is consequently localized and uncoordinated, conflicts arising from disputes over water access are virtually nonexistent. The type of conflict found in Tapay differs radically from that encountered elsewhere in the Colca Valley and the Peruvian highlands. Upstream-downstream disputes are avoided because downstream users of the Seprigina River in Urinsaya have access to independent spring sources that supplement the downstream river flow after upstream users in Hanansaya have taken their due. The only coordinated activities of individuals occur in the upper reaches of the village oriented to agriculture. In this area, the gates giving access to field divisions are opened successively in one field division after another, at the end of the respective harvests, to allow animals to graze on the stubble. They are closed at the beginning of the agricultural cycle. This coordination, however, does not affect water distribution.

If conflict has no basis in the systems of water transport and water distribution, is the irrigation system conflict-free? My material suggests not. Conflict, rather, emerges in the system of beliefs and ritual attached to water. It is here that the dispersed nature of water transport and distribution is unified in space and time, through beliefs in the importance of ritual offerings to the sources of water and their incorporation into a wider, villagewide scheduling. The majority of Tapeños believe water sources are linked through spiritual powers affiliated with the feared ancestors (*gentiles*) who control access to water. The only way to ensure continued water is by making offerings to these powers in a complex schedule of space and time. All local and independent sources of water are thus united through a village cosmology, superficially categorized as Catholic because of the Christian overlay of saints introduced by the colonial Spanish clergy, but heavily embedded in pan-

Andean origin myths associated with mountains, earth, and sources of water.

When Protestant Evangelicals, a newly emergent group in Tapay, attempted to take control of the institutions of power in the village, they chose the arena of the beliefs and rituals surrounding water. They sought to impose their worldview of hard work and punctuality on the domain of corvée labor, counterpoising it to the traditional festive nature of these occasions. The clash they provoked highlights the importance attached to the beliefs and rituals of public work groups and irrigation. Thus, what appears to be a system of irrigation free from conflict and centralized control hides an arena for competition between systems of belief. Evangelicals have done the field worker a great service by throwing light on the nature of community in a decentralized irrigation system.

NOTES

Acknowledgments. Fieldwork in Tapay was carried out in 1986 for one year with a fellowship from the University of Copenhagen and travel expenses financed by the Danish Research Council for the Humanities. During my stay in Peru I was associated with the Catholic University of Peru in Lima. I am indebted to David Guillet and William Mitchell for inspiration, comments, and criticism while writing this chapter.

1. The district is the lowest rung of the administrative hierarchy to have a full complement of municipal officials. Internally, Tapay is organized into a district capital of the same name and a number of smaller administrative units including annexes (*anexos*), small house clusters (*caserios*), and herding ranches (*estancias*). Annexes nominate their own authorities and *caserios* and *estancias* are administered directly by the district capital. In spite of these divisions and their administrative subleties, we consider Tapay an independent coherent community characterized by sets of clustered agricultural hamlets and a dispersed population of herders.

2. The regional bartering system includes all the villages on the north side and some villages on the south side of the Colca Valley and herding populations in the Caylloma puna region. Fruits grown in Tapay are bartered for corn from Cabanaconde; broadbeans, barley, alfalfa, quinoa, and potatoes from the rest of the valley; salt from Huambo and Lluta; and frozen, dehydrated potatoes (*ch'uño*) and meat from the puna. One of the most popular annual trading excursions is to travel in June, July, and August from Tapay to Yauri in Cuzco Department to exchange fruit for *ch'uño*. Participating in the regional bartering system is of crucial importance to the economy of Tapay. It permits villagers to obtain all basic Andean staples and a series of industrial and commercial products. The only common cash transactions involve the purchase of alcohol, which enters Tapay through migrant networks and the sale of cochineal dye.

3. Most of the Colca villages are divided into moieties (Benavides 1988). In two cases, however, this dual division differs from the others. In Chivay a third sector is added to the two original moieties, while in Cabanaconde the only recognized members of the moieties are the water authorities (Gelles 1988:18). Another interesting case is Yanque, where the dual division has caused a split in the communal organization of the village. Although the village constitutes one single district, each moiety has formed its own recognized peasant community (*comunidad campesina*), a formal communal organization that is independent of the political-administrative division of the Andean districts (Valderrama and Escalante 1988:29).

4. Until recently, this fiesta was the occasion for violent combat between young men from Hanansaya and Urinsaya. These clashes represented ritual battles between the two moieties rather than an expression of personal rivalry between individuals. The dual structure of Tapay, however, is changing. Ritual battles between Hanansaya and Urinsaya have been abolished, and the rivalry is taking new forms. Twenty years ago Urinsaya began to celebrate the Candelaria fiesta independently in Cosñihua, thus challenging the established hierarchy of the moieties. Moreover,

when Tapay was designated a microregion (*microregión*), entitling it to government aid in a new government program, Urinsaya's more dynamic leadership enabled it to benefit to a greater degree than Hanansaya, causing envy and resentment. Competition between the moieties–traditionally expressed through symbols and ritual–is reemerging in new guises. Moiety competition has also occurred among out-migrants. Tapeños living in Lima have formed two soccer teams, Hanansaya and Urinsaya, that participate in the matches organized by the club of Colca migrants.

5. All Tapeños, even the herders of Tocallo, are considered fruit growers, lending a sense of unity that transcends their internal divisions. Villagers acquire fruit for barter outside Tapay by exchanging their own products for fruit, or by getting the fruit on consignment from the fruit growers in lower hamlets. When fruit is sold on consignment the profits are split in half.

6. The descent is about 1,100 meters, and the ascent up to the central hamlet, about 700 meters.

7. In the Andean highlands, the organization of irrigation varies considerably from one region to another (Gelles 1986:101, 129, 132). In some communities, control over water is centralized (Fonseca 1983:64-65; Gelles 1986:121); in others it is decentralized (Mitchell 1976:40; Bunker and Seligmann 1986:162; Fock 1981a:324-25, 1981b:406).

8. Tapay's 25 reservoirs are found in Tapay, Puquio, Chugcho, Cosñihua, Malata and Paclla hamlets.

9. *Pagos* (literally, payments) are also called *t'inka*, a generic term for all sacrifices. In addition to offerings to water spirits there are *t'inka* to Mother Earth (*pacha mama*), to the ancestors (*gentiles*), to the mountain deities (*apu* or *machu*), to animals, and to houses. Other Colca villagers make similar offerings to the water and the earth. See Guillet (1987:412) for Lari; Valderrama and Escalante (1988:129-150) for Yanque; and Gelles (1988:20) for Cabanaconde.

10. The hierarchy of saints is similar to Ossio's description of water ritual in Andamarca (Ossio 1978:384-386).

11. For further information, see Ansión (1987:83-114) and Paerregaard (1987:33-34, 1989).

12. The 52 Evangelicals out of a total village population of 983 are distributed as follows: central hamlet, 11; Puquio, 10; Chugcho, 15; Cosñihua, 1; Paclla, 7; and Llatica, 8. No Evangelicals live in Malata, Pure, Tocalla hamlets nor in the *puna estancias*. They are found in Hanansaya (36), Urinsaya (1), and in the two adjacent hamlets of Llatica and Paclla (15), which are closely linked by kinship ties. New converts enter the congregation individually rather than as entire families. Few Evangelicals are related by kinship bonds (although they address each other as brother) and Catholic spouses or children rarely follow a family member into the congregation. Thus, although Evangelicals are concentrated geographically, they are dispersed socially.

REFERENCES CITED

Ansión, Juan

1987 *Desde el rincón de los muertos: El pensamiento mítico en Ayacucho*. Lima: GREDES (Grupo de Estudios para el desarrollo).

Benavides, María A.

1988 "La división social y geográfica Hanansaya/Urinsaya en el valle del Colca y la provincia de Caylloma (Arequipa, Peru)." *Boletín de Lima* 60:49-53.

Barthel, Thomas S.

1986 "Agua y primavera entre los atacameños." *Allpanchis* 28:147-184.

Bastien, Joseph W.

1985 "Qollahuaya-Andean Body Concepts: A Topographical-Hydraulic Model of Physiology." *American Anthropologist* 87:595-611.

Bunker, Stephen and Linda Seligmann

1986 "Organización social y visión ecológica de un sistema de riego andino." *Allpanchis* 27:149-178.

Cáceres, Efraín

1986 "Agua y vida en mitos andinos." *Allpanchis* 28:99-122.

Fock, Niels
1981a "Ecology and Mind in an Andean Irrigation Culture." *Folk* 23:311-330.
1981b "Ethnicity and Alternative Identification: An Example from Cañar." In *Cultural Transformation and Ethnicity in Modern Ecuador*, Norman E. Whitten, editor, pp. 402-419. Urbana: University of Illinois Press.
Fonseca, Cesar M.
1983 "El control comunal del agua en la cuenca del rio Cañete." *Allpanchis* 22:61-73.
Gelles, Paul
1986 "Sociedades hidráulicas en los Andes: Algunas perspectivas desde Huarochirí." *Allpanchis* 27:99-147.
1988 "Irrigación, comunidad y la frontera agrícola en Cabanaconde." Paper presented at the 46th Congress of Americanists, Amsterdam.
Guillet, David
1987 "Terracing and Irrigation in the Peruvian Highlands." *Current Anthropology* 28:409-418.
Hunt, Robert C., and Eva Hunt
1976 "Canal Irrigation and Local Social Organization." *Current Anthropology* 17:389-398.
Isbell, Billie Jean
1978 *To Defend Ourselves: Ecology and Ritual in an Andean Village*. Austin: University of Texas Press.
Kelly, William W.
1983 "Concepts in the Anthropological Study of Irrigation." *American Anthropologist* 85:880-885.
Lansing, J. Stephen
1987 "Balinese 'water temples' and the management of irrigation." *American Anthropologist* 89:2:326-341.
Mitchell, William P.
1976 "Irrigation and Community in the Central Peruvian Highlands." *American Anthropologist* 78:25-44.
Ossio, Juan M.
1978 "El simbolismo del agua y la representación del tiempo y el espacio en la fiesta de la acequia de la comunidad de Andamarca." *Actes du XLIIe Congrès International des Américanistes*, pp. 377-396.
Paerregaard, Karsten
1987 "Death Rituals and Symbols in the Andes." *Folk* 29:23-42.
1989 "Exchanging with Nature: T'inka in an Andean Village." *Folk* 31:53-75.
Sherbondy, Jeanette
1982 "El regadío, los lagos y los mitos de origen." *Allpanchis* 20:3-32.
1986 "Los ceques: Código de canales en el Cusco Incaico." *Allpanchis* 27:39-74.
Skar, Harald O.
1982 *The Warm Valley People: Duality and Land Reform among the Quechua Indians of Highland Peru*. Oslo: Universitetsforlaget.
Treacy, John
1987 "Comment on Terracing and Irrigation in the Peruvian Highlands by David Guillet." *Current Anthropology* 28:425.
Valderrama, Ricardo F., and Carmen Escalante Gutierrez
1986 "Sistema de riego y organización social en el valle del Colca–caso Yanque." *Allpanchis* 27:179-202
1988 *Del Tata Mallku a la Mama Pacha: Riego, sociedad y ritos en los Andes peruanos*. Lima: DESCO (Centro de Estudios y Formación del Desarrollo).
Zuidema, Tom
1986 "Inca Dynasty and Irrigation: Another look at Andean Concepts of History." In *Anthropological History of Andean Polities*, John V. Murra, Nathan Watchel, and Jacques Revel, editors, pp. 177-200. Cambridge: Cambridge University Press.

CHAPTER NINE

An Andean Irrigation System: Ecological Visions and Social Organization

Linda J. Seligmann
James Madison University

&

Stephen G. Bunker
University of Wisconsin-Madison

INTRODUCTION

Felipe Guaman Poma de Ayala, an indigenous Quechua commentator on the massive changes he witnessed in the transition from Incaic to Hispanic domination of his native land, in the following text refers to a centralized authority whose customs and laws were essential both to the coordination of the labor to build irrigation ditches and to the restrictions on other economic activities that might cause these structures to deteriorate. This text ends with an appeal to reestablish the same kind of authority and to authorize local-level bureaucratic officials to enforce the old laws and customs:

Consider that in a village, they have dug out irrigation canals from the rivers or wells, from the lakes or reservoirs. In ancient times, they dug them out with such great labor that were they to have paid for them, they would have spent ten or twelve thousand pesos or twenty thousand pesos. Consider that before there was an Inca, as there was such a great number of Indians and there was but one king and master, they opened and dug out the ditches and all the cultivable fields, terraces that they call *pata*, *chacara* [field], *larca* [irrigation canal]. And they dug them out by hand without tools, with the greatest ease in the world; it appears that every Indian raised a stone. That was sufficient given the number of people there were.

. . . [Consider] the bridges and rivers, ditches, lakes, reservoirs and bogs which were dug out on orders of the first kings and master of the kingdom and afterwards, he ordered that the Inca kings enforce the custom and law not to disturb the said ditches, the irrigation water. The said fields, even the grazing lands, were irrigated in the heights and the valleys . . .

. . . And thus he established a penalty without appeal, a sentence that no one could damage or disturb a single stone and that no livestock could enter the said ditches. And they enforced this law and ordinance for the service of God and his Majesty and the welfare of the republic of this kingdom . . .

. . . And thus, this law has not been enforced. And so, all the cultivated fields perish for lack of water. Because of this, the Indians lose their lands and his Majesty loses his royal fifth and the Mother Church the tithe she is owed. And thus, in this time, the Spaniards let loose their livestock and their herds of cattle and mule trains and the goats, sheep cross [the ditches] and do grave damage [to them]. And they take out the said waters and they break the ditches that they are unable to repair without money. And the little water [there is], they take away solely from the poor Indians. And thus, the Indians absent themselves from their villages . . .

And because of this, there ought to be a judge of water . . . in every village who distributes the water and punishes and penalizes and throws out the cattle from the said ditches and fields (Guaman Poma de Ayala 1980:958 [f.944]).

The resonance between Guaman Poma's description and appeal and Wittfogel's notions of the relationships between centralized authority and irrigation systems contrasts strongly with Mitchell's (1976) and Netherly's (1984) assertions that their own studies of past and present irrigation systems in Peru falsify Wittfogel's (1957) assumptions that hydraulic systems engender despotic bureaucracies. If Wittfogel is read as unequivocally associating irrigation systems with large-scale bureaucracies and state formation, clearly the Peruvian cases invalidate his claims. In fact, in Wittfogel's (1985 [1928]) earlier work on irrigation systems, he constructed typologies based on varied combinations of differents kinds and sizes of drainage basins, social and political organization, and irrigation systems. He distinguishes, for example, between the "large scale Egyptian type, which included ancient Babylon and China as nearly pure forms of Asiatic despotism" and the Japanese type, which had "no extensive sphere of irrigation and drainage construction. The rivers could be handled locally. Thus one finds many isolated centers of production" (1985:56). Systems of governance similarly vary.

In his earlier work, Wittfogel was far more concerned with the general question of the relationship between nature and society, and particularly with disentangling the role of the natural and the social in the historical development of different modes of production. Insisting that materialist explanations are possible only if we understand that natural conditions or the conditioning of nature is the starting point of historical analysis, Wittfogel explains, "Naturally conditioned powers of production were crucial for establishing the direction of development of the production of food, soil, fertility, irrigation, et cetera . . . These naturally conditioned powers of production were decisive in the beginnings of cultures" (1985:57). Neither the natural nor the social alone can bring social production into being, but when they interact, their influence soon becomes dialectical.[1] According to Wittfogel:

> It is self-evident that in the course of history man effects changes in his natural environment by constantly modifying it and that these modifications have repercussions in man himself. He must accordingly modify the way in which he effects changes in the natural environment which he himself has modified. There are, of course, a great number of secondary factors which play a role in the formation of the social processes of production–the social conditions of the process, political and legal forms, the profusion of "higher ideologies" (altogether, a possibly tremendous weight of tradition), the effects of other organisms of production–what Marx called the "interplay of international relations" (1985:55-56).

Wittfogel's point throughout is that the natural conditions of production set the initial directions of the social determinations of modes of production. Subsequently, there is ongoing interaction between the natural and the social, both of which change dialectically over time (dialectically in precisely the sense of interaction with, and response to, each other).[2] In this article, we are not so much interested in determining the validity of the "Wittfogel hypothesis" that hydraulic systems engender despotic states. Rather, we explore the relationship between the physical configuration of irrigation systems (size, scale, and topography) and the organization of labor necessary for their construction and maintenance. Specifically, we want to determine if the systemic interaction between the physical structure of the irrigation system and the social organization of labor causes predictable changes over time in maintenance, water allocation, and the capacity of social groups to construct new canals. Different topographic configurations create differing problems for an irrigation system; these require different means of mobilizing topographic skills, materials, and labor. An established canal system imposes additional organizational requisites for maintenance of infrastructure and the distribution of water. Events in the larger socioeconomic and political system can also disturb relationships between humans and land, individuals and community, and between and within distinct corporate groups.

Fonseca (1984) has called our attention to the paucity of information concerning functioning highland irrigation systems in the Andes, particularly information concerning the tension between individual and communal control over land, labor, and water. He, along with Gelles (1984), Guillet (1985), Mayer (1977, 1979, 1985), Mayer and Fonseca (1979), Mitchell (1976), and Montoya, et al. (1979), has begun to provide us with important data about contemporary irrigation systems.[3] Most of these scholars correctly observe that for Andean highland peoples, their cosmology is inseparable from their technological control of their environment and their sociopolitical organization. However, largely missing from these studies is a recognition, fundamental to Wittfogel's work, that indigenous topographic perceptions and geographic and geological knowledge are essential to their attempts to modify the environment. This is especially true of the construction and maintenance of irrigation systems.

In this article, we analyze the organization of a complex and extensive irrigation system located in the village of Huanoquite (department of Cuzco).[4] We are particularly concerned with the discrepancies we observed between the labor and skill incorporated into the contemporary maintenance of ditches and the labor and skill needed to design and construct them in the first place. This comparison of past and present requires us to combine very different kinds of evidence. First, we must deduce the topographic knowledge, construction skills, and nature of control over labor for the original construction from such physical evidence as measurements of canal trajectories, the angles and conditions of soil and rock through which the canals traverse, and the topographic obstacles that had to be overcome in building them. Second, we explore the social organization such construction would have required. Evidence of these relations between modifications of the physical environment and Huanoquite's social organization is difficult to find, but early chronicles, surviving documentary evidence, and contemporary oral traditions give us indicators of how ditch builders perceived their physical environment and how they organized their labor to convert topographic knowledge to useful changes in that topography. We complement these physical observations with general knowledge of the history of labor relations and land tenure for the region as a whole.

Despite the great importance of their irrigation system, Huanoquiteños have abandoned some canals and are allowing the rest to deteriorate. Because they lack a cohesive sociopolitical organization, few villagers can imagine the reconstruction of the system. The cumulative effect of political and economic changes brought about by, *inter alia*, two land reforms, individualized land tenure, participation in cash markets, and migration in search of wage labor have caused this loss of collective action. In turn, this fragmentation has led to changes in the kinds of geographic concepts, topographic knowledge, and understanding of natural cycles that farmers have relied on to maintain and manage their canal system. Huanoquiteños thereby have become increasingly limited in their ability to organize irrigation, construct and maintain canals, and distribute the water. The consequent deterioration of the irrigation system, in turn, affects the productivity of labor, political relations between different groups in Huanoquite, relations between Huanoquite and agencies of the state, and the condition and erosion of soils. Oral traditions about water, mountains, rulers, and deities mirror this social and ecological fragmentation. They also tell us about the ideal conditions under which Huanoquiteños believe canal construction and water management should take place. These beliefs about the environment, moreover, affect the Huanoquiteños actions in that environment. The oral traditions structure perceptions of the possible, thereby affecting their work on the ditches. Huanoquiteños are responding to environmental conditions which are themselves partly determined by their beliefs.

ENVIRONMENTAL OBSTACLES AND ADVANTAGES

Huanoquite, a nucleated village of more than 900 inhabitants, is located 22 kilometers south southwest of Cuzco at 3,383 meters above sea level. Historically, it has been an extremely fertile agricultural region, attracting first the Inca nobility, then Spanish *encomenderos* followed by *hacendados*, and finally the government-operated cooperatives (CAPS) that were established after the 1969 Agrarian Reform. Before the Spanish Conquest, the population lived in dispersed *ayllus*,[5] which in turn were organized into moieties. Each of the *ayllus* conformed roughly to a distinct canal system. The canal systems of all the *ayllus* were linked to one another within the same drainage system, with the exception of one in which a ditch carried water out of its original drainage system and into another. The organization of this particular ditch was consequently fraught with political conflict. The political structure of these *ayllus* is not entirely clear, but it appears that the larger units were led by a local lord called a *kuraka* and his second-in-command. The ethnic lords, in turn, paid obeisance to the imperial Incas, who were represented by a provincial governor.

During the colonial period, these dispersed populations were reduced to what is now known as Huanoquite, a single village laid out in a typical Spanish grid pattern. Nevertheless, it is likely that small hamlets continued to survive, surrounding the center. Large agricultural landed estates which gradually usurped much of the indigenous land, maintained and even extended the irrigation system by harnessing the labor of their own peons. Today, Huanoquite corresponds to an administrative district composed of different communities, radiating from the colonial center. The center is primarily inhabited by *hacendados* and their heirs. Surrounding it is the population of the *ayllu*/community Maska and the cooperative of Inkaq Tiyanan de Tihuicte. Originally a hacienda, the cooperative became an agricultural experimental station in the early 1960s. It was subsequently organized into a government-run cooperative (CAP) after the 1969 Agrarian Reform. Beyond Maska and Tihuicte lie the maximal *ayllu*/community of Tantarcalla (with three minimal *ayllus*), the annex of Llaspay, and the maximal *ayllu*/community of Chifia.

The most recent of the major ditches is more than eighty years old; the oldest are Incaic or pre-Incaic.[6] The Huanoquiteños have abandoned the oldest canals, but the more modern ones continue to make a critical difference to the productive capacity of Huanoquite farmers, permitting them to harvest both subsistence and market crops. Because of their irrigation water, Huanoquite farmers can sow early potatoes (*papa maway*) and maize before the rainy season begins, and thereby receive more favorable market prices for the crops. Another advantage to sowing maize before the rainy season is that farmers can extend its cultivation into zones where maize cultivation without irrigation would not be viable because of destructive frosts.

Huanoquite's irrigation sustains an agricultural economy based on a wide range of crops oriented both to subsistence and to the market economy. The steep, broken topography allows Huanoquiteños access to multiple ecological zones and permits them to harvest a great diversity of crops (see Map 9.1 of Huanoquite's production zones).[7] To take advantage of this ecological diversity, however, they must maintain a complex labor schedule, which exacerbates enormously the problems of constructing, maintaining, and using their irrigation canals.

MAP 9.1 PRODUCTION ZONES IN HUANOQUITE

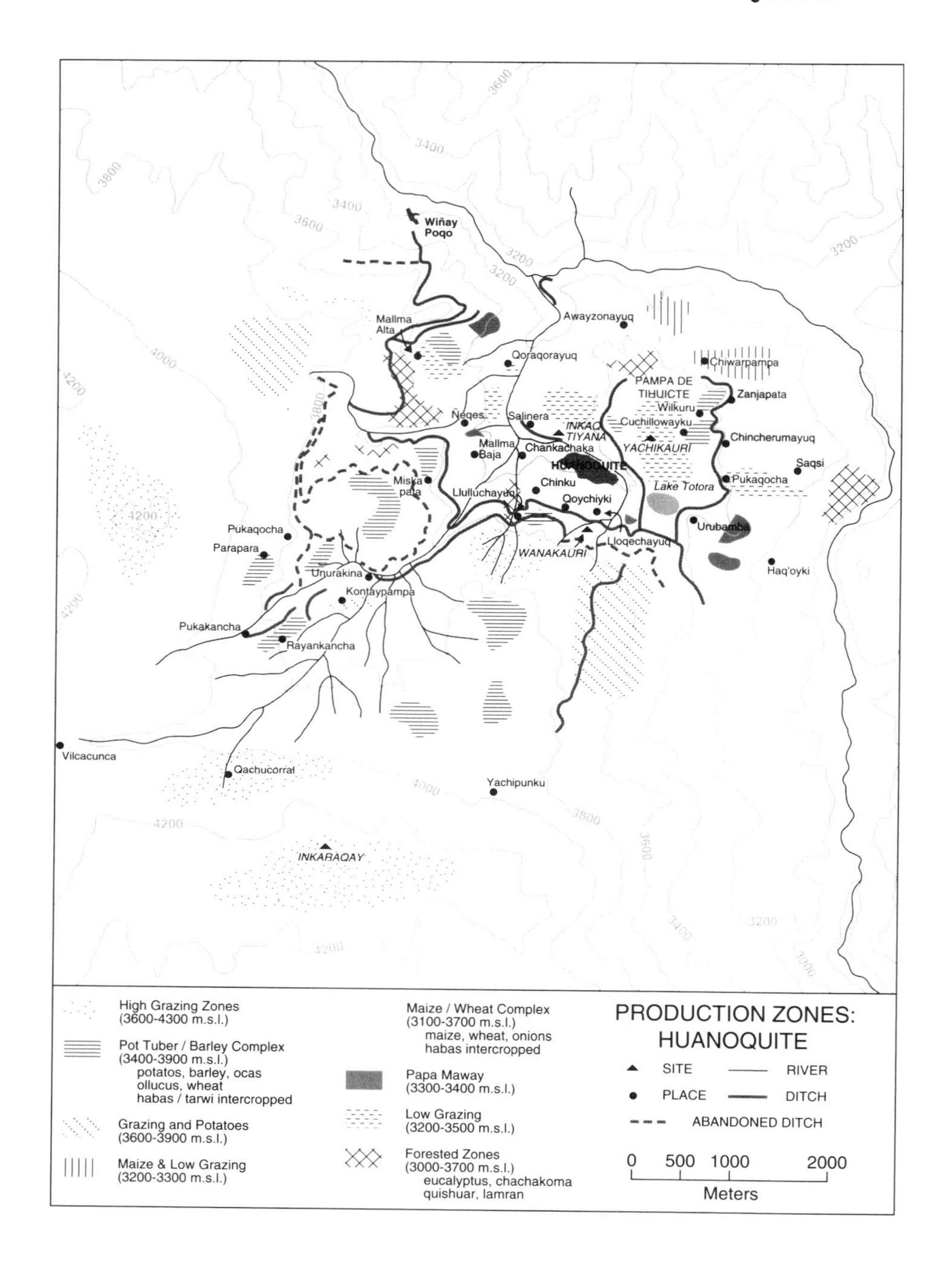

THE DRAINAGE SYSTEM OF HUANOQUITE

Unlike such large-scale irrigation systems as those in Egypt where ditches flow along relatively flat terrain, Huanoquite's ditches must take into account multiple steep slopes that change grades within short distances, varying amounts of water available from different rivers and springs, multiple valley systems, and a wide range of soil types, some of them highly unstable.[8]

The construction of the Huanoquite canals required impressive social and technical skills. The ditch builders first had to conceptualize the geographic relations between complicated systems of rivers, valleys and mountains. Second, they had to imagine the possibility of bringing water over these difficult geographic formations. Finally, they had to mobilize the labor and technical skill required to construct ditches with a minimum of fall, sometimes over slopes whose incline exceeds seventy degrees. (See Map 9.2, "Huanoquite Base Map" for all the topographic features mentioned below).

One topographic feature, more than any other, has influenced the conceptualization of Huanoquite's irrigation system. The spring-fed Qewar River sustains a flow ranging from 335 liters per second in February, the wettest month, to 447 in June and 414 in August, two of the driest months.[9] Not only is the Qewar by far the most copious river in the region, its volume is greatest when there is least rainfall, thus providing water when it is most needed. Yet it is separated from the best agricultural areas by the large mountain called Ch'uspi Grande and a deep gorge formed by the other main river, the Qolla Wayku. The Qolla Wayku's flow peaks at 265 liters per second during the rainy season, when it receives runoff from several small tributaries, and diminishes to less than half of that during the dry season. Although conveniently located in proximity to the best agricultural areas, the Qolla Wayku River is more of an impediment to irrigation than an asset. Together with Mountain Ch'uspi Grande, it poses a major barrier to the construction of canals from the more abundant Qewar River.

The initial idea to carry water over these two obstacles required a precise understanding of the relative levels of different valleys and of the hills that separated them, an understanding extremely difficult to achieve because there is no single site where one can simultaneously see all the relevant topographic features. To implement this idea required the construction of high and long retaining walls, of wooden trestles wedged into sheer cliffs, of weirs over which the ditches could cross washes and river beds, and of controlled waterfalls at the bottom of canal courses. Irrigation enhanced the carrying capacity of the land by artificially extending cultivable zones and agricultural seasons. The irrigation system, in turn, required new organizational forms to maintain the ditches and associated infrastructure and to distribute irrigation water. These achievements constituted part of the legacy that today's Huanoquiteños have received from their ancestors and that, for reasons we will explain below, are barely being conserved. Both the transformations of Huanoquite's natural topography and the complexity of the organizational forms to manage irrigation are not immediately apparent. Only by carefully scrutinizing the lay of the land, following the trajectories of canals from

MAP 9.2 BASE MAP OF HUANOQUITE

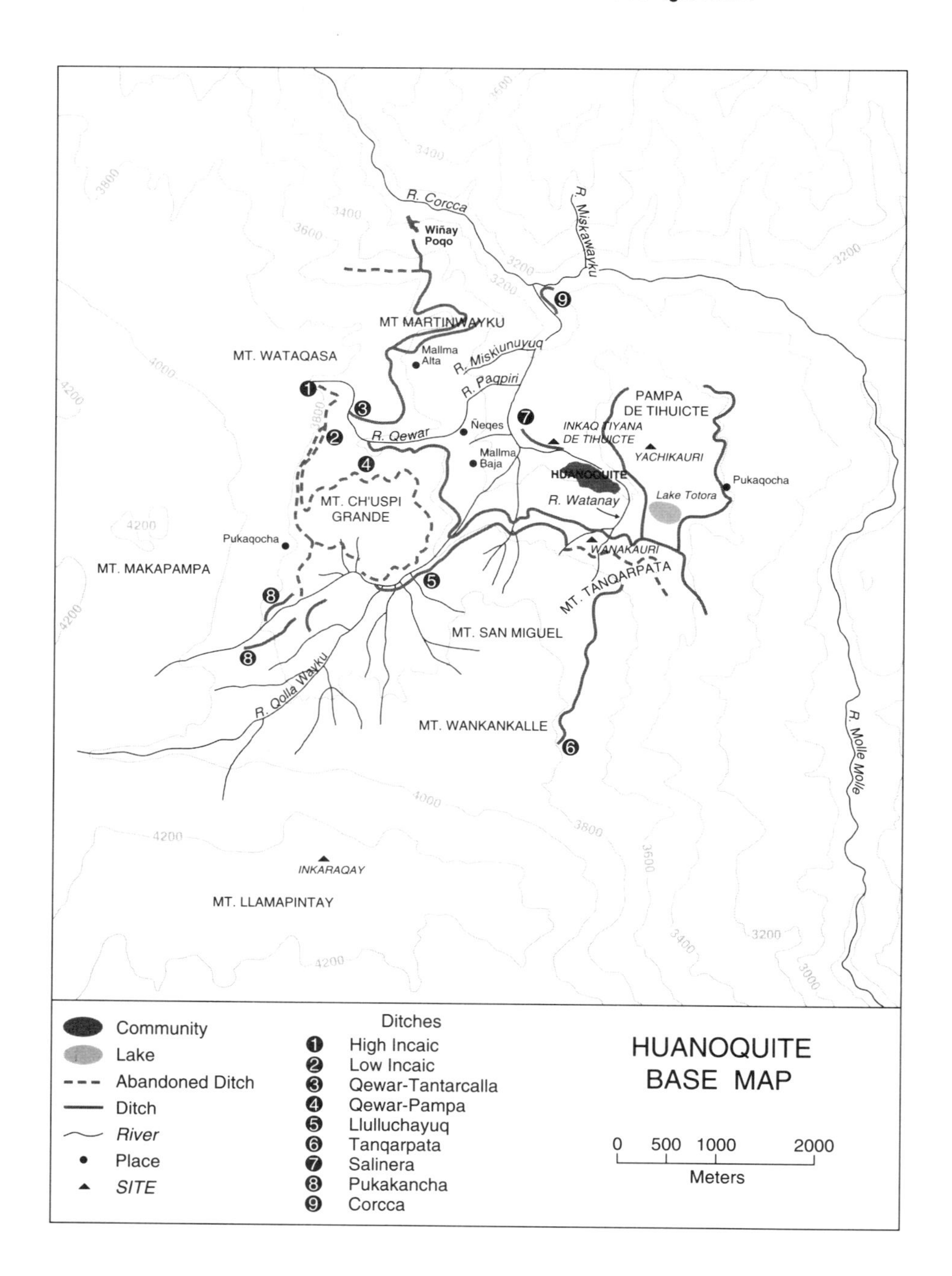

their destination to their source, and by asking Huanoquiteños about their irrigation system were we able to begin to understand the challenge of maintaining this system and why it had deteriorated.

The agricultural zone in which the village of Huanoquite lies corresponds neatly to the drainage system of the two rivers mentioned above, the Qewar and the Qolla Wayku. This drainage system and the agricultural area it defines extend six kilometers from north to south, seven kilometers from east to west, and more than one kilometer vertically, from less than 3,000 meters to more than 4,200 meters above sea level. The mountains of Makapampa and Llamapintay form the western and eastern limits, respectively, of the drainage system. Wataqasa, a pyramidal mountain of limestone, and Martinwayku, a long ridge that projects from Wataqasa, create another wall to the north. Finally, the Corcca River and its deep valley form the lower border of the drainage system, curving widely around the great 800-hectare pampa of Tihuicte. This pampa, or high plateau, formed by a massive uplift of underlying sedimentary strata to a nearly vertical plane, constitutes one of the most favorable areas for agriculture in this region. Such extensive flat areas are seldom found in the highlands. Unfortunately, this fertile pampa has no water source of its own, an environmental fact that constitutes the second major influence on the design of Huanoquite's hydraulic system.

Topographic features within the drainage system constitute additional physical determinants of the system's design. As Map 9.2 shows, the Qewar Valley–the one with abundant water and little cultivable land–descends from the northwest. It is separated from the pampa by the Qolla Wayku Valley, which descends from the southwest, and by the two mountainous formations around it. The first is Ch'uspi Grande, connected to Makapampa by a land bridge called Pukaqocha. The second formation, Wankankalle, extends from the easternmost part of Llamapintay to Ch'uspi Grande, thus forming a wall at the bottom of the Qolla Wayku Valley and separating this valley from the pampa of Tihuicte and the village settlement. Rising above the flat top of Wankankalle are the twin peaks of San Miguel. Their slopes descend directly to the village. A series of shelves on these slopes provide stable and fertile soils for agriculture. They lie conveniently close to the homes and storage areas of farmers. Thus, Ch'uspi Grande divides the narrow and relatively infertile Qewar Valley from the wide and flat upper Qolla Wayku valley. The lower Qolla Wayku Valley, which narrows into a deep gorge, itself becomes an obstacle, separating the Qewar Valley from the slopes of San Miguel and the pampa of Tihuicte. The Incaic and more recent ditch builders used different trajectories to overcome these obstacles, but their solutions resembled each other in the ingenuity of their conceptualization and in the scale and skill of the labor force their projects demanded.

CONSTRUCTION AND MAINTENANCE OF THE HYDRAULIC SYSTEM

The Ancient Ditches

There are two abandoned canals, presumably Incaic in origin, whose traces are still visible. The first (1 on Map 9.2) comes directly from the Qewar's source, where the river gushes forth with more than 400 liters per second from beneath an imposing crag at the foot of Makapampa. This canal crosses a slope so steep that a retaining wall had to be built for more than half of its first hundred meters of flow. For almost 300 meters it continues with a minimum drop above the Qewar. It then leaves the Qewar Valley, following the contour of Makapampa in a south-southwesterly direction while the river flows north, already far below the ditch. The ditch flows atop long retaining walls, which range up to four meters in height and thirty meters in length in a single stretch. Two kilometers from its origin, it encounters the slope of Pukaqocha, the land bridge between Makapampa and Ch'uspi Grande. In this entire trajectory, despite having to cross extremely steep slopes as well as a series of washes on the slopes of Makapampa, the ditch only falls twenty-two meters–about one meter of drop for every one hundred meters of run. The ditch builders' success allowed them to open a pass approximately four meters deep on the ridge of Pukaoqocha, thus giving them access to their ultimate objective, the Qolla Wayku Valley. The ditch follows the high part of the valley at the foot of Makapampa. There is evidence it may have circled the entire valley at one time, continuing below San Miguel until arriving just above the village, and traversing three separate drainage systems in the process.

The other ancient ditch (2 on Map 9.2) departs from the first approximately one-half kilometer from its source at the Qewar. It descends the slopes of Makapampa more steeply, arriving at Pukaqocha thirty meters below the higher ditch. It then doubles back and begins to follow first the western, then the northern, and finally the eastern slopes of Ch'uspi Grande, effectively circling its entire crown. In this way, the Huanoquiteños succeeded in irrigating all the slopes of the mountain except for that part nearest the other side of Pukaqocha, which received irrigation water from the first ditch. Thus, the ditch builders, in the process of overcoming the obstacles they confronted, simultaneously took advantage of them. The ditch that circles Ch'uspi Grande did not need as many retaining walls since it was able to follow a lower trajectory and avoid the major physical obstacles. Nevertheless, this ditch is impressive because of the creativity of its design and the sophisticated geographic perceptions that its construction required.

The Qewar-Pampa Ditch

The more recent canals still in use follow lower but no less difficult trajectories. The principal canal (4 on Map 9.2), from the Qewar to the pampa, takes water around the same obstacles as the two ancient ditches but irrigates less ground. It carries more than 110 liters of water per second and flows for more than seven kilometers. It irrigates fields over a range of almost one-half vertical kilometer–from 3,570 meters to less then 3,100 meters above sea level–crossing above areas where early potatoes grow well along with others that favor maize.

This canal takes its water from the Qewar in its second, or middle, valley, which lies between Ch'uspi Grande and Wataqasa. The slope of Ch'uspi Grande, which is seriously eroded by breaks in the old ditch, is extremely steep until it reaches this valley. The modern canal crosses a series of deep, narrow and unstable washes created by this erosion. Therefore, those who built the canal had to put up retaining walls for a good part of the first 800 meters of its course. The concentration of rainwater in these steep washes makes the maintenance of these walls extremely difficult.

The canal flows from the Qewar valley, curving to follow the eastern slope of Ch'uspi Grande, which is far less steep. Here, the canal construction was far easier, and contemporary maintenance less problematic. Numerous feeder ditches begin in this area, irrigating some fields located immediately below the canal and others up to 200 meters below and one-half kilometer from the main canal. This land is held in a few large holdings, most still the private property of *hacendados* who managed to keep these fields despite the 1969 national Agrarian Reform. The remainder of the fields in this stretch were turned over to the community of Maska, one of the two main social and political units of the village. They are controlled by the six families who received them from the landowner for whom they had worked.[10]

After crossing the eastern slope of Ch'uspi Grande, the canal encounters the most difficult obstacle in its entire trajectory. Only one place is suitable for crossing the Qolla Wayku gorge while keeping the canal high enough to reach the best soils in the area. The canal follows the curve of Ch'uspi Grande toward the southwest, entering the Qolla Wayku Gorge and flowing against the current of the Qolla Wayku River. It flows above high retaining walls constructed on very unstable foundations that hug the almost vertical cliffs of the gorge. It continues along the cliffs until it reaches the riverbed. It crosses the riverbed on a trestle made of rocks covered with sod. The canal then leaves the gorge along retaining walls that are as high as those on the other side. More than two kilometers from its source, the canal begins to cross the slope of San Miguel Mountain, which at this point is quite steep and eroded. Its trajectory remains difficult for another kilometer.

The course of the canal, and thus its maintenance, becomes far easier when it reaches the shelves that break up the slope of San Miguel above the village. Here, the canal follows a long terrace about four meters high. The fields above the ditch are almost flat, probably the result of villagers throwing the soil uphill over years of ditch cleaning. The rains in this area do not course down with the force that causes frequent ruptures in the embankments along other sectors of the canal.

Multiple ditches irrigate not only the fields on this slope but also vegetable gardens in the village and fields below it. The ditches are also used to provide water for the construction of adobes. At certain times of the year, the streets of the village are transformed into a series of streams, sometimes purposely and sometimes accidentally, because of carelessness in closing the ditch outlets. Skill and patience are needed in the construction not only of these major canals but also of the minor canals that farmers make to channel the water down to their particular fields or patios.

After flowing for almost one kilometer along these gentle and fertile shelves, the canal descends abruptly from Wanakauri into a gorge that begins at the peak of San Miguel. Wanakauri and its valley separate the fertile shelves of San Miguel from the steep hill of Tanqarpata, whose soils are extremely poor and unstable.[11] After leaving Wanakauri, the ditch builders' tactics changed dramatically. Here, they abandoned their efforts to achieve minimum drop, a sensible choice given the poor quality of the soils that could have been irrigated by keeping the ditches more level. Instead, the canal drops precipitously over a series of small, controlled waterfalls until reaching the floor of the valley between San Miguel and the pampa of Tihuicte. There, it divides into two branches. Both flow across low ridges to the pampa. The first branch courses along the western side of the pampa, while the second follows the eastern side. The second branch of the canal is the most important because it feeds the ditches that irrigate the extensive slopes below the pampa.

Although irrigation continues to permit Huanoquiteños to sustain a varied and relatively prosperous economy, they no longer manage the canals as well as they once did. In its first three kilometers of flow, repeated breaks in the ditch have undermined the foundations of the retaining walls. The retaining walls in the first 800-meter stretch at the foot of the face of Ch'uspi Grande are already in an advanced state of deterioration. Similar erosion has weakened the foundations of the walls that support the canal as it makes its difficult crossing through the Qolla Wayku gorge. The multiple falls after Wanakauri have weakened the walls that hold the water up along the valley wall. People have organized public work projects (*faenas*) at all these crucial points to clean the ditches, but they have not been able to deal with the structural problems that increasingly threaten to undermine the viability of the canal system itself. Even on the pampa, where farmers do not have to devise solutions to steep grades, there are problems of erosion. This is the lowest, warmest, and most favorable area in Huanoquite for maize, and irrigation is essential for its cultivation. The maintenance of this part of the canal is a difficult and expensive task because the pampa is composed of a limestone base mixed with veins of chalk, a geological formation highly susceptible to erosion. The auxiliary canals are also seriously eroded, at times to such an extent that it is extremely difficult to get water to the fields.

The Qewar-Tantarcalla Ditch

The second canal still in use is the Qewar-Tantarcalla ditch (3 on Map 9.2). It carries just over half (58 liters per second) the water of the Qewar-Pampa canal, and is far less elaborate. It branches off at a higher point from the Qewar, curves around Mount Wataqasa, and continues to the northeast on a long ridge called Martinwayku. It crosses above Mallma Alta, an extensive stretch of property that was not expropriated during the Agrarian Reform. It doubles back at the north end of Martinwayku, leaving the Qewar-Qolla Wayku drainage. It then traverses a deep, narrow gorge, before arriving at a point from which it drops 200 meters to the hamlet of Wiñay Poqo.[12] It has retaining walls along much of its course around Mount Wataqasa. When it enters the gorge, it crosses a cliff on an ingenious trestle of sod and tree trunks. Given that its trajectory is far higher than the lands it must irrigate, the ditch builders had a greater margin of error in avoiding the worst obstacles. In recent times, however, maintenance has been insufficient. Serious conflicts between the community of Tantarcalla and the *hacendado* of Mallma Alta, who owns the land the ditch crosses in the Qewar Valley, have persisted over the years and periodically erupted into violence. Even though Tantarcalla is responsible for cleaning the entire canal, a task that takes several days (Reforma Agraria n.d.), the *hacendado* has repeatedly tried to prevent Tantarcalla from using the water by either blocking or destroying the ditch. Maintenance is made even more difficult because the slopes that the ditch traverses are extremely susceptible to erosion wherever there is a break in the ditch. As a result, the community receives only a small fraction of the water that enters the canal system from the Qewar.

There are already serious problems with the irrigation system. Its collapse would create a crisis for Huanoquite. Nonetheless, as we will explore below, sociopolitical fragmentation impedes the development of the coordinated efforts needed for the conservation of the canal system.

HYDRAULIC BELIEFS AND PRACTICES

In their beliefs about water and in their ritual practices, Huanoquiteños recognize some of the current sociopolitical problems that impede ditch maintenance. These beliefs portray an understanding of the relationship between topography and canal structure, and the need for appropriate sociopolitical mechanisms to manage the canal system. They suggest that canal construction and organization should be managed by an overarching authority and ritual specialists with supernatural powers. Unfortunately, however, these oral traditions also reinforce Huanoquiteños' sense of powerlessness in the face of their own sociopolitical fragmentation.

It is not that the beliefs and ritual practices have brought about the actual degradation of the canal system. Rather, these traditions, because they have become integral to the fabric of daily life, are taken as given, reinforcing a view of the world in which Huanoquiteños lack the powers of the past and are unable to solve contemporary problems. They believe that canal construction was the work of their ancestors and that they can no longer replicate such feats. Consequently, even while fascinated by where their water comes from, Huanoquiteños do little in the way of long-term ditch maintenance. At the same time that they believe their water is a gift from nature, they fight bitterly over it during irrigation season. Huanoquiteños attribute the origin of water and canals to supernatural forces and the powers of the Incas. Oral traditions collected during the colonial period (dating most probably from a much earlier time) emphasize this: Incas and deities make water appear; animals with supernatural powers determine the trajectory of canals; water guardians take care of special lakes and springs; priests supplicate water sources and decide agricultural scheduling; Inca officials calculate land and water boundaries; water judges resolve conflicts over water; if water judges cannot resolve these conflicts, ethnic polities will fight to do so (see Avila 1966 [1598]; Hernández-Príncipe 1923 [1621]).

These oral accounts are preoccupied with water availability, canal trajectories, and the location of retaining walls. They also focus on increasing the size of the territory to be irrigated, as well as on controlling floods through the building of dams. Human specialists, whose authority is validated supernaturally, mediate between the physical and the social. They have the potential to organize labor and ensure the continuity of specialized knowledge and the investment of individual knowledge in the realization of collective public works projects.

Two principal myths in Huanoquite offer explanations of the origin of the springwater that feeds the Qewar River. In these myths, the Qewar is thought to originate in a small lake (called Qaranqa) in the barren, wild, and distant puna (*sallqa*). An eerie crag that rises above the lake bears a startling resemblance to a bearded visage, a parallel noted by villagers. Beards are used to characterize both Viracocha (a deity) and the Spanish invaders. In Quechua, *viracocha* is a compound word meaning "fat/foam of the sea/lake." As a founding deity, Viracocha is believed to have carved out land/water districts and created distinct ethnic polities, assigning each of them to different geographic regions. He carved these ancestral peoples out of clay and painted them in different types of dress.

These ancestors subsequently traveled to different homelands by way of subterranean canals. Emerging from natural openings, springs, mountains, caves, and rivers, they founded the indigenous *ayllus* and then transformed themselves into sacred sites (*huacas*), often with human characteristics and personalities. At times, these sites are seen as exceedingly abstract forms such as streams of light; at other times, they are manifested in powerful and unusual rock formations (see Allen [1988] and Sherbondy [n.d., *intra* for a discussion of these myths and sacred sites).

In both myths, the water descends from the lake by way of subterranean channels constructed by the Incas. About 80 years ago, according to the myths, a priest threw red *kantuta* flowers (*Cantua buxifolia*) into Lake Qaranqa and waited to see where they would go. In the first myth, the flowers appeared far below in the Qewar and Qhashwa (another spring that irrigates the nearby community of Llaspay, also a former sociopolitical unit of Huanoquite). In the second myth, the flowers appeared in yet another community, Huanca Huanca (another former *ayllu* of Huanoquite). The Huanoquiteños therefore say that the Qewar and Qhashwa are "brothers."[13]

A number of similar oral traditions refer to subterranean canals that connect the distant ocean (Lake Titicaca) to Lake Qaranqa and then to the Qewar, which then gushed forth from beneath a hole in the form of a full-blown river.[14] One man from Tantarcalla whispered that the Qewar descended from its source after a human couple emerged from the same site. Several Huanoquiteños also told us that the Incas, after building their canal, climbed to a cavernous hole into which they threw red *kantutas*. They then returned to their thrones (*tiyana*) far below . These thrones, still revered by the Huanoquiteños, are carved into stone and located where the Inca road enters Huanoquite. There, the Incas awaited the water. The water that filled the canal appeared in the form of the Qewar River.

These myths emphasize a sanctified connection between irrigation use and the sociopolitical linkages among different *ayllus*. In addition, they assign supernatural origins to the irrigation system. Water sources are seen as created by either supernatural powers or the Incas, who themselves are given divine characteristics. The water sources, however, become socially useful only through the application of human labor.

In the myths, Huanoquiteños retain bits and pieces of the geological and topographic knowledge that had been integral to precolonial canal management. At the present time, this fragmentary information continues to serve a purpose by assigning names to distinctive topographic sites, thereby perpetuating some of the environmental particulars that were used in precolonial irrigation management. Nevertheless, due to the fragmentary nature of the contemporary sociopolitical order, this knowledge is no longer systematic or coherent enough to be used successfully to manage the irrigation system.

Four sacred sites in Huanoquite (Quypan, Yachikauri, Wanakauri, and Inkarakay) form part of a *ceque* line that extends from Cuzco and passes through Huanoquite. Polo de Ondegardo (1917 [1571]:37) specifically lists Cuipancalla (that is, Quypan), the highest point on the Inca road to Cuzco through Huanoquite, as one of the sacred sites. The *ceque* system was an imperial Inca model that fused space and time. It organized the Inca empire into hierarchically and socially differentiated land and water districts, and it also marked the place and timing of periodic ritual occasions (Sherbondy *intra*; Zuidema 1964). The *ceque* lines were identified by sacred sites that radiate from Corikancha, the Temple of the Sun in Cuzco. Beyond Cuzco, the mountain shrines marking the *ceque* line belonged to the local populations. These sacred sites remain significant in local environmental knowledge and have direct associations with water and terraces. At the sacred site of Wanakauri, where the Qewar-Tihuicte ditch intersects a *ceque* line, for example, the main canal leaves the gentle slopes and begins to drop rapidly through very steep slopes, along which one today finds terraces and enormous boulders that are scattered on the slopes.

Huanoquiteños tell us that the Incas were herding these boulders to build a bridge across the Corcca River but that the work was interrupted when a rooster crowed. In some other versions of the story, the ancestors from beneath the ground announced, "our time has come" (*tiemponincisña*), when the Spaniards arrived. These huge boulders, formed of magma and embedded sandstone fragments, are completely different in composition from other rocks in the zone, except for a few that lie precisely where the bridge was to have been built. This oral tradition thus communicates important knowledge about the geological characteristics of Huanoquite as well as information about transformations in political authority from the age of the Incas to the age of colonialism. The people we talked with did not interpret this story but told it in answer to our queries about why the higher canal system had been abandoned. Our interpretation of both stories is that native authority's ability to command the local labor force for ongoing infrastructural projects was not only challenged by the conquering Spaniards but also undermined to such a degree that the projects of empire simply came to a halt.[15]

The site of Wanakauri is on a direct line with the site of Yachikauri (a rocky formation resembling a lizard that is located at the highest point of the pampa), with the site of Quypan,[16] and with the temple of the Sun in Cuzco (Corikancha) from which all the *ceque* lines radiated. This particular *ceque* line continues beyond Wanakauri to Inkarakay, also on the outskirts of the Huanoquite region, where a group of carved rocks and subterranenan channels may have served some purpose in the Incaic irrigation system. This site is similar in appearance to the Incaic ceremonial site of Qenko near Sacsayhuaman in Cuzco, though on a much smaller scale.

What is extraordinary about the *ceque* system is that it used natural geological formations to create an overarching administrative and religious system. The *ceque* system, as a unifying imperial device, was based on unusual geological formations and the location of mountain peaks. The permanent nature of these formations, in contrast to water whose course may change over a brief period of time, meant that they continued to serve as mnemonic devices for storing local environmental knowledge even after the Inca State was dismantled. Today, even though the *ceque* system is no longer maintained, it continues to provide spatial, historical, religious, and ecological orientation to Huanoquiteños because of its embeddedness in physical features of the environment itself (Seligmann 1987a).

The explanations of the abandonment of the high canal system tell us that the Incas (in one version, an *hacendado's* peons) building the canal sent three men back to its source to release the water. The men opened the canal and the irrigation water began to descend, guided by an *amaru* (a boa constrictor) with golden braids. Another group of men waiting below killed the *amaru*. The water retreated to its source, never to return. Others say that after the Incas had built the canal, the water gushed forth from a spring farther below, and thus the canal became useless. Still others suggest that the same thing happened when the *amaru* was killed. Some recount that a rooster crowed and the water dried up; others say that the water disappeared when the people killed a frog living at the source.

The Andean belief system associates *amarus* with political power and royal dynasties. The canal system, indeed, takes the form of a snake as it crosses mountains and valleys. More dramatically, when Huanoquiteños open the ditch after cleaning it during the public works project, the descending head of water picks up the debris left on the ditch floor and looks very much like a huge, wriggling, muddy snake. In the late afternoon light, the foam and the flotsam carried on top of the water appear to be the snake's golden braids.

These beliefs structure the Huanoquiteños' perceptions of their irrigation system. They tell Huanoquiteños that water has supernatural origins. In addition, the image of the segmented boa constrictor (*amaru*) reflects an ideal sociopolitical organization in which distinct segments of the canal system originating from a common source are controlled by different *ayllus* but are part of the same "organism" or sociopolitical system.

The mythical explanations for the abandonment of the high canal system provide plausible reasons for irrigation mismanagement. First, they acknowledge that accurate environmental knowledge is necessary to ensure the functioning of canals. Second, they portray the need for cooperation in successful irrigation management. The version of the myth in which one group of men kills the snake released by the first group can be interpreted in several ways. It is possible that both social units had not cooperated in building the ditch and that they therefore began to fight over the water. It is also possible that one group wanted to deprive the other of water. Finally, killing frogs or *amarus* in the Andes is akin to killing guardians of natural and social resources. Frogs protect water sources; *amarus* protect political power.[17] The death of the *amaru*, the crowing rooster and the frog are polysemic metaphors of the dissolution of political power and the fragmentation of the hydraulic cycle (see Earls and Silverblatt 1978; Seligmann 1987b; Whitten 1985) that communicate the loss of the personal power *(sami)* that would allow the Huanoquiteños to control their own political destiny (Delran 1974). The Huanoquiteños are aware of their subordinate status in contemporary Peru, and they clearly see the deleterious consequences of injustice and discord in the operation of their irrigation system.

In the next section, we turn to the ways in which the Huanoquiteños' contemporary sociopolitical status has affected the use of their canal system. Here we are able to draw on more reliable evidence concerning the roles of authority and labor organization in irrigation management. Interview data and direct observation of contemporary irrigation practices support our empirical data on the environment and our interpretations of ethnohistorical and contemporary oral traditions.

SOCIAL ORGANIZATION AND TOPOGRAPHIC KNOWLEDGE

The Huanoquite canal system was probably constructed over a long period of time, a project that must have required considerable labor. In addition, Huanoquiteños would have had to mobilize labor at different times of the year, whenever problems with the ditches arose, a point that both Wittfogel and the native Andean chronicler, Guaman Poma, have made. Huanoquiteños, however, are no longer able to muster this labor, a condition that has caused the decay of the present-day canal system. For example, they are unable to obtain sufficient workers or organize them for the annual ditch cleaning or for carrying out short-term repairs to the canal system. (See Map 9.3 for an example of the most recent effort by the mayor in 1983 to organize irrigation cleaning by sectors.) At another level, the need for overarching authority also becomes apparent in the endless disputes that arise between canal users. Efforts to establish a system for water distribution have repeatedly failed. In one case, in 1959, the Water Judge himself was accused of corruption and ignominiously thrown into a ditch.

Changes in land tenure patterns, political organization, and labor relations have weakened Huanoquite's collective control over water distribution and canal maintenance. These changes have occurred in other communities, but their effects have been more severe in Huanoquite in part because of the more difficult topography of Huanoquite, particularly when compared with the less difficult

MAP 9.3 SECTORAL IRRIGATION SCHEDULE OF HUANOQUITE (TANDA SYSTEM)

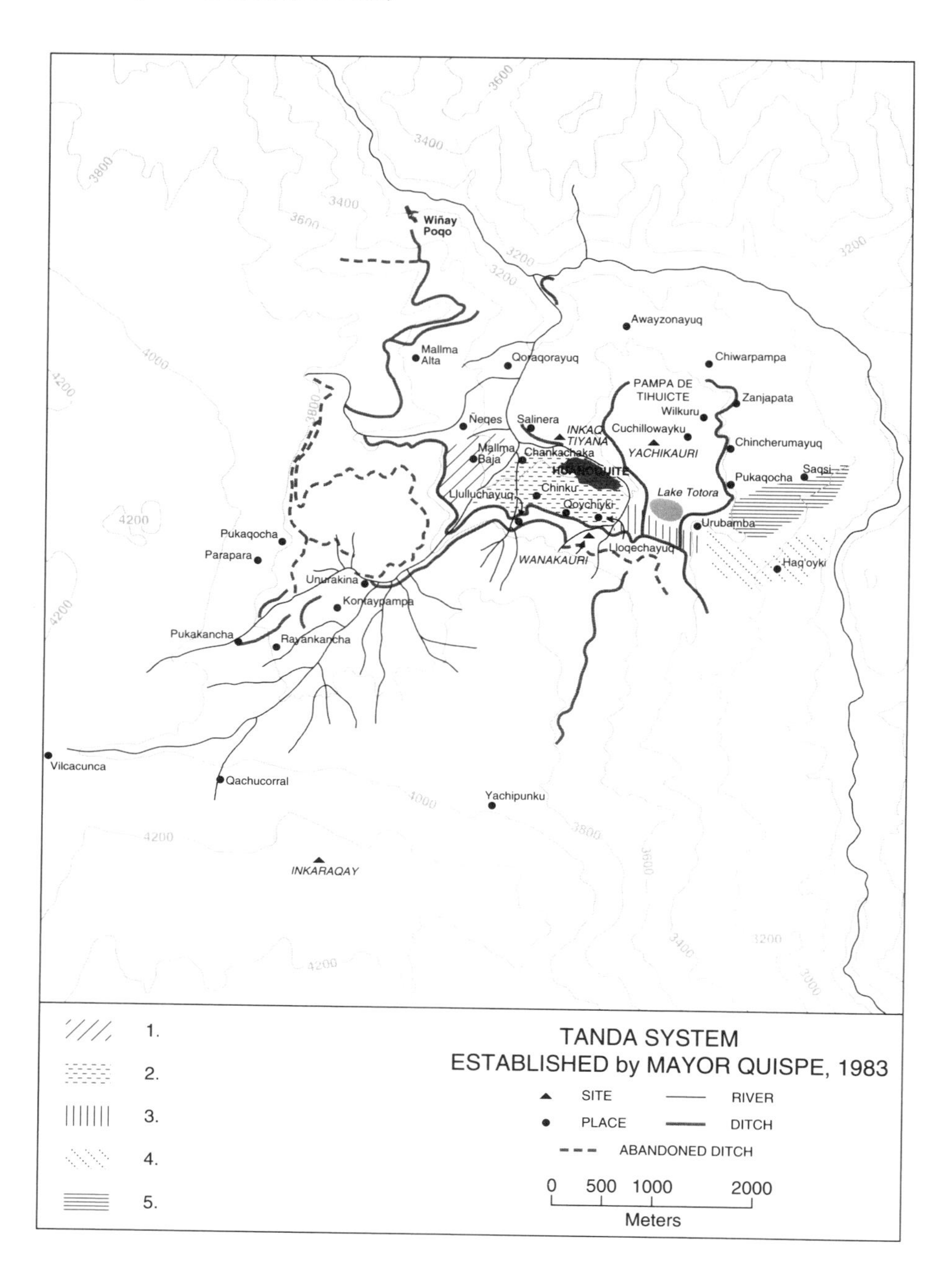

Note: The sequence of irrigation is ordered by sector, beginning with the sector number 1 and ending with number 5 on the legend.

terrain in the communities described by Fonseca (1984), Golte (1980), Mayer and Fonseca (1979), and Mitchell (1976).[18]

One reason for the poor management of Huanoquite's irrigation system lies in the fragmentation of its *ayllus*. *Ayllus*, indigenous units of social organization, permitted households to gain corporate rights to different kinds of resources, especially land and water. *Ayllus* were linked to one another through the authority of a principal lord, a linkage that facilitated the mobilization of labor for suprahousehold tasks (see Guillet 1985). These ties to the principal lord were thus essential to the coordination of tasks and the capacity to harness labor from among the *ayllus*. The process of *ayllu* fragmentation began in the late sixteenth century following the Spanish Invasion and the Toledan Reforms. At that time, dispersed social units were concentrated in a single site. Indigenous households were forced to abandon residences and lands so that colonial bureaucrats could better control and collect tribute from them. Even when households retained access to their lands, they found it more difficult to manage them because they now lived far from their fields and pastures.

Today, Ayllu Maska (formerly Urinsaya of Huanoquite) and the cooperative of Tihuicte (formerly part of Maska and then peons on the hacienda of Tihuicte) are the only formally recognized social units of the village of Huanoquite. The Ayllu Inkakuna withdrew from Huanoquite and attached itself to the community of Chifia as a result of a land dispute between Ayllu Maska, Ayllu Inkakuna (formerly Anansaya of Huanoquite), and a local *hacendado*. Ayllu Chanka became a *hacienda* during the colonial period and is now an immense cooperative. Finally, Ayllu Tantarcalla has distanced itself from Maska because Maska refuses to return the image of their patron saint. Because of this sociopolitical fragmentation the three *ayllus* that use the same water source for irrigation cannot easily collaborate to clean the entire irrigation system.

The fragmentation of Huanoquite's *ayllus* is in part caused by the preponderance of *haciendas* in the community, as well as the failure of the Agrarian Reform to address the inequalities in land ownership or to create mechanisms for coordinated irrigation maintenance.[19] Before the 1969 Agrarian Reform, a small group of *hacendados* competed among themselves over land and water but shared a fundamental interest in maintaining their hegemony over labor and in directing that labor in the construction and repair of ditches. The Reform abolished this brutal system of authority without instituting an effective substitute. The Reform also failed to abolish completely the unequal access to land and water. The technical reports prepared by Agrarian Reform personnel before expropriation proceedings reveal gross disparities between the actual amount of land owned by *hacendados* in Huanoquite and the reported amounts that were subject to expropriation. Collusion between Agrarian Reform agents and *hacendados* is apparent from these documents. *Hacendados*, whose holdings were only partially expropriated, remained with the best of their lands, those under irrigation.[20]

In consequence, the agrarian reform exacerbated conflicts over both the maintenance of the ditches and the distribution of water. Although post-Reform access to irrigation water is supposed to be egalitarian, large property owners not only own the most irrigated land but they also gain first access to irrigation water because they are located at the beginning of the canals. Even so, they contribute no labor to the canal cleaning other than a *ch'akipa* (a "thirst quencher" of alcohol) for the workers. Montoya et al. (1979) have remarked that a hegemonic ideology obscures the reality of unequal distribution of irrigation water in the central highland community of Puquio. Huanoquiteño villagers, in contrast, recognize the injustice of their situation.

To clean the canal as thoroughly as possible without damaging the ditch, no one is supposed to irrigate for at least four days before the cleaning. Despite this customary law, a few large landowners continue irrigating their properties, thus weakening the ditch walls even further.

The fragmented system of land tenure also impedes the effective mobilization of labor for work on irrigation maintenance. The heirs of *hacendados*, along with community and cooperative members who have acquired private lands through rent, purchase, or inheritance, farm smaller irrigated parcels interspersed among community and cooperative lands expropriated from the larger *haciendas*. Additionally, four large properties remain in private hands interspersed among Maska's community lands. It is almost impossible to organize the cleaning of this portion of the ditch or to allocate access to water rights judiciously given the multiple, coexisting tenure patterns and administrative structures (see Map 9.4).

The Qewar-Tantarcalla ditch disputes provide a poignant example of how pre-Reform inequities have persisted despite changes in the laws.[21] The canal passes above the lush estate of Mallma and drops down to the *ayllu* of Wiñay Poqo. Farmers in Wiñay Poqo speak bitterly of the three generations of the Paz family and their in-laws, who have owned Mallma. "From time immemorial," Perpetuo Socorro Cabeza told us, the Paz family has taken Wiñay Poqo's water. They are "malicious and wasteful with it," overwatering their fields and even watering uncultivated lands. The records of the Ministry of Agriculture, Comunidades Campesinas (Reforma Agraria, n.d.), show that Humberto Paz's grandfather, Cirilo, opposed Tantarcalla's recognition as a formal community in the late 1920s. As a result, a complex history of attempted murders, brandings, and beatings entered the legal record of the Ministry concerning the litigations over water and land. In 1938, Cirilo Paz wanted to charge Wiñay Poqo one *sol* per *topu* of land they irrigated. When they did not pay him, he destroyed the ditch. Tantarcalla's headman accused Paz of trying to murder one of their members by throwing him into the river. In 1941, the spokesperson of Tantarcalla described Paz as "the terrible enemy of the people." That same year, Paz whipped the spokesperson and threatened to murder four other "indians" from Tantarcalla.

The irrigation canal, after it reaches Wiñay Poqo, waters a tiny portion of the ayllu's thirty *topus* of cultivable land. The peasant farmers clean the entire length of the canal in a two to three day labor party, including the portion that passes above Mallma. After they return to their village, located four kilometers away, they "look up, waiting for the water to come," hoping that Humberto Paz has not cut it off for use on his own fields, which he often does.

Even where the Agrarian Reform effectively redistributed land, it failed to

MAP 9.4 LAND TENURE STRUCTURE OF HUANOQUITE

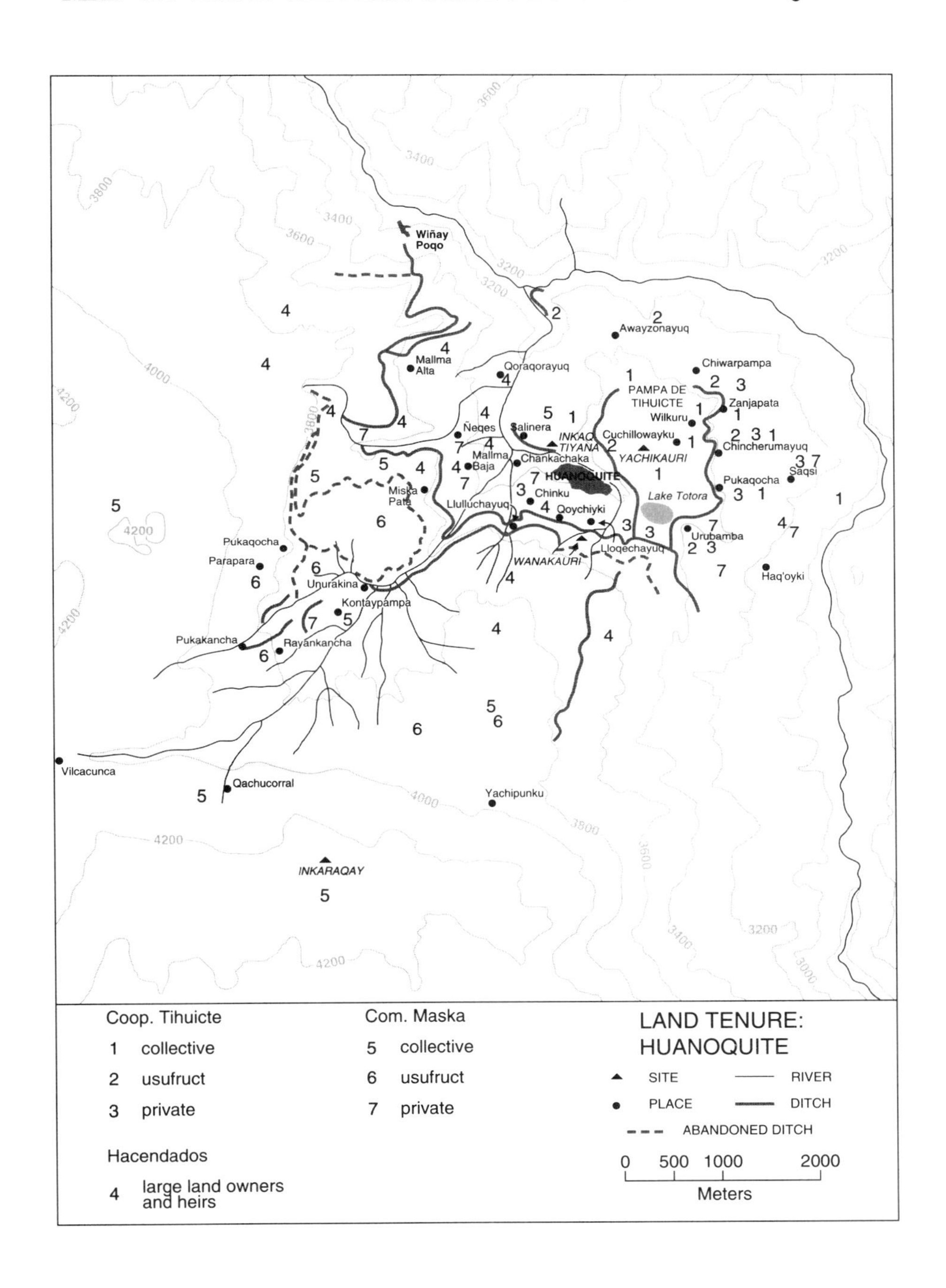

develop a coherent policy of land tenure and irrigation use for the entire community, so that the Reform actually exacerbated conflicts over water and land. The hacienda Tihuicte, which controlled most of the pampa, for example, was transformed into a cooperative. The independent community of Maska gained lands from other haciendas. Even though they share an irrigation ditch, the community and the cooperative have different juridical and organizational bases so that they are often unable to act jointly for the common good. Since most of Maska's lands lie along the upper course of the ditch and Tihuicte's lands are near the end of the ditch, the two entities are unable to agree about how much labor each owes to the collective endeavor. Coordinating ditch repair between the two units is thus extremely difficult. These tensions over ditch cleaning, moreover, carry over into conflicts over water distribution between upper and lower stretches of the ditch.

The establishment of the cooperative also incited conflicts between the cooperative and community over access to lands in different production zones. Today, members of the cooperative control the majority of lands in the *qheshwa,* or lower valley zones, where they cultivate maize and wheat, while the members of the community control the majority of the *puna* lands where they grow potatoes, barley, and broad beans.[22]

The lack of a central authority severely impedes coherent water management. The jurisdictions of the community of Maska, the cooperative of Tihuicte, and the District Council, which is supposed to represent the entire district of Huanoquite, are not clearly defined and in many instances overlap. The factionalism caused by these artificial administrative structures is exacerbated by the scarcity of economic resources in the region, deepening further the existing cleavages between and among villagers and their administrative representatives. As farmers have become increasingly tied to the market economy, moreover, they have to work harder and harder on their own farms. Because of the Reform, each household now has to make far more of its own farming decisions and organize its own transport for the sale of its products in the city of Cuzco. As most households now have a greater land base than they did before the Reform, they must spend more time meeting their own labor needs at the same time that authorities are demanding more of their time to work on public projects. Getting agricultural credit and information also takes a great deal of a farmer's time. Given the problems that peasant households face in juggling their household labor needs, it is not surprising that political authorities are unable to harness labor for the canal cleaning. Many households adhere to the "least-effort" principle, exerting minimal collective effort in the short-term at the risk of a long-term collective economic collapse in the future. The best that they can achieve under these circumstances is irrigation maintenance so superficial that it is insufficient to conserve the essential infrastructure.

Compounding these political and economic problems is the apparent loss of the topographic knowledge and skills necessary to maintain the ditches. The land tenure structure that has developed since the Agrarian Reform has diminished the general topographic knowledge of most farmers. Members of each community have tended to become intimately familiar with only a relatively small portion of the environment over which the canal traverses, and each Huanoquiteño finds it more difficult to be knowledgeable about the distinct soil conditions, slopes, geological and geographic conditions, and the relations among different topographic features in the areas where they no longer walk or work. It may be the case that a few people from different social and territorial units control enough environmental knowledge so that if they pooled their knowledge, they could create a general understanding of Huanoquite's regional topography. It is highly unlikely that they would be able to do so because of the fragmented sociopolitical order of the region. Consequently, Huanoquiteños are unable to create a coherent vision of necessary environmental knowledge for the needed maintenance of their irrigation works, further jeopardizing the infrastructure.

The growing number of children who attend school has exacerbated the decline in general topographic knowledge. Children rarely accompany their parents to work in their fields or to herd livestock, the way they once did. Older Huanoquiteños, even though they recognize the value of formal education, have told us that they acquired their environmental knowledge by helping their parents in the fields or herding. Young men who migrate to urban centers and return to Huanoquite as adults experience similar difficulties in learning how to farm the lands they receive from their parents.

Huanoquiteños are much more oriented to the market and individual productive strategies today than they were previously. Because of this they are much more careful in the ways they use irrigation water in their own fields than they are in the collective care they give the system. Because they are concerned with how frequently they should irrigate their crops, they consider carefully weather conditions, kinds of cultigens, and the quality of the soil. They plow vertical furrows for nonirrigated crops, which suffer from waterlogging and root rot. They use contour plowing for early potatoes. To prevent the loss of previous fertilizer, they irrigate their maize only once. They irrigate their early potatoes several times. They complain about the haciendas whose sandy soils require far more irrigated water than their own fields of clayey soil.

Here, one clearly sees the differences between collective and individual care of irrigation water. Individuals are far more conscious of the best way to use irrigation water in their own fields. In contrast, although the water is destined for individual use, because farmers perceive the community's canal as public property, they do not preoccupy themselves with its maintenance. Even though water is supposed to be used for the general welfare of both the community and cooperative, the two political groups compete for it.

CONCLUSION

The inability to mobilize labor to maintain the irrigation system because of the process described in this chapter is illustrated by the story of one moderately well-off *hacendado* who lives in an isolated area of Huanoquite near Wataqasa. To irrigate, he constructed with some difficulty, his own ditch. An old man from Tantarcalla had told him that he would not be able to move the rocks that lay in the path of the ditch because sorcerers had put them there. When the workers tried to move them, their arms went numb. However, because the spell was ancient, it finally moved to another place, so that the workers were able to break up the rocks and build the ditch.

What this *hacendado* does not believe, and what no one in the village, as far as we could ascertain, believes, is that this *hacendado* is using the water that formerly supplied the now abandoned canal. He, like many others, believes the abandoned canal's original water source either dried up or dropped to a lower source because of the death of the golden serpent, and that the water he is now tapping comes from the lower source. One day, however, we followed the traces of the abandoned canal to its origin and discovered that water continued to flow from it and that this water eventually reached the ditch built by the *hacendado*. This meant that the entire trajectory of the abandoned canal could be repaired, a task that would require coordinated labor power.

If the abandoned canal were resucitated, it would irrigate at least 80 hectares of land belonging to the community in the Qolla Wayku Valley where it used to flow. Nevertheless, Huanoquiteños have not been able to mobilize the labor to build it. Their oral traditions have become incorporated into their worldview, even to the point of distorting the empirical evidence that the original water source is still available and that the abandoned canal could be repaired. The story of the killing of the snake with golden brades evokes the Huanoquiteños' experiences of conquest and their effects–the destruction of empire and the transformation of sociopolitical organization that makes it far more difficult for Huanoquiteños to coordinate their labor. These stories also depict the challenges that repairing the canal would represent for a fragmented polity.

Clearly, it would not be a simple task to repair the abandoned canal. Oral traditions incorporate the ideology of subordination that Huanoquiteños have absorbed over the centuries. These myths also accurately portray the problems Huanoquiteños would have to surmount to revitalize their environmental understanding and to coordinate their labor to maintain their existing canals, let alone construct and maintain one that has been abandoned. Huanoquiteños have lost much of their prior technological knowledge and rarely experiment with new ditch technologies. For them, the canals were not constructed by common men but rather by their more powerful and capable ancestors who knew how to control and tame the savage forces of nature and political society. Huanoquiteños also pay heed to a taboo that prevents them from tampering with their ancestors' projects.[23]

Although Huanoquite continues to retain an identity it has forged over centuries, the Agrarian Reform has divided Huanoquite into two competing political organizations at the same time that it has turned over to them their own lands and given them a political legitimacy to challenge the power of the landed class. The Agrarian Reform alone, of course, is not responsible for Huanoquite's political fragmentation. Huanoquite, as is true of many other colonized regions, has suffered irrevocably from the earlier impact of the Spanish Invasion–demographic collapse, the usurpation of indigenous lands, the exploitation of indigenous labor, and the extirpation of idolatries that fragmented a coherent religious belief system by destroying the sacred landmarks that embodied signficant indigenous environmental knowledge. Nonetheless, the Agrarian Reform has accelerated Huanoquite's sociopolitical fragmentation.

The case of Huanoquite illustrates one of Wittfogel's fundamental observations: the construction and maintenance of an irrigation system requires knowledge that is transmitted from generation to generation, technology to implement that knowledge, and the authority to coordinate a labor force and manage material resources within the specific environment in which it is found. An environment as complex and mountainous as Huanoquite's imposes greater constraints than those found in less difficult geographic surroundings. The existing chaos in irrigation maintenance, water allocation, and canal reconstruction can only be eliminated by some form of overarching authority. In other parts of the Andes, this authority has been placed in the community itself, which continues to be strong enough to force members to comply with their collective obligations. Huanoquiteños have been unable to vest such authority in the community.

Golte (1980) observes that the dynamic persistence of Andean organization is founded on the flexibility of socioeconomic relationships within Andean collectivities that permit farmers to manage multiple cycles of agricultural production in conjunction with market demands, migration patterns, and other activities that intervene in production and exchange. Huanoquite's sociopolitical fragmentation, the continued power of *hacendados*, and the multiple demands on villagers' labor make collective and authoritative control over irrigation management almost impossible.

Huanoquiteños repeatedly tell us that their infrastructure can be expanded or maintained only through the intervention of "*el poder*," authority, whether in the form of a centralized state, a development agency, or some other social or supernatural authority. Their belief that they need such an outside power is reinforced by the visible deterioration of their irrigation system, which endangers their already fragile collective control over the environment. The basis for Huanoquiteños' social reproduction, therefore, is seriously jeopardized. Although the canal system continues to function, its survival remains tenuous. Eventually, another story may become part of the record, giving us bits and pieces of a major environmental disaster that has led to the collapse of Huanoquite's canal system and its varied agricultural economy.

NOTES

Acknowledgments. Grants from the Wenner Gren Foundation for Anthropological Research, Fulbright-Hayes, and the Nave Bequest of the University of Wisconsin-Madison have helped fund this research. We are grateful to the editors of this volume, William P. Mitchell and David Guillet, for their critical comments. Special thanks to the people of Huanoquite who took the time to tell us about the ways of water and the lay of the land.

1. By dialectical, we are not suggesting that the environment is sentient but rather that the cause and effect relationships between human beings and their environments are complex, multidirectional, and cumulative. The dramatic changes to environment that human beings unexpectedly bring about may require human beings to modify their productive and exchange relations. Likewise,

transformations in social organization may alter the environment in ways that prevent a community from depending on it for its subsistence. Well-known examples include soil and water depletion and deforestation attendent on sugar plantation and milling, salinization and erosion in certain irrigation systems, and contamination of soil and water from the mining or processing of many minerals.

2. Reichel-Domatoff (1976), Taussig (1987), and Whitten (1985) have also recently explored the idea that ecological systems are constituted in dialectic fashion through the interaction of society and nature in the Amazon, Colombia, and among the Quichua of Ecuador, respectively.

3. A two volume issue of the Peruvian journal *Allpanchis Phuturinqa* (1987) reflects a new interest in the topic of Andean irrigation. The research of Sherbondy (1982, n.d.), Isbell (1976, 1977), Ossio (1978), Zuidema (1964, 1978) and Bastien (1985) complements our growing knowledge of the sociopolitical organization of irrigation in the Andes. This body of research analyzes Incaic and contemporary rituals and beliefs associated with water, especially the festive cleaning of irrigation canals that takes place in the central highlands. Sherbondy has also traced and reconstructed theoretically the Incaic canal system of Cuzco, meticulously documenting how the technology of Incaic agriculture was linked to carefully prescribed rituals marking the annual hydraulic cycle under Inca hegemony.

4. Our findings are based on a yearlong period of fieldwork from December 1983 through December 1984.

5. The definition of *ayllu* changes according to time and place in the Andes. It can refer precisely to a social unit composed of related kin within three generations; on the other hand, it may be a more general kind of social unit composed of families related by ethnicity, common land holdings, rank, or class. A woman's *ayllu* might be her extended family, or her lineage, probably her bilateral kin, and even the members of her community or province. *Ayllu* members usually have obligations in the form of labor to their *ayllu* in return for the usufruct of corporate lands and any other resources that are considered collective. The members of an *ayllu* claim common shrines and sacred places. In contemporary Huanoquite, *ayllu* generally corresponds to community with a bias toward endogamy. Some flexibility has allowed outsiders occasionally to be incorporated into an *ayllu*.

6. Dating canals is difficult. Most villagers agree on the approximate age of the most recent ditches. That there were colonial and Incaic canals has been ascertained through oral tradition, archival references, and an analysis of canal construction materials and techniques, confirmed by an archaeologist working in the area (Brian Bauer, personal communication). What is not clear is whether functioning and nonfunctioning canals have pre-Incaic origins.

7. These production zones lie along a vertical gradient where distinct crops can be grown or where specific activities such as herding take place. The lower valley, or *qheshwa*, ranges from 3,100 meters above sea sea level to 3,600 meters above sea level, where farmers specialize in maize, early potatoes and wheat. During the dry season, livestock are brought down from the puna to pasture in this area. The upper valley, or puna, ranges from 3,800 meters above sea level to 4,200 meters above sea level. At the higher reaches only herding occurs, whereas lower down, barley, tubers, some broad beans, and potatoes are cultivated. Huanoquite is located in the *chaupi*, midway between these two general ecological zones.

8. Kosok (1965) was aware of the importance of these typological differences. The frontispiece of his book *Land and Water in Ancient Peru* consists of two air photos of irrigation systems, one from the Nile, the other from coastal Peru. They are dramatic examples of the distinctions among hydraulic systems that Wittfogel was trying to make.

9. Flow rates are all based on measures taken in 1984.

10. See Seligmann (1987a) for an analysis of the ways that earlier land holding and land use patterns affected the implementation of the Agrarian Reform, maintaining an unequal distribution of land.

11. The severe erosion of Tanqarpata is probably the result of the collapse of another more ancient canal that carried waters to the precolonial center of the area, known today by Huanoquiteños as Ñawpaq Llaqta (Ancient Town).

12. Wiñay Poqo is one of three former sociopolitical units, or minimal *ayllus*, that were part of the maximal *ayllu* of Tantarcalla, which belonged to Huanoquite before the Spanish Conquest.

13. Another myth known by fewer Huanoquiteños than the two principal myths speaks of a smaller lake located at a higher altitude, Lake Qompu. The same priest tried to plumb the depths of this lake without success. Finally, he filled a bucket with water and red *kantuta* flowers and let it down by a rope. The force of the subterranean waters swept away the *kantutas*, and they reappeared in the sites mentioned above. Both lakes are located more than ten kilometers and two steep valleys away from the village.

14. Many repeat the story of the son of an *hacendado* who faints after lifting a rock under which he sees water gushing and boiling. He returns with his father but can no longer find where the place is. Ardiles (1986) has done archaeological surveys at a site near Cuzco and has found evidence of existing subterranean canals. Knapp (1987) has also found extensive archaeological evidence of such tunnels in Ecuador and has been able to obtain ethnographic data on how they were built and cleaned. Many references to such channels are made in the chronicles as well.

15. See Seligmann (1987b) for a detailed discussion of how contemporary Andean highland oral traditions interpret the historical experience of the Spanish Conquest, particularly the inversion of pre-Hispanic notions of power and authority, and how these interpretations, in turn, have led to the development of messianic beliefs and movements.

16. Other Huanoquite sites also resemble those in imperial myths and traditions, being called by the same name as the imperial sacred place and containing similar characteristics. For example, the Inca site of Wanakauri figured prominently in the origin myth of the Inca state (see Urbano 1981) and was the site where royal initiation rites took place. The Huanoquite place with the same name is located near a lake, filled with *totora* reeds. Zuidema and Urton (1976) have pointed out the importance of *totora* reeds in initiation rites. Perhaps the Inca nobility residing in Huanoquite performed their initiation rites at this local Wanakauri site. It would be fascinating to determine the reasons why identical toponyms are commonly found in different and distant regions of the Andes. It may be the result of similar geography, migration, political propaganda, and/or common historical traditions.

17. Our debt to Durkheim in this and other parts of our argument should be fairly clear. The authority vested in and maintained by a community can be as despotic as Wittfogel's Asiatic Mode of Production; Durkheim saw socially beneficial results in this.

18. In this article, we provide a general sketch of these changes. For a more detailed analysis, see Seligmann's (1987a) *Land, Labor and Power: Local Initiative and Land Reform in Huanoquite, Peru*; and Bunker and Seligmann's (n.d.) *The Huanoquite Waterworks*.

19. By 1689, ten substantial estates had been established in the Huanoquite area (Villanueva 1982:433-440), and by 1786, there were at least fifteen *haciendas*, a higher number than that found in the other six parishes (*doctrinas*) that then comprised the province of Paruro (Mörner 1977).

20. To determine the amount of land that would be expropriated from an *hacienda*, Reform agents evaluated the *hacienda's* different kinds of land–forest, pasture, irrigated, nonirrigated, and fallow–by differentially weighting them and converting them to a single measure. To satisfy expropriation demands, many land owners willingly gave up great portions of forested, nonirrigated, and pasture lands to remain with a much smaller amount of productive irrigated land.

21. Guillet (1985) discusses how changes in water laws affected efforts by communities at the local level to regain access to irrigation water. The litigations in Tantarcalla during the early twentieth century may well have been provoked by new laws that were passed concerning the return of water management to local communities and the rights of Indian communities to gain formal recognition. As Guillet points out, after the Spanish Conquest, increasing legal emphasis was placed on the management of land rather than water; in addition, the indirect rule of the Incas through local ethnic lords was replaced by a Spanish peninsular model of local government, the cabildo or town council, which became responsible for irrigation management. In the late nineteenth century, the French Municipal Code was introduced. Local control over irrigation legally became the responsibility of the sub-prefect, located in the provincial capital. Although theoretically this meant that control had shifted away from a local to a regional level, in reality, disputes continued to be settled locally by a variety of means since the distance of most communities from the provincial capital meant that these laws could not really be enforced.

22. A degree of coordination has evolved between the community and the cooperative.

Cooperative members need labor power from the community. Community members need pasture lands that the cooperative controls. Furthermore, community members covet maize to make the beer (*chicha*) needed for reciprocal labor exchanges (*ayni*). These interdependencies moderate the conflicts between the community and the cooperative, but the overriding relationship is one of competition.

23. In other areas, it may very well be that tectonic uplift (Mosely 1983), demographic collapse (Cook 1982; Denevan 1976; Treacy 1987), extreme climatic shifts (Guillet 1987), migration, which brings a better return than farming in some cases (Mitchell *intra*), or some combination of these events (Denevan 1987) may be responsible for the abandonment of canals or terrace systems. The Conquest caused depopulation in Huanoquite, which probably contributed greatly to difficulties in labor management, particularly when it involved large-scale infrastructural projects. However, the fact that the abandoned canal's water continues to be tapped, that land pressure has increased dramatically in Huanoquite, and that Huanoquiteños insist the canal is irreparable do not point so much to environmental causes for the canal's abandonment but to sociopolitical and cultural causes.

REFERENCES CITED

Allen, Catherine
1988 *The Hold Life Has*. Washington, DC: Smithsonian Institution Press.

Ardiles, Percy
1986 "Sistema de drenaje subterraneo prehispánica." *Allpanchis Phuturinqa* 27:75-97.

Avila, Francisco de
1966[1598] *Dioses y hombres de Huarochirí*. José María Arguedas, translator. Lima: El Museo Nacional and Instituto de Estudios Peruanos.

Bastien, Joseph
1985 "Qollahuaya-Andean Body Concepts: A Topographical-Hydraulic Model of Physiology." *American Anthropologist* 87(3):595-611.

Bunker, Stephen G. and Linda J. Seligmann
n.d. "The Huanoquite Waterworks." Unpublished manuscript.

Cook, Noble David
1982 *The People of the Colca Valley: A Population Study*. Boulder, CO: Westview Press.

Delran, Guido C.
1974 "El sentido de la historia." *Allpanchis Phuturinqa* 6:13-28.

Denevan, William M.
1976 *The Native Population of the Americas in 1492*. Madison: University of Wisconsin Press.
1987 "Terrace Abandonment in the Colca Valley, Peru." In *Pre-Hispanic Agricultural Fields in the Andean Region*. William M. Denevan, Kent Matthewson, and Gregory Knapp, editors, pp. 1-43. Oxford: British Archaeological Reports.

Earls, John and Irene Silverblatt
1978 "La realidad física y social en la cosmología andina." *Actes du XLII Congres International des Americanistes*, volume 4, 1976, pp. 299-326. Paris.

Fonseca M., Cesar
1984 "El control comunal del agua en la Cuenca del Rio Cañete." *Allpanchis Phuturinqa* 22:61-73, Cusco, Peru.

Gelles, Paul
1984 "Agua, faenas y organización comunal: San Pedro de Casta, Huarochiri." *Revista Antropológica del Departamento de Ciencias Sociales* 2:305:334.

Golte, Jürgen
1980 La racionalidad de la organización andina. Lima: Instituto de Estudios Peruanos.

Guaman Poma de Ayala, Felipe
1980 [1615] *El primer nueva corónica y buen gobierno*. John V. Murra, Rolena Adorno, et al., editors. Mexico: Siglo veintiuno.

Guillet, David
1985 "Irrigation Management Spheres, Systemic Linkages and Household Production in Southern Peru." Paper presented at the Annual Meeting of the American Anthropological Association, Washington, DC.
1987 "Terracing and Irrigation in the Peruvian Highlands." *Current Anthropology* 28(4):409-430.

Hernandez Principe, Rodrigo
1923 [1621] "Mitología Andina." *Inca* 1:25-78.

Isbell, Billie Jean
1976 *To Defend Ourselves: Ecology and Ritual in an Andean Village*. Austin: University of Texas Press, ILAS.
1977 "Those Who Love Me: "An Analysis of Andean Kinship and Reciprocity Within a Ritual Context." In *Andean Kinship and Marriage*. Special Publication Series, 7. Ralph Bolton and Enrique Mayer, editors, pp. 81-105. Washington, DC: American Anthropological Association.

Knapp, Gregory
1987 "Riego precolonial en la sierra norte." Ecuador *Debate*: 14:17-45.

Kosok, Paul
1965 *Life, Land and Water in Ancient Peru*. Long Island, NY: Long Island University Press.

Mayer, Enrique
1977 *Tenencia y control comunal de la tierra: caso de Laraos (Yauyos). Cuadernos*, 24-25:59-72.
1979 *Land Use in the Andes: Ecology and Agriculture in the Mantaro Valley of Peru with Special Reference to Potatoes*. Lima: International Potato Center, Social Science Unit Publication.
1985 "Production Zones." In *Andean Ecology and Civilization: An Interdisciplinary Perspective on Andean Ecological Complementarity*. Shozo Masuda, Izumi Shimada, and Craig Morris, editors, pp. 45-84. Tokyo: University of Tokyo Press.

Mayer, Enrique and Cesar Fonseca M.
1979 *Sistemas agrarios en la Cuenca del Rio Cañete*. Lima: Instituto Indigenista Interamericano & ONERN.

Mitchell, William P.
1976 "Irrigation and Community in the Central Peruvian Highlands." *American Anthropologist* 78:25-44.

Mörner, Magnus
1977 *Perfíl de la sociedad rural del Cusco a fines de la colonia*. Lima: Universidad del Pacífico.

Montoya, Rodrigo, María J. Silveira, and Felipe J. Lindoso
1979 *Producción parcelaria y universo ideológico: el caso de Puquio*. Lima: Mosca Azul Editores.

Mosely, Michael
1983 "The Good Old Days Were Better: Agrarian Collapse and Tectonics." *American Anthropologist* 85:773-799.

Netherly, Patricia
1984 "The Management of Late Andean Irrigation Systems on the North Coast of Peru." *American Antiquity* 49(2):227-254.

Ossio, Juan M.
1978 "El simbolismo del agua y la representación del tiempo y el espacio en la fiesta de la acequia de la comunidad de Andamarca." *Actes du Congres XLII International des Americanistes*, v. 4, 1976, pp. 377-396. Paris.

Polo de Ondegardo, Juan
1917[1571] *Relación de los fundamentos acerca del notable daño que resulta de no guardar a los Indios sus fueros*. Horacio Urteaga, editor. Colección de Libros y Documentos Referentes a la Historia del Perú, 3. Lima: Sanmarti.

Reichel-Dolmatoff, Gerardo
1976 "Cosmology as Ecological Analysis: A View from the Rain Forest." *Man*, New Series II:307-318.

Reforma Agraria, Oficina de Comunidades Campesinas, Tantarcalla.
n.d. Cusco, Peru: Ministerio de Agricultura, Oficina de la Reforma Agraria, Zona Agraria XI.

Seligmann, Linda J.
1987a *Land, Labor and Power: Local Initiative and Land Reform in Huanoquite, Peru*. Ph.D. dissertation, University of Illinois, Urbana.
1987b "The Chicken in Andean History and Myth: The Quechua Concept of *Wallpa*." *Ethnohistory* 34(2):139-170.

Sherbondy, Jeanette
1982 *The Canal Systems of Hanan Cusco*. Ph.D. dissertation, University of Illinois, Urbana.
n.d. "Water and Power in Inca Peru." In *Andean Kaleidescopes*. Billie Jean Isbell, editor. In press.

Taussig, Michael
1987 *Shamanism, Colonialism, and The Wild Man: A Study in Terror and Healing*. Chicago: University of Chicago Press.

Treacy, John
1987 "An Ecological Model for Estimating Prehistoric Population at Coporaque, Colca Valley, Peru." In *Pre-Hispanic Agricultural Fields in the Andean Region*. William M. Denevan, Kent Mathewson and Gregory Knapp, editors, pp. 147-162. Oxford: B.A.R.

Urbano, Henrique
1981 *Wiracocha y Ayar: héroes y funciones en las sociedades andinas*. Cuzco: Centro de Estudios Rurales Andinos 'Bartolomé de Las Casas.'

Villanueva Urteaga, Horacio
1982 *Cusco 1689: Informes de los párrocos al Obispo Mollinedo*. Cuzco: Centro de Estudios Rurales Andinos 'Bartolomé de Las Casas.'

Whitten, Norman E., Jr.
1985 *Sicuanga Runa: The Other Side of Development in Amazonian Ecuador*. Urbana: University of Illinois Press.

Wittfogel, Karl
1957 *Oriental Despotism*. New Haven: Yale University Press.
1985 [1928] "Geopolitics, Geographical Materialism, and Marxism." Translated by G. L. Ulmen. *Antipode* 17(1):21-72.

Zuidema, R. T.
1964 *The Ceque System of Cusco: The Social Organization of the Capital of the Incas*. Leiden: E.J. Brill.
1978 "Lieux sacres et irrigation: tradition historique, mythes et ritueles au Cusco." *Annales* 33(5-6):1,037-1,056.

Zuidema, R. T., and Gary Urton
1976 "La constelación de la llama en los Andes Peruanos." *Allpanchis Phuturinqa* 9:59-120.

CHAPTER TEN

Channels of Power, Fields of Contention: The Politics of Irrigation and Land Recovery in an Andean Peasant Community

PAUL H. GELLES
University of California, Riverside

INTRODUCTION

This chapter analyzes the politics of irrigation and land recovery in the highland peasant community of Cabanaconde, located in the Colca Valley of southern Peru. I argue that the organization of irrigation is determined by social and cultural models and by the historical and political processes–local, regional, and national–that have given form to these models.[1] To invoke Geertz (1980), irrigation systems are "texts to be read." However, the reading of these systems should allow us to understand not only the cultural models that they embody but also the power relationships that underwrite the use of these models. The analysis presented here takes account of the culturally plural nature of Peruvian society, of the ways that models for irrigation management are negotiated and manipulated by individuals and groups, and of the transformation of these models over time.

I use two cases to elucidate these processes. The first concerns attempts to recover a large number of abandoned terraced fields. Although water recently became available to cultivate these fields, political conflict in the community has interfered with their complete recovery. These power struggles, which have a long history, are exacerbated by political forces at the regional and national levels. These combined forces limit water availability and consequently the viability of land recovery within Cabanaconde's territory.

The second case concerns an important feature of irrigation politics in Cabanaconde and elsewhere in the Andes: the tension between local and state models of irrigation management. The local model is based on a system of dual organization in which peasant water mayors (*yaku alcaldes*) distribute the water. The water mayors and the dual model are a legacy of both Inca and Spanish hegemony. Today, however, they constitute the local "indigenous" form of irrigation management, one that is used to ritually obtain abundant water and to resist state interference in local affairs. This local model is contested by the supposedly more "rational" model of the Peruvian state's irrigation bureaucracy (see Bolin *intra*; Guillet *intra*). This section of the paper demonstrates that these local and state models have resulted from different historical processes and that they embody different cultural rationales concerning power, authority, resource management, and ethnic identity.

In the concluding discussion, the two sets of conflicts–the politics of water availability and land recovery, and state versus local models of water distribution–are examined together in terms of the larger cultural politics found in Peruvian state and community relations. Both cases demonstrate that the micro-politics of irrigation and community are embedded in regional and national contexts. Conversely, they also show that the social and cultural forms of the community, as well as local initiatives, determine the extent of state intervention. The two cases also reveal that in addition to ecological and technical considerations, we need to look more closely at the political forces that may inhibit–or encourage–the recovery of abandoned terraced fields for agriculture.

THE COMMUNITY OF CABANACONDE

Located at 3,270 meters above sea level (m.a.s.l.) on the semiarid west slope of the southern Peruvian Andes (Map 10.1), the town of Cabanaconde was established as a nucleated settlement (*reducción*) in the 1570s. It was made a district capital in the early 19th century and became an officially recognized peasant community with corporate legal status in 1979. It has two annexes, Pinchullo and Acpi. The people of Cabanaconde today are bilingual Quechua and Spanish speakers. They distinguish themselves ethnically from other groups in the southern Peruvian Andes and even from other groups of the Colca Valley.[2]

The population of Cabanaconde has more than doubled over the last century, growing from 1,796 inhabitants in 1876 to 2,960 in 1940, 3,421 in 1981, and 4,000 in 1987 (Cook 1982:41, 84; Denevan 1987:17; Ministry of Agriculture 1987). Although it is difficult to get a precise figure, there were at least 600 households in the community in 1988.[3] Population growth has placed more and more pressure on land, water, and other productive resources. The demographic expansion, combined

MAP 10.1 MAP OF CABANACONDE AND SURROUNDING AREA

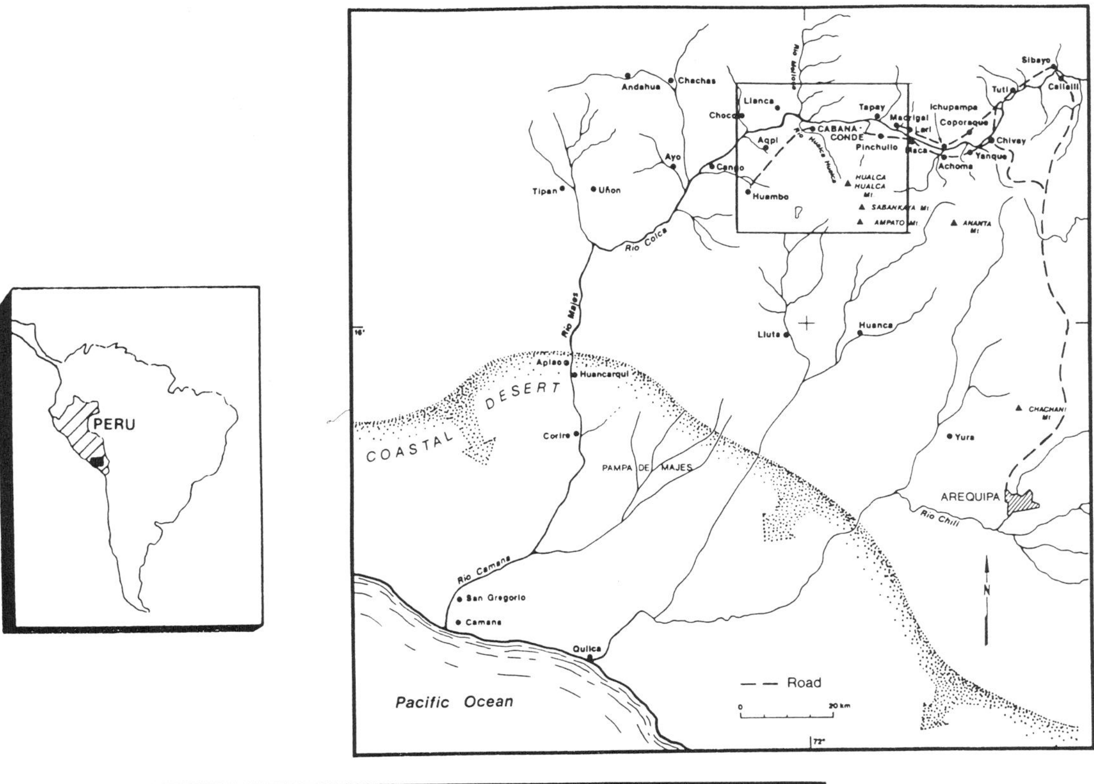

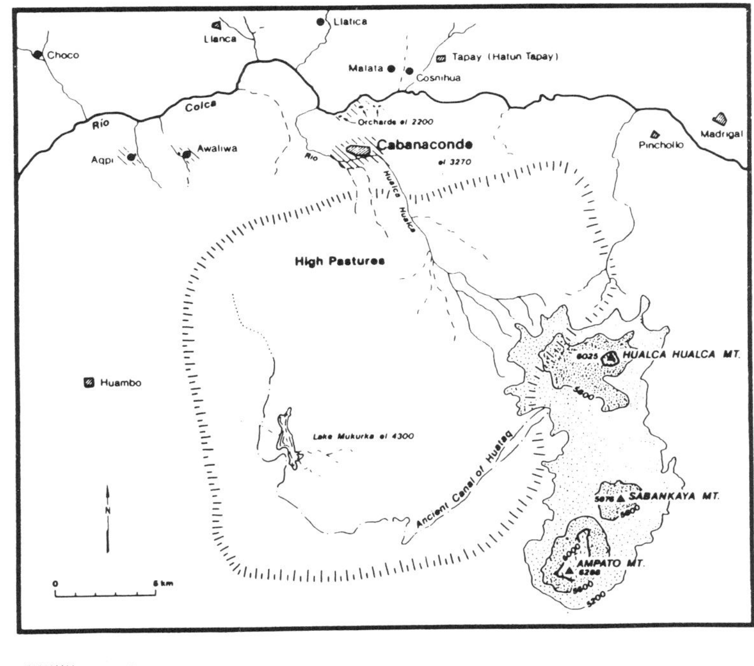

Area of high pastures
Cultivated fields

with a bilateral inheritance pattern, creates rampant partitioning of land holdings (*minifundismo*). Because of this and because Cabanaconde is economically differentiated (although not as much as many other communities in the region), many farmers must rely on sharecropping arrangements and land rental to make ends meet. A road linking Cabanaconde to the city of Arequipa was built in 1965, increasing seasonal migration and participation in the market economy. Population pressure has also been an important factor in permanent outmigration. Today, there are large migrant colonies of Cabaneños in Arequipa and Lima, as well as a smaller colony in Washington, D.C.[4]

The environmentally diverse "production zones" (Mayer 1985) of Cabanaconde range from 2,000 to 4,500 m.a.s.l. (Figure 10.1).[5] These zones are all within a day's walk of the community, constituting a classic case of "compressed verticality" (Brush 1977). In the Colca canyon at 2,000 m.a.s.l., the farmers cultivate fruit, alfalfa, maize, and prickly pear cactus. Prickly pear cactus, in addition to producing a delicious fruit, also serves as a habitat for cochineal (small insects that contain a red colorant that is sold commercially as dye). The high pastures, located between 3,800-4,500 m.a.s.l., are home to a number of alpaca, llama, sheep, and cattle herds. The bulk of agricultural production takes place between approximately 3,000 and 3,350 m.a.s.l., where the famous Cabanita maize grows. There are more than 1,200 hectares of irrigated fields, approximately three-quarters of which are dedicated to maize. Peasants from many parts of the southern Andes come to barter for this product.[6]

GENERAL FEATURES OF IRRIGATION IN CABANACONDE

The Hualca-Hualca River gathers water from supplementary sources, including the Majes canal, as it winds its way down to the town's fields from the snowmelt of the looming Hualca-Hualca Mountain (6,025 m.a.s.l.).[7] The ritual complex associated with Hualca-Hualca Mountain is part of the local model discussed below. It is quite ancient. In 1586, a Spanish Crown official described the use the town of Cavana made of Hualca-Hualca's snowmelt to irrigate its fields (Ulloa Mogollón 1965 [1586]). The same document also reveals that the Cavana people worshipped Hualca-Hualca and believed that their ancestors emerged from this mountain. Beliefs about origins in, and worship of, mountains and other sources of water is an ancient and widespread characteristic of many ethnic groups in the Andes (Reinhard 1985; Sherbondy 1982b, *intra*). Although today the people of Cabanaconde are influenced by many diverse ideologies, Hualca-Hualca Mountain continues to be a principal deity and the object of much ritual activity.[8]

Irrigation is an important arena of social interaction. There were 865 irrigators (usually heads of households) in Cabanaconde in 1980 according to the Ministry of Agriculture. Although the heads of households and the water authorities described below are predominantly men, both men and women are skilled and active in irrigation from a young age.[9] Family ties and alliances are often expressed in water use. In disputes over land, for example, water is sometimes used as a weapon: a

FIGURE 10.1 CABANACONDE PRODUCTION ZONES

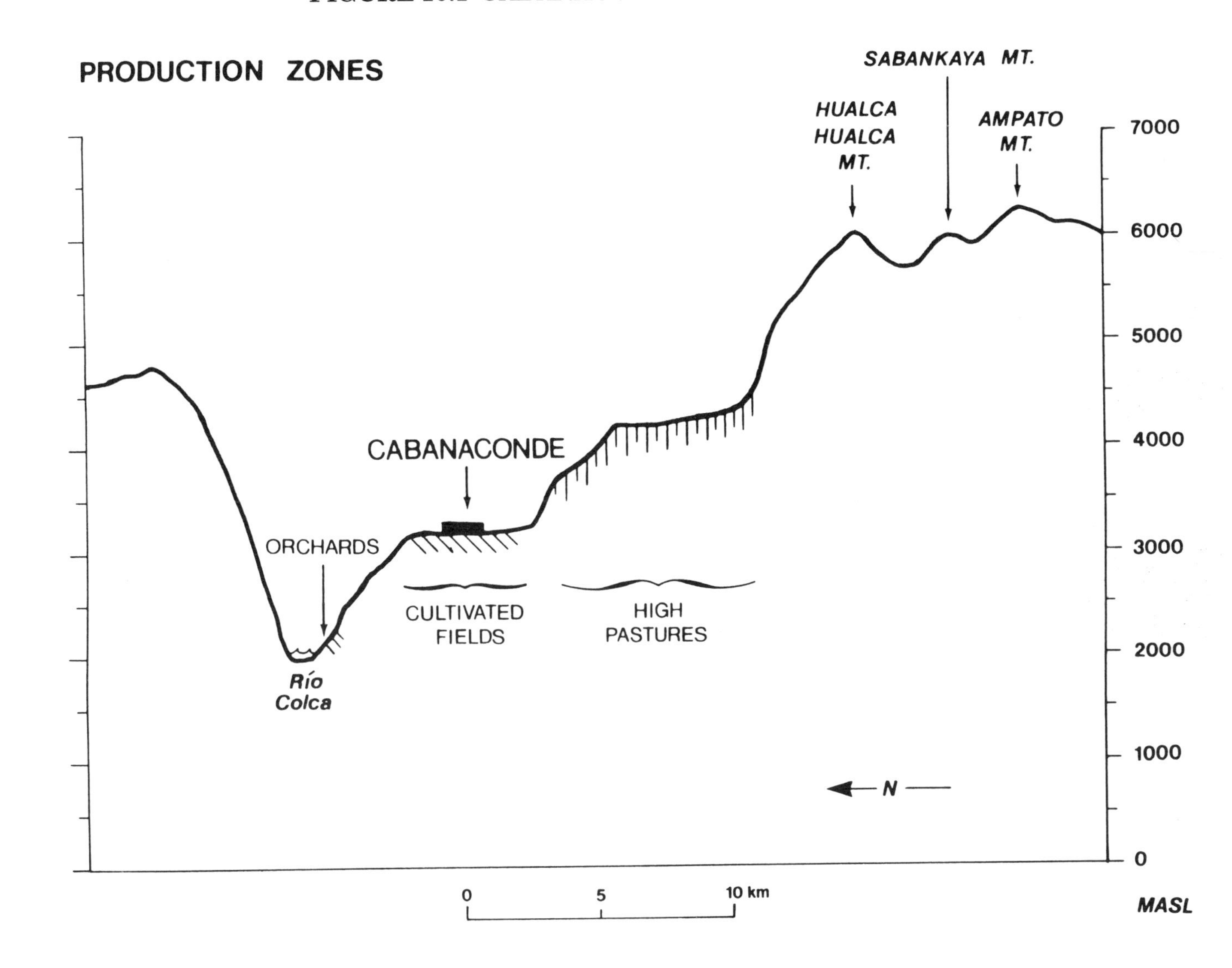

slight turn of a rock can send water streaming into the already irrigated plot of an enemy, destroying the small plants. The importance of irrigation is also demonstrated by the fact that the irrigation cycle structures the private and collective calendars of the Cabaneños during much of the year. It gives life to their terraced fields and to many of their conversations. As evidenced in the centuries-old cult to Hualca-Hualca, water is also a basic part of the Cabaneños' ethnic identity.

At present, the waters of Hualca-Hualca and the Majes canal are used without letup from the first of June until the rains begin, usually in December or January. River water in Cabanaconde passes directly and continuously to the fields through an intricate canal system. People irrigate both day and night. Reservoirs, used for storing water at night in other regions, are not important to this system (Map 10.2), a feature that also distinguishes Cabanaconde from other communities in the Colca Valley, such as Coporaque (Treacy 1989b), Lari (Guillet 1987), and Yanque (Valderrama and Escalante 1988), as well as nearby Tapay (Paerregaard *intra*). The Hualca-Hualca River is estimated to supply Cabanaconde's agriculture with between 75 and 150 liters per second of water during the dry season months of May to December (Abril Benavides 1979; ORDEA 1980). Before the Majes water became available in 1983, each complete cycle or round of irrigation water through all the cultivated fields lasted 90-120 days. Majes has doubled the amount of available water, and today each round lasts 45-50 days. The additional water has intensified existing agriculture and allowed Cabanaconde to expand its area of cultivated land.

Direct and heavy rainfall from January through March is important for the proper maturation of maize. This precipitation is also needed to increase the snowpack on Hualca-Hualca and the volume of melt for irrigation.[10] Annual rainfall, however, is extremely variable in Cabanaconde, and drought occurs periodically in this area of the Andes (see Figure 10.2 and also Guillet 1985). The people of Cabanaconde irrigate to complement the low levels of rainfall throughout the growing season. They also irrigate before the rains begin to extend the growing season of maize, permitting its sowing four months before the rains and thereby allowing its harvest before frost sets in. In this way the people of Cabanaconde–like those of Quinua described by Mitchell (1981)–are able to plant maize at higher altitudes than would otherwise be possible.

CONFLICT AND THE POLITICS OF LAND RECOVERY

The first case to be examined in this essay concerns the impact of communal and regional conflict on land recovery. Municipal records show that communal efforts to tap new water sources, and thereby rehabilitate large tracts of land, date back to at least 1916 (BMC). These efforts, which continued throughout the century, bore fruit in 1988 and continue to do so today. Large tracts of fertile land, however, are not being recovered because of local and regional politics.

MAP 10.2 THE CANAL SYSTEM OF CABANACONDE

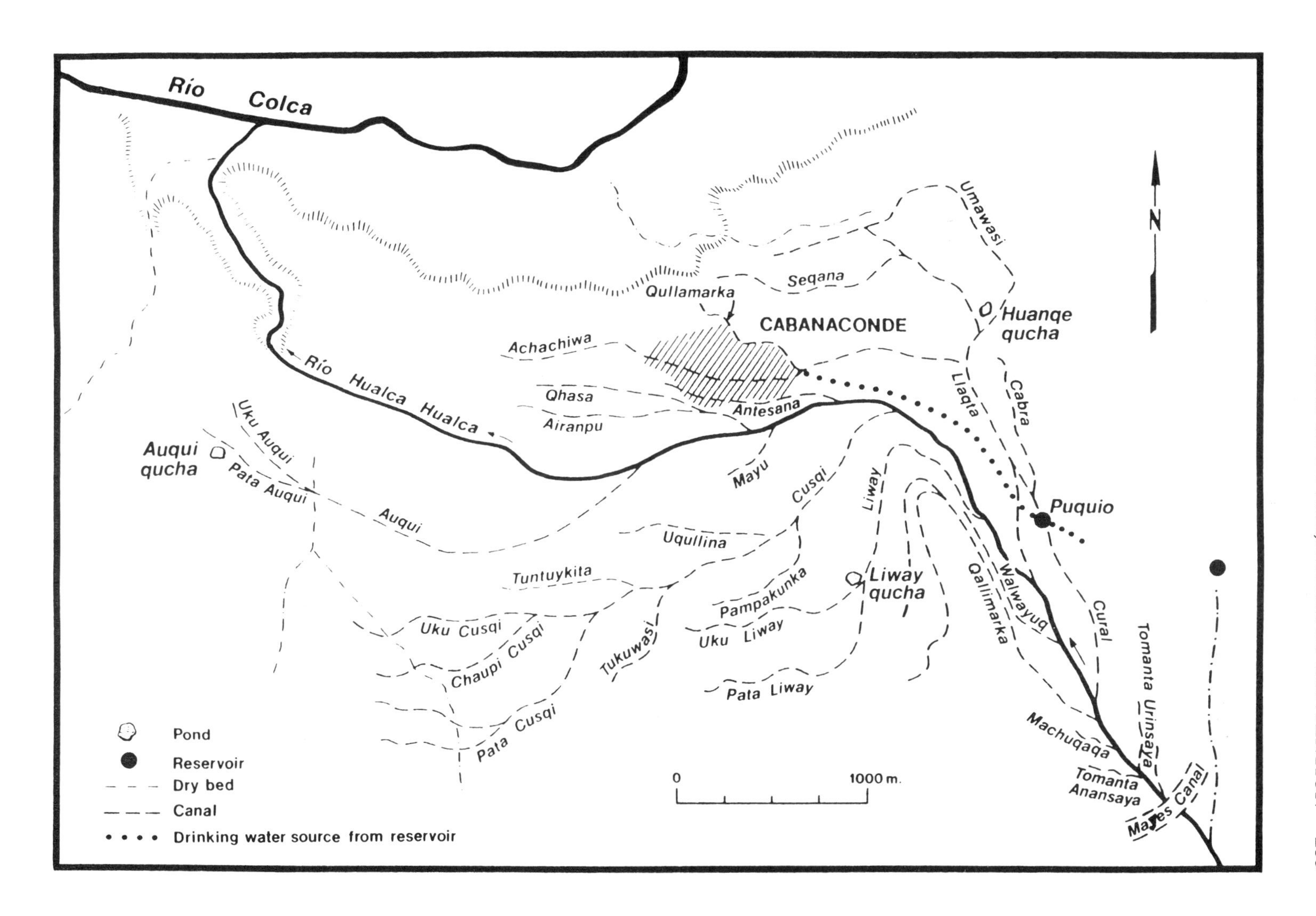

FIGURE 10.2 RAINFALL IN CABANACONDE

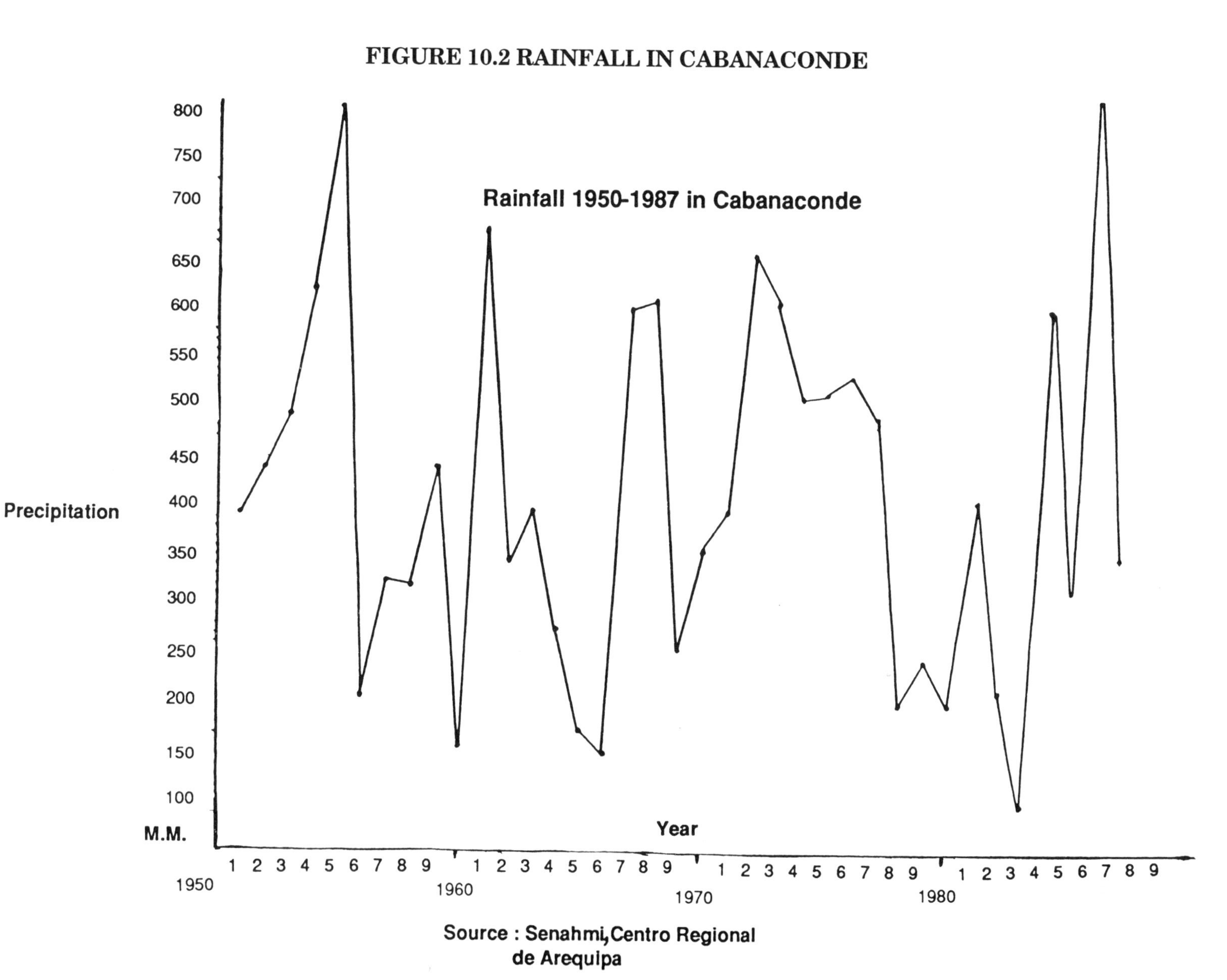

Communal Conflict

Cabanaconde is economically differentiated, and people have conflicting interests. Different types of assets (for example, land and cattle holdings, social networks, godparenthood ties [*compadrazgo*], migrant remittances, access to market opportunities) vary greatly from family to family. Competition, factionalism, and envy, therefore, are part of community life and play an important role in the political processes of the community.

The official political structure of Cabanaconde, like many of the recognized peasant communities (*comunidades campesinas*) in Peru, consists of a municipal council (*concejo municipal*), a governor (*gobernador*), a peasant community (*comunidad campesina*), and an irrigators commission (*comisión de regantes*). These political institutions often cooperate on projects of mutual interest, such as the construction of a school or the organization of a civic event. At the same time, they also compete for control over natural resources, personal loyalties, and money (funds that arrive from government and private development organizations). In Cabanaconde the different political offices are often filled by individuals known for their competence and probity; such people have sometimes been compelled to take office by community consensus and pressure. Nonetheless, other individuals seek office to fulfill personal or political party ambitions.

According to Cabaneños, political parties first appeared in the late 1970s, greatly changing the political dynamics of the community.[11] Different parties often control the different political institutions, making it difficult for the community to create a unified policy. The governor is appointed by the prefect of the province and is usually a partisan of the party in power. Municipal council and peasant community officials, on the other hand, are elected by the townspeople. In 1988 the municipal council and governor's office were controlled by APRA (Alianza Popular Revolucionaria Americana), the political party then in power in Peru, whereas the peasant community was affiliated with United Left, an opposition party. Sometimes party allegiances reflect adherence to political ideology, but often they represent self-interest. Families and other groups, for example, often use political party affiliation to secure power and control community decision-making processes, sometimes even using their power to resolve personal conflicts. Such manipulation of political offices has a long history. Cabanaconde gained official recognition as a peasant community with corporate legal status in 1979. Before that date the central political forces were the municipal council and the governor, nonrepresentative offices that were easily controlled by powerful individuals within and outside the community.

Until the 1960s, the important political posts of mayor, the trustees of the municipal council, the governor, the president of the "irrigators council" (*junta de regantes*), and the local water administrator (*subadministrador de aguas*) rotated among the members of five powerful families.[12] These were some of the only literate families in Cabanaconde during the first half of the twentieth century. Although some of these elites used their position to benefit the entire community, they often used political office for personal gain. When a growing number of community members became literate, the power of this elite group was challenged and positions of authority began to rotate among a wider group of people. However, several people refused to relinquish control and in some cases even imprisoned opponents. Through family alliances within Cabanaconde and contacts with powerful officials in the province of Caylloma and the city of Arequipa, these powerful families continued to dominate the political life of the community until the 1960s. These families, many of which traced direct descent to Spanish families, formed part of the larger regional elite.

Most central to the question of irrigation politics and land recovery is the fact that from the 1920s through the 1950s these powerful families took advantage of their positions within the community's political structure to sell–mostly among themselves–untitled fields in an extremely fertile area that were not authorized for any type of cultivation (*terrenos eriazos*). Some built stone walls around these large extensions using communal labor (some of these fields were as large as 15 hectares). Plots of land, promised in return for this labor, were never given to the majority of peasants. Instead, a few influential peasants were coopted through the "gift" of small plots in a less fertile area. However, when it came to the question of irrigating the unauthorized lands–which would have meant prolonging the round of water for the other crops–the community refused. After a legal battle that lasted many years, and in which the local elites pressed their contacts in Arequipa and even Lima, they received water for less than a hectare of land. Their other fields have remained uncultivated.

These elites also attempted to secure water for their unauthorized fields by introducing new forms of water distribution. As described below in the section on state and local models of irrigation management, these efforts were also thwarted. Nevertheless, the conflict between the community and local elites over the unauthorized fields continues to militate against the recovery of abandoned land.

These same powerful families opposed Cabanaconde's recognition as an indigenous community (*comunidad indígena*), the institutional precursor to the peasant community. Arguments such as "why should we call ourselves *'indígenas'* when we are more than that" masked their fear of a competing power base, one that the majority of peasants could use to oppose the abuses of this elite group. During the 1970s the Velasco regime made it easier for communities to gain official recognition. At the same time, the need for corporate legal status grew as the community became increasingly threatened by the interests of powerful community members and by external forces, such as the Majes conflict described below. One particular arena of conflict will illustrate the need that existed for a representative body.

Until the late 1970s, the Church owned thirty-three fields (*chakras cofradías*), each dedicated to a different saint. These fields, ranging between one to three hectares each, rotated for three-year periods among community members, often among the poorest. These beneficiaries, who previously had to sponsor the *fiestas* associated with each saint, were (and are) called *mayordomos*. Purposely misinterpreting the cry of the revolutionary military government (1968-75)– "the land belongs to those who work it!"–many of the thirty-three *mayordomos* tried to usurp these lands in the mid-1970s. The local priest helped them in their attempts to claim permanent rights to the Church lands. Since he was afraid that the lands would be expropriated by the military government, he sold them to many of the *mayordomos* (BPA).

In 1979 the newly formed peasant community, supported by the majority of community members, engaged the *mayordomos* in a prolonged legal battle and finally took the *chakras cofradías* by force, destroying the crops that had been planted. The priest and several of the *mayordomos* left the community, and the fields came under the community's jurisdiction, rotating as in the past among the poorer peasants. This battle established the peasant community as a strong, representative force within the town.[13] It soon turned its efforts to the defense of other community interests, among which was the recovery of abandoned terraced fields. The peasant community has been successful in expanding the cultivated area, thereby realizing the many aspirations and efforts to do so that had begun in the last century. Nevertheless, the contested lands held by elite families continue to hamper recovery efforts in certain areas.

Regional Conflict: The Huataq and Majes Canals

Conflicts between the community and regional forces have also had great impact on the availability of water for irrigation and on the viability of efforts to recover lost agricultural lands. An example of this is the case of "Huataq," a spring that is located in the high pastures of the community at approximately 4,500 m.a.s.l. (see Map 10.3) and that produces more than 600 liters per second. Since the early part of this century, Cabanaconde has attempted to rebuild the 35 kilometers of abandoned canal that, during Inca times, had apparently brought water from this spring to what were then cultivated lands. In the early 1980s, this project was renewed in earnest and showed every sign of success. However, competition among communities over highland sources of water is often intense (Gelles 1984). Powerful individuals in Arequipa, who were (and are) landowners in an adjacent valley that uses the waters of Huataq, opposed this project and enlisted the help of Ministry of Agriculture officials to stop it. The neighboring community of Lluta, which uses the Huataq waters, also opposed the project. This community has fought Cabanaconde over these waters legally, and sometimes physically, since at least 1933. Thus far the Cabaneños have been unable–because of political influence, the conflict with Lluta, and the need for the logistical support of the Peruvian state–to complete this difficult job.

The community did prevail in a different case, however. Another source of water, as well as of contention, is a large canal built by the Majes Consortium in the late 1970s. This canal, which pipes highland water to the arid coast, passes through Cabanaconde's territory. The official map of this billion dollar development project neglected to show that more than a dozen communities and tens of thousands of peasants lay in the path of the proposed canal (see Map 10.4). This is symptomatic of the lack of regard that the Majes Consortium had for the inhabitants of the Colca Valley, even though the project brought widespread social and environmental problems to the area (Benavides 1983; Hurley 1978; Sven 1986). Until 1983 the Colca Valley received few benefits from the project, gaining only an improved road

MAP 10.3 LAND AND WATER IN CABANACONDE

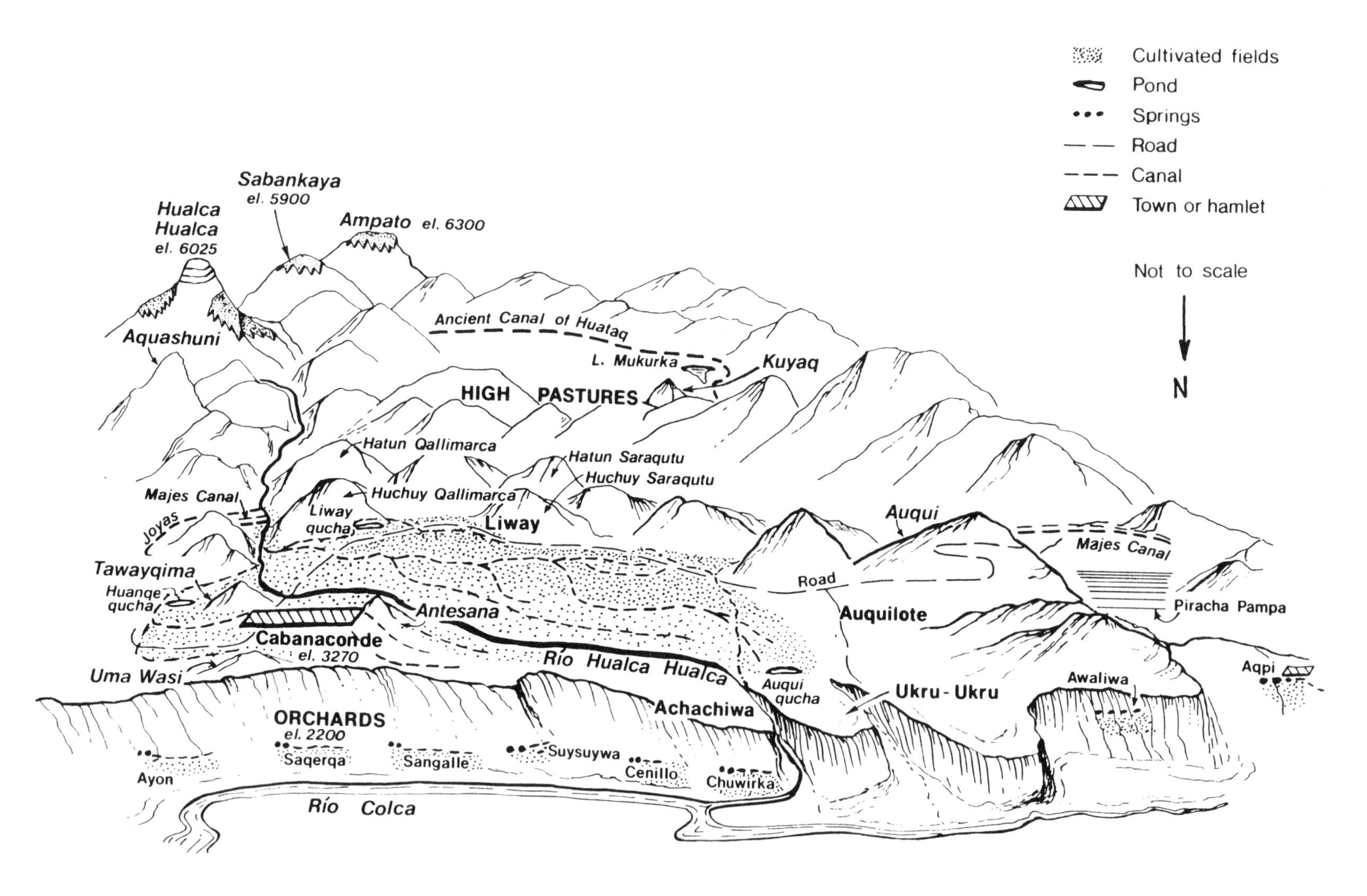

MAP 10.4 THE MAJES PROJECT

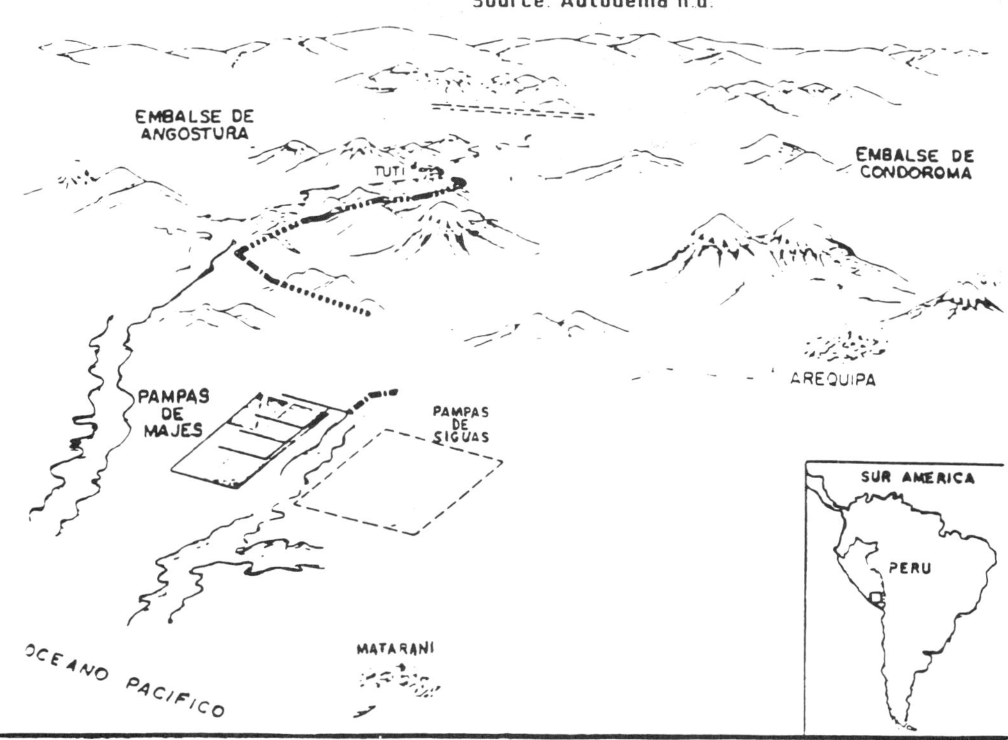

PROYECTO

MAJES

PRESA DE CONDOROMA

and poorly paid, dangerous temporary jobs.

In 1983 the Cabaneños finally acted against the injustices committed against them by opening a hole in the canal. This demonstration of peasant resistance was primarily stimulated by the 1983 drought–the most serious one in the last 30 years (see Figure 10.2), and one that had created a real possibility of famine. More out of fear of further conflicts than any real concern for the community, the Majes Project Administration (*Autoridad Autónoma de Majes*) patched the hole and installed an outlet, ceding 150 liters per second (l.p.s.) to the community. This heroic feat of the Cabaneños set an example for the other communities of the left bank of the Colca Valley; they threatened to take similar action and were soon given access to the "Majes water."[14]

The Land Recovery Project in Cabanaconde

Once the flow of 150 l.p.s. from the Majes canal was secured in 1983, attempts to increase the area of cultivated land began almost immediately. This took the form of using the increased irrigation water to cultivate more land on the average plot than before, especially on the fringes, and to expand the total area of cultivable land. Canals in the lower part of the agricultural lands were extended by means of communal labor, and abandoned terraced fields in an adjacent area called Auquilote (see Map 10.3) were distributed through lottery.

To decide which families could participate in the lottery for the thirty-six hectares being recovered (one hectare per family), the peasant community held a communal assembly. A list of the community members was read. As each name was presented, the public decided whether the person met the criteria established by the peasant community: one had to be a full-time and responsible farmer, a household head with dependent children, a permanent resident in the community, and a small landholder. More than 200 people qualified to draw numbers out of a box. The thirty-six winners quickly organized into an association, elected a president, and began to rehabilitate their lands through cooperative labor.

Although land recovery in Auquilote has had many problems, more than half the newly recovered fields yielded good harvests in 1988 and 1989.[15] Auquilote was the first step in a very ambitious plan to recover more than 1,000 hectares of agricultural land in Cabanaconde. This plan was predicated on an increase in the flow of water from the offtake of the Majes canal. The Majes Project Administration promised an additional 350 l.p.s. if the community could provide a suitable plan for the use of the water. The peasant community commissioned an engineer to study and delimit 340 hectares of agricultural lands in three areas: "Ukru-Ukru," "Pirachapampa," and "Liway" (Map 10.3). This study, which took several months, cost the community more than U.S. $2,000.[16]

A "qualifying commission" was named to evaluate eligibility for the 340 one-hectare lots. Using similar criteria to those in Auquilote, a list of eligible community members was drawn up and later approved in a communal assembly; another lottery was then held. This time, all who qualified obtained lots, but the quality of the land they received was a matter of chance.

This latest land recovery project has met with problems resulting from the communal and regional conflicts discussed above. First, the grown children and grandchildren of the elite families have entered into an intense legal battle with the peasant community over the lands of Ukru-Ukru and Liway, many of which had been illicitly purchased and appropriated by these families earlier in the century. This legal battle has brought the land recovery process in these areas to a standstill. The peasant community cannot deliver titles to the lottery winners until the problem is resolved by a land judge. In 1988, moreover, the elite families used their influence to harass local officials. They convinced the police to detain the president of the peasant community. Although a throng of more than 300 angry peasants secured his release half an hour later, such overt and covert threats from the powerful families keep the community divided and impede the recovery of some of the lands. Even if the peasant community wins the legal battle, it is possible that many individuals, fearing reprisals, will not cultivate their new lands.

In 1988, Auquilote was the only area in which the recovery of lands had been successful. At that time, community infighting further endangered land recovery, as the Majes Project Administration would not release the promised water until the land judge reached a decision on the contested fields. Release of more Majes water was also contingent on the community signing a document in which they promised not to touch the waters of Huataq in the future. The real reason for the Majes Project Administration's generosity comes to light: by releasing 350 l.p.s. to Cabanaconde, they retain the 600 l.p.s. of Huataq as a backup for the Majes canal and for the powerful Arequipeños who currently use Huataq water for their fields.[17] In 1989, the additional water was granted to the community, and many plots in Pirachapampa yielded a good harvest in 1990 and 1991. However, the lands of Liway and much of Ukru-Ukru are still in litigation and remain uncultivated.

In 1988, groups within the community other than the powerful families opposed the allotment of new lands by the peasant community. One group pushed for increasing the flow of water to land already under cultivation. They represented a loose coalition of relatively large landholders and others who did not qualify for the new lots. Although few in number, they raised yet another issue related to the increased availability of water: the betterment of existing agricultural lands versus the recovery of new lands. Community members who had been disqualified from the lotteries because their children were already adults and therefore had not been considered dependents created additional problems. Their discontent resulted in a promise that they would receive land in the next allotment. This has come to pass. By February 1990, several hundred more hectares of land in the areas of Joyas, Huanque, and elsewhere had been allotted. Although these fields have yet to obtain irrigation water from the Majes canal, many farmers have cultivated this land with only rainfall.

In sum, after a century of efforts to expand its hydraulic resources and cultivable area, Cabanaconde has been able to recover a considerable amount of abandoned terraced fields. The lands that remain uncultivated, however, are testimony to the difficulties that still must be overcome. At the communal level, political and economic divisions militated against the land recovery program. In the cases of the

Huataq and Majes canals, regional and national interests greatly influenced water availability and the possibilities of land recovery. Thus, the expansion of water supply systems and the recovery of abandoned agricultural terraces is not only a technical and ecological problem but a social and political one as well.

STATE VERSUS LOCAL MODELS OF IRRIGATION MANAGEMENT

The second case to be examined concerns the conflict between state and local models of irrigation in Cabanaconde. At first sight, contemporary local and state models of water distribution–*anan/urin* and *de canto* respectively–appear to be merely different options for the spatial distribution of water. In fact, they embody significantly different historical processes as well as widely diverging cultural rationales. Foley and Yambert have recently asserted that the state provides "the arena and the rules of the game in which more local political struggles can be carried out" (1989:67). In contrast, it is argued here that local cultural frameworks also provide "rules of the game" for these struggles.[18]

The Local Model of Irrigation: Dual Organization

Although irrigation is important throughout the year in Cabanaconde, the months of June through December are crucial, as irrigation is necessary for the planting and proper maturation of maize, Cabanaconde's major crop. At this time, water is distributed by means of dual organization, a cultural and social model based on dual divisions (Maybury-Lewis 1989), which in this case is a remnant of ancient moieties known as *anansaya* and *urinsaya*. Variants of this model can be found in other Colca Valley villages (see, for example, Guillet *intra*; Paerregaard *intra*; Treacy 1989b; Valderrama and Escalante 1988).

Water mayors (*yaku alcaldes*) are entrusted with overseeing the actual distribution of irrigation water.[19] There are two water mayors at any one time. One is responsible for providing water to the fields classified as *anansaya,* the other is responsible for those classified as *urinsaya*. With the help of his wife and other family members, each water mayor works a shift of four consecutive days and nights, after which he is relieved by the other water mayor. The water mayor is one of many rotating political offices that townspeople assume as a form of community service. Called a *cargo*, this type of political post contrasts with the formal religious *cargos* of the community, which today are voluntary and sometimes have long waiting lists. The *cargo* of water mayor lasts an entire round of water distribution (at present about 45-50 days) and is considered to be the most onerous of the *cargos*. People try to avoid being named to this office because they fear the dangers of nocturnal irrigation, the diversion of almost two months of labor from their own agropastoral activities, and the responsibility of managing such a politically and spiritually charged resource.

There is an established order for the distribution of water: by canal, by the different sectors along each canal, and by the plots of land within each sector. The irrigators, either by watching the course of the water during the day or through word of mouth, anticipate what time of day or night the water will reach their particular fields. They usually arrive a few hours beforehand to prepare their plots of land and feeder canals. They then await the commands of the water mayor. He makes sure that water follows the established order and that intakes are opened and closed correctly. He also mediates any conflicts and guards against water theft, a chronic problem. Although officially condemned, water theft is a socially accepted practice. However, there are "correct" and "incorrect" ways to steal water. The latter are wasteful, cause the water mayor to suffer, and are detectable. People generally steal water at night in the upper reaches of the fields. For this reason the water mayor is usually accompanied by his son or son-in-law at night. One of them oversees distribution while the other patrols the length of the canals. In return for all this work, the irrigators give the water mayor coca leaves, alcohol, and, occasionally, food and small amounts of money.

Evidence suggests that an earlier system of distribution based on socially and spatially localized dual divisions has been transformed into one in which the only remaining expression of affiliation to the old moiety divisions is the water mayors themselves. Such a transformation is also supported by the fact that in many parts of the Andes today, the *anansaya/urinsaya* division creates distinct spatial and social groups. In these situations, people often live in *anansaya,* have their fields there, and receive water from the *anansaya* irrigation system (for example, Mitchell 1981). The same is true for the people of *urinsaya,* who constitute a distinct social group within the community. This is not the case in Cabanaconde.

In Cabanaconde, almost every individual possesses both *anansaya* and *urinsaya* lands. The water mayors are the only people designated as belonging to *anansaya* or *urinsaya,* and then they are so designated only during the time of their *cargo.* Water is distributed according to whether land is classified as *anansaya* or *urinsaya,* but unlike many other parts of the Andes, these divisions are not geographically exclusive. A fairly clear axis dividing the two types of fields, however, does indicate that a stricter division probably occurred in the past.[20] This division has now disintegrated. Although there are large areas where *anansaya* fields are geographically contiguous, a few *urinsaya* fields are found among them. The opposite is also true: the areas that are predominately *urinsaya* contain pockets of *anansaya* fields. Therefore, along any one canal some of the fields are *anansaya,* others are *urinsaya* (Figure 10.3).

The *anan/urin* system of dual organization gives the local model of water management its general form and is semantically related to a wider set of dualisms that are part of the Cabaneño worldview. The dual system also encompasses a set of religious ideas and practices associated with irrigation water, mountains, and the earth. Together, dual organization and ritual remain a strong structuring force. Because the sacred water from Hualca-Hualca Mountain not only nurtures fields and people but can also bewitch and kill, the water mayors must perform the proper rituals to guarantee abundance of water, fertility, and the personal safety of their families.[21] The offerings by the two water mayors not only reproduce an ancient cosmology and ethnic identity but also communicate many meanings that are of

FIGURE 10.3 STATE AND LOCAL MODELS OF IRRIGATION

State Model of Irrigation Distribution: De Canto
(fields are watered sequentially without distinction of the dual divisions)

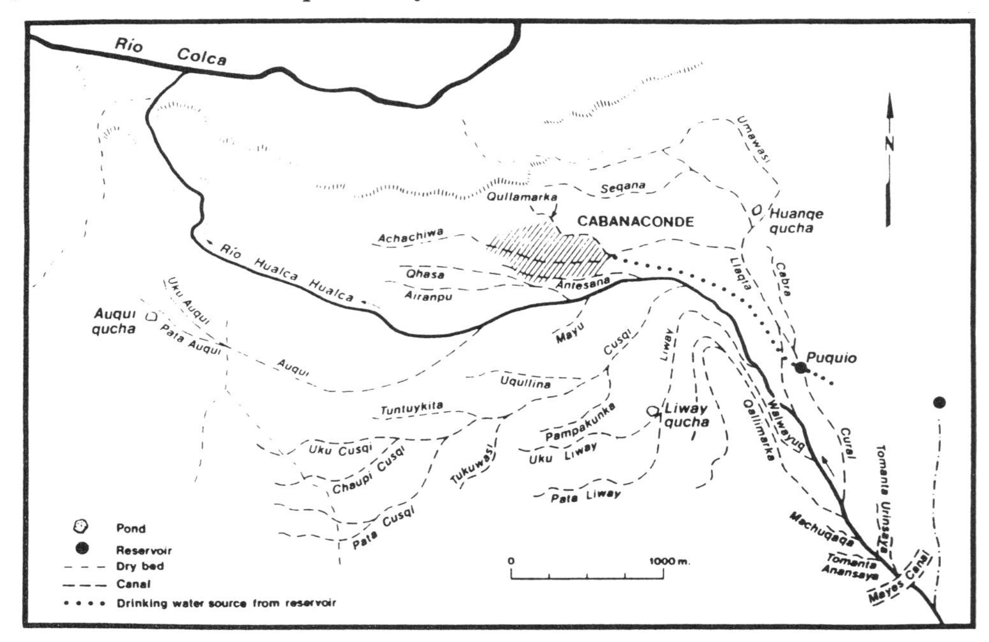

Local Model of Irrigation Distribution: Dual Organization
(fields are watered according to the dual divisions)

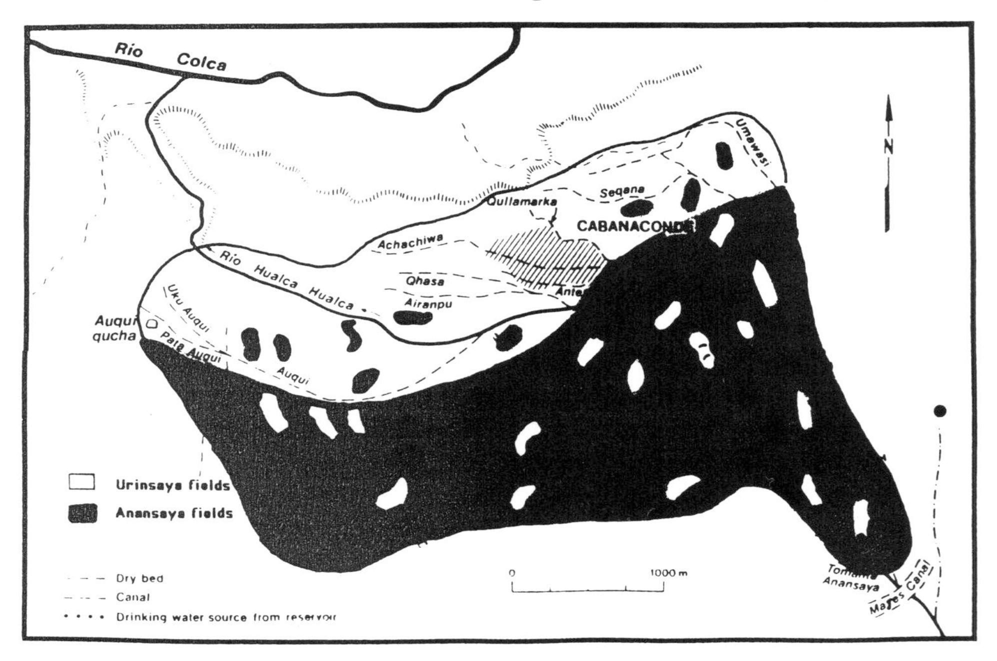

transcendental importance to the lives of the Cabaneños. The rituals have "social efficacy" and "actively produce practices and policies that constitute social reality" (Tambiah 1990:2).

On the first day of his 45-to-50-day round, each water mayor, assisted by a ritual specialist (*paqu*), must make offerings (*q'apa*) to the mountain Hualca-Hualca and to the irrigation water itself. These offerings assure that water is abundant and runs swiftly and that no accidents befall the water mayor or his family. In this elaborate ritual, coca leaves, alcohol, and cigarettes are consumed, and the fetus of a llama, alpaca, or vicuña is offered to the mountain Hualca-Hualca. The water mayors continue to make offerings, or at least pray with incense, at the beginning of each four-day period or at the intake of each major canal. These and other periodic rituals are viewed as necessary for the proper completion of a round of water.

The rituals of the first morning are followed in the afternoon by a large social event called *tinkachu,* sponsored by the water mayor and his family. Family members, neighbors, and irrigators gather to drink, eat, and pledge support to the water mayor, his wife, and other family members who help him during this important and hazardous *cargo.* Here the power of the water mayors is manifest: not only family and friends attend, but also other community members who want to assure that their fields receive plenty of water.

The two water mayors compete with each other in the performance of their *cargo.* Townspeople comment on the quality of the *tinkachus,* noting the number of family members and friends (*munaqkuna*) that each water mayor is able to mobilize. Water mayors also compete to be the first to complete the water round. To do so, they employ many techniques, including magic, to hurry their water along and to slow down that of the competition. Similarly, ritual offerings during the rounds are used to secure safety for the water mayor and to create greater water volume.

Although the present system of dual organization encompasses many domains, such as competition and spirituality, and is related to a wide range of conceptual and social dualisms within the community, its origins exhibit an entirely different dynamic. Fuenzalida (1970) and Benavides (1988) have asserted that under Spanish rule, dual divisions were used by the colonial administration to organize labor and tribute. This is certainly the case for Cabanaconde, where dual organization served the interests of the Inca empire, Spanish colonists, and later local elites (Gelles 1990). Dual organization, one of many pre-Inca institutions that the Inca empire used, took on a strongly political character under Inca rule.[22] Two ethnic lords (*kurakas*) were appointed to each conquered group, the leader of *anansaya* being superior and commanding the leader of *urinsaya.* Dual organization was the conduit through which great revenues and surpluses were extracted from peasants by the state.

This dual administrative system–like other Inca institutions such as labor service (*mit'a*)–was appropriated and redirected by the Spaniards to serve the ends of the Spanish Crown and *encomienda* system.[23] *Encomiendas* were grants of Indian labor given to certain Spaniards during the sixteenth century for their service to the Crown; there are references to the *encomiendas* of Cabana *anansaya* and Cabana *urinsaya* as early as 1549 (Cook 1982:4). Labor and tribute to the *encomienda* owners were extracted through this renovated form of dual organization. When Cabanaconde was established as a nucleated settlement (*reducción*) later in the century, these endogamous groups became spatially localized within the town. After the *encomiendas* faded, dual organization continued to serve the interests of the Crown and of the large population of Spaniards who had settled in the town.

As Benavides has noted, during the nineteenth and twentiethth centuries the dual divisions (often called *parcialidades*) were "a mechanism of control exercised by the local authorities in the province of Caylloma and in other regions of the Andes" (1988:51). In Cabanaconde these divisions continued to serve administrative purposes even after independence: there were tribute collectors for each *saya* or *parcialidad* well into the mid-nineteenth century. Although the principle of endogamy appears to have disappeared before the turn of the century, the dual divisions continued to be used to organize varied activities, such as the sponsorship of fiestas and "public works," well into the 1920s. These public works often provided private benefits for the local priests and for the descendents of the powerful Spaniards who used to rule over the community–the local elites mentioned in the first section of this essay.[24] Clearly, then, the present division of plots of land into *anan* and *urin,* each with a corresponding water mayor, was at one time part of an all-encompassing system of dual divisions used for extractive purposes by the Inca and Spanish states, one that had social and spatial correlates within the community itself.

Thus, the physically demanding, time-consuming, and dangerous job of water mayor is a carryover of an exploitive system that required poor peasants, formerly illiterate "indians," to oversee the physical distribution of water. The exploitative nature of this *cargo* is all the more apparent if we consider that before the 1970s, one served for an entire year, not for seven weeks, as is the case today. Today, the water mayors are sometimes jokingly referred to as the "ragged mayors" (*saqsa alcaldes*), partly because of their tattered appearance after several consecutive days and nights on the job, and partly as an ironic comparison to the much more powerful town mayor. Until the 1960s, there was a fairly rigid social divide between town officials (largely local elites) and the water mayors (primarily poor peasants). Even today powerful individuals often manage to avoid serving as water mayor by occupying more prestigious offices or by paying poor peasants to take their place.

One of the paradoxes of this system is that the people who assume the office of water mayor, and consequently must divert their energy from their own agropastoral duties, become spiritually, socially, and materially empowered during their tenure. This is indicated by the important rituals they are entrusted with, by the social events they sponsor, and by their ability to provide additional water to the fields of family and friends.[25] Although the water mayor's power is slight compared to that of many other communal authorities, these staff-bearing officials are nonetheless respected as important ritual actors and as water managers during their *cargo.*

The Informal System

The *anan/urin* system of water distribution ends with the first heavy and sustained rains, usually in late December or January. During the rainy season (approximately December through March), water management is less ritualized, involves fewer social activities, and either follows the *de canto* state model or an informal system without any formal political authority. The latter system is called the *sistema informal* by the Cabaneños. This latter informal system is, in fact, a third model that begins with the first strong rains of late December or early January. In this informal system of distribution (also known as *"el que pueda"*) water goes to the person who gets it first or the person with the most social or physical muscle.

With the exception of the heaviest and sustained downpours, Cabaneños believe that most rainfalls are insufficient to soak the fields thoroughly. Farmers often irrigate their fields on rainy days to ensure maximum saturation of the soil. If the rains let up even temporarily, a growing number of conflicts emerge. Families and groups of irrigators band together to take the water. Water guards (*tiyaq*) and patrollers (*muyuq*)–who are family members, hired day laborers, or irrigators taking turns among themselves–guard each of the major intakes. Nevertheless, the absence of any water official results in many conflicts and even physical violence. This conflict has plagued the community since at least the 1920s, but the repeated attempts to do away with the informal system have been unsuccessful.[26]

The State Model of Irrigation

During prolonged dry spells in the rainy season, the state model of irrigation distribution is initiated. The irrigators commission, the local manifestation of the state's irrigation bureaucracy, is technically operative throughout the year as it has legal authority over the water mayors and the traditional *anan / urin* system. Yet it is only during the dry periods in the rainy season that the state model, based on the General Water Law (Ley General de Aguas), is used to distribute water.

The Irrigators Commission (see also Bolin *intra*) is a villagewide water management association with an elected governing board of president, vice-president, secretary, treasurer, two aldermen (*vocales*), and two controllers (*controladores*). It holds its meetings in a small dusty room, which has a desk, five chairs, some weathered wheelbarrows, a few bags of cement, as well as accounting and minute books that date from the 1940s. In the provincial capital Chivay (a two-hour drive from the community) a Ministry of Agriculture hydraulic engineer (*ingeniero de aguas*) represents the state at the provincial level. He and an extension agent (*sectorista*) of the ministry supposedly oversee the activities of the irrigators commission. At times they intervene in positive ways, curbing abuses and preventing powerful peasants from irrigating unauthorized lands (*terrenos eriazos*). They do not always have such a positive impact, however, as they often provide corrupt access to "officially approved" water.

The local irrigators commission oversees the distribution of water, the cleaning of canals, the naming of water mayors, and the application of fines to those who steal water or evade canal cleanings. In practice, the efficacy of the commission varies from year to year. Commission members frequently take advantage of their positions to irrigate more than their share, to enrich themselves (pocketing fines and bribes), and to favor friends and family with water. This is not always the case, however. When a good commission is installed, the water mayors and the distribution system benefit greatly.

During the prolonged dry spells in the rainy season, the commission names *controladores* to oversee water distribution. Although these controllers alternate in four-day cycles as do the water mayors, they do *not* follow the *anan/urin* system. Rather, they ignore the dual classification of the plots and distribute water sequentially (*de canto*) from one plot to the next (Figure 10.3). Many of the norms that accompany the state model of *de canto* distribution are different from those used in the local model. For example, the controllers receive money from the irrigators instead of coca leaves and liquor. Unlike the water mayors, they are not fulfilling a major *cargo*, but a minor civic duty. Instead of a snake-headed staff to legitimate their authority, the controllers have an official decree from the irrigators commission. They neither perform elaborate rituals nor sponsor large social events as do the water mayors. Moreover, passing the office of controller does not exempt them from assuming the office of water mayor at a later date.

Like its divergent cultural rationale, the historical roots of the state model differ from those of the local model. The origins of Peruvian water legislation are found in Spanish practices and their antecedents in the Arab world, and the *de canto* form of distribution is no exception (Glick 1970; see also Lynch 1988:77). Colonial water legislation, which dates at least as early as 1556 (Costa y Cavero 1939:155), was directed toward the coast and highland cities. Even when the Water Code of 1902 declared all water the property of the state (Andaluz and Valdez 1987), Peruvian authorities rarely intervened in the water practices of the rural highlands.

It was only in 1969, when the Velasco regime initiated its general reordering of agrarian social structure, that the state began to intervene directly in the water management of highland villages (Bolin *intra*; Guillet 1985, *intra*; Lynch 1988; Seligmann 1987). The water laws and other policies the regime put into effect–what Lynch has called the "bureaucratic transition"–incorporated institutions, norms, and values similar to state-financed irrigation systems throughout the world (Lynch 1988). Nonetheless, the water-user associations and new forms of distribution mandated by Velasco's reforms have not been fully incorporated into the life of many Andean communities, including Cabanaconde (Gelles 1990).

Cabanaconde has had a special purpose irrigation association, the irrigators commission, composed of community members and linked to the state since the 1940s. This institution, however, continues to follow to a large degree the dictates of the local model of irrigation.[27] It was only in the 1970s, moreover, that the state began its attempt to directly alter the community's irrigation cycle, doing so by means of an extension agent overseeing the controllers and the *de canto* system. Since the early part of this century, the Cabanaconde elites attempted to institute the *de canto* form of distribution. But because these efforts were associated with their attempts to irrigate unauthorized lands, the community resisted and returned to the *anan/urin* model. Since the 1970s the Peruvian state has insisted, as part of its program to "rationalize" the hydraulic resources of the highlands, that communities adopt the *de canto* model of distribution. Nonetheless, except for a limited portion of the rainy season in which the state model is preferred to the informal system, the Cabaneños have resisted this.[28]

Unlike many other communities of the Colca Valley and elsewhere (Guillet *intra*; Treacy 1989b; Valderrama and Escalante 1988), Cabanaconde has no current list of water users and does not pay a water tariff. Resistance to such state control is partly explained by the fact that throughout this century the state has been unresponsive to requests for help and has even worked against community interests. In addition, some people benefit from the lack of state control: the absence of a water users' list allows wealthy people to shift the burden of canal cleaning onto poorer peasants. Finally, the state's failure to impose the *de canto* model also stems from local resistance to the cultural domination of coastal society.

Rationality and Ritual Assurance

The water mayors and the *anan/urin* model of dual organization embody a fundamentally different rationale concerning power and authority from that of the controllers and the *de canto* model of water management. The state model legitimates its authority and attempts to control irrigation primarily through the written word, which until the 1960s was the province of a small elite. Today the seals of authority and the minutes of the irrigators commission are treated as powerful objects; the written word is still a symbol of power. However, this symbolism yields to the ultimately more powerful authority represented by the snake-headed staffs and offerings of the water mayors. The main reason for this is that the irrigators commission is made up of community members who, like other townspeople, participate fully in the religious practices of the local model.

Adherence to the local model must also be seen as resistance to the hegemonic ideology implicit in the state model. The state model incorporates a colonial mind-set that views the human and natural resources of the Andean highlands as inferior to those of the Hispanic coast. Such sentiments are expressed in behavior and even documents, such as that of the Ministry of Agriculture elaborated in 1980, which states that the problems of Colca Valley villages (which most surely applies to the highlands as a whole) are due to "the low cultural level of the irrigators . . . [and] a certain resistance to rational work methods" (ORDEA 1980). This prejudiced view completely ignores the state's own shortcomings in helping to improve and expand the hydraulic resources of the region. It also ignores the indigenous knowledge systems and cultural rationales found in local models of irrigation.

This lack of understanding is exhibited in the state's attempt to "rationalize" the distribution process itself. As previously noted, during the rainy season the controllers do *not* follow the *anan/urin* system; rather, they irrigate continuously from one field to the next (*de canto*). For many years the state has attempted to institute the *de canto* system for the entire yearly cycle, which would do away completely with the *anan/urin* model of distribution. The *de canto* model is supposedly more rational and efficient since it saves a considerable amount of water that is lost when one water mayor has to return to a few plots of land along a canal in which the majority of plots have already been irrigated by the other water mayor.

Although townspeople are well aware of this water loss, they say that the state model is less efficient. Since they receive hourly wages, the controllers slow the irrigation process down to collect more money on each plot. Moreover, the dual model provides a cognitive map and fixed sequence of distribution not found in the state model. Under the state model, as evidenced in the previous attempts by local elites to use the *de canto* system to irrigate unauthorized fields, decisions regarding distribution are changeable and therefore subject to manipulation by powerful individuals. The nature of the offices of controller and water mayor differs significantly, and there is greater pressure for the water mayor to be conscientious and fair in distribution.

The local and state models also embody different conceptions of efficiency and availability. The water mayors prepare ritual offerings to secure an abundance of water and ensure the fertility of the fields. These notions of abundance and fertility are part of larger semantic and ritual domains and constitute an important component of Cabanaconde's religious tradition. Taken together, the actions of the two water mayors constitute another ritual level which seeks equilibrium through the complementarity of opposites and which is related to many other conceptual and social dualisms. In this sense, the local model of dual organization not only establishes a pace of distribution along a set path but also provides the gameboard for the water mayors. The ritual competition between them, in which their social and spiritual selves are publicly displayed, advances the water quickly. The entire community observes and participates in this other ritual level. The local model, then, has its own "communicative rationality," one that suggests "designs for living" and that structures social practice.[29]

The power of the local model, and its capacity for resisting and absorbing the state model and other ideological intrusions, is quite remarkable.[30] Over the last fifty years there have been repeated attempts–first by local elites and later by state officials–to replace the *anan/urin* system with the *de canto* model. During an irrigators' meeting in 1988 the hydraulic engineer, supported by a member of the irrigators commission, once again made the irrevocable decision to abolish the local model. As in 1947, 1955, 1960, 1971, 1980, and 1984, to make it official, this decision was inscribed in the minutes of the irrigators commission.[31] The minutes were signed by all present, including the water mayors. Yet the water mayors later ignored the decree.

Despite pressure to adopt the *de canto* system and to charge for their services as the controllers had done during the rainy season, the water mayors continued to follow the dual model and to receive only coca leaves, alcohol, and pocket change. Reluctance by the irrigators to pay money, which is difficult to obtain, is one reason for the success of the local model. People are also concerned about the possible negative consequences of abandoning the tried-and true-formula of the *anan/urin* system. Social pressure to fulfill this *cargo* in traditional fashion is great; as one of the water mayors told me, "If I try to charge, people will talk." Yet, according to a member of the irrigators commission, it was the water mayors themselves who were most opposed to changing the system, a system that assures not only a fairly equitable distribution of water and the continued fertility of the fields but the safety of the water mayors as well. As of July 1991, the dual system continued in practice.

To summarize the second case presented in this essay, state and local models of irrigation have different histories and represent two distinct ways of conceptualizing and implementing water management. Whereas the state model takes a secular, bureaucratic, and supposedly more "rational" approach to water management, the local model is focussed on ritual assurance and views water as part of a larger social and symbolic universe. Each model has different historical and cultural roots, and they are both products of supralocal political systems mapped onto local beliefs and local social forms. Most of the social, spatial, and political correlates of the moiety system have withered over the last century and a half, but they persist in the local model.

Today, irrigation is the primary social domain in which the dual system continues. Yet there are many other conceptual dualisms, as well as a few social dualisms, that continue to structure community life. Some of these dualisms are residuals of the defunct moiety system, which was used to organize people and resources for administrative purposes. Others are the result of the cultural categories of "high" and "low" and "upper" and "lower," from which institutionalized duality in pre-Hispanic times probably originated. These cultural categories remain a strong structuring force independent of the dual divisions instituted by the moiety system.[32] It is the strength of the larger semantic fields of duality and mountain worship that has conditioned the transformation of the dual model from an exploitive system of extraction by Inca and Spanish states to a form of resistance against the contemporary Peruvian state.

DISCUSSION AND CONCLUSIONS

In this paper I have explored two sets of interrelated conflicts that provide insight into the culturally and politically charged nature of irrigation and land recovery in the Andes. The first case demonstrated that the recovery of abandoned terraced fields and the acquisition of an increased supply of water in Cabanaconde is not simply a technical and ecological problem. Rather, political forces at the communal and regional levels have determined the availability of water and the viability of land recovery.

The second case explored the diverging cultural rationales and sets of social relations found in the local and state models of irrigation management. Water availability is conceived by Cabaneños as something that can be enhanced by ritual and by adherence to the institutions of water mayor and the dual model of water distribution. These timeworn vestiges of Inca and Spanish hegemony now constitute the local "indigenous" model of irrigation, a model that differs from that promulgated by the Peruvian state. Although the *cargo* of water mayor is dangerous and burdensome, the individuals who fill this office derive social, material, and symbolic power during their tenure. The dual model and the rituals associated with the water mayors are linked to the cultural and ethnic identity of the Cabaneños. This local model constitutes a center of resistance to the secular, intrusive, and supposedly more "rational" model of the contemporary Peruvian state, a model that is in fact easily manipulated by local elites. In sum, state and local models of water distribution embody different historical processes as well as different cultural rationales concerning power, authority, resource management, and ethnic identity.

In both sets of conflicts–the politics of water availability and land recovery and state versus local models of distribution–extracommunal forces are a critical part of local politics. In the following discussion I will situate these cases in the larger cultural politics that conditions the relationship between coastal and highland society.

Local Models, Hegemony, and Cultural Resistance in the Andes

Conflict between state and local models is not restricted to Cabanaconde. It is found in many Andean communities. For instance, in the highland community of San Pedro de Casta, province of Huarochirí (on the arid west slope of the central Peruvian Andes [see Map 10.5]), one again finds differences between the *de canto* model implemented by the state through the irrigators commission and the local form of water distribution. Traditionally, the eldest persons in each sector had been given priority in the distribution of water. The distribution process today is a compatible mixture of the two systems.[33] There is, however, conflict between the traditional authorities and those of the irrigators commission. In the years 1982-84, the irrigators commission attempted, unsuccessfully, to supplant the traditional staff-bearing authorities (*varayuq*).

In San Pedro de Casta, as in Cabanaconde, water is sacred and water management is highly ritualized. Unlike Cabanaconde, however, ritual activity focuses on one eight-day "water fiesta" (Spanish: *fiesta de agua;* Quechua: *yarqa asp'iy*). Work and religion are fused through the physical and spiritual cleansing of the irrigation canals. The staff bearers and ritual specialists make repeated offerings to the canals and reservoirs, thereby placating the hero-deities who constructed these works and assuring an abundance of water and fertility. Their actions are overseen by a council of elders (Gelles 1984, 1986).

This ritualized communal work project is found in many other parts of the highlands (Arguedas 1964; Fonseca 1983; Isbell 1974; Mitchell 1981; Ossio 1976), and even in Colca Valley communities (Treacy 1989b; Valderrama and Escalante 1988). It has a wide geographic distribution (Map 10.5). However, there are subtle and not so subtle differences in the way the water fiesta is realized from one community to another. The difference between these communities, in which water-related ritual is concentrated in an annual week-long period, and Cabanaconde, where ritual activity is implemented throughout most of the distribution cycle, is even more marked. As we saw in the cases of San Pedro de Casta and Cabanaconde, there are also well-defined differences in the way water is distributed from one community to the next.

It is evident, then, that local beliefs, rituals, and distribution practices associated with irrigation vary considerably from one Andean community to another. Nevertheless, local models of irrigation share important features: water is sacred

MAP 10.5 DISTRIBUTION OF THE WATER FIESTA IN PERU

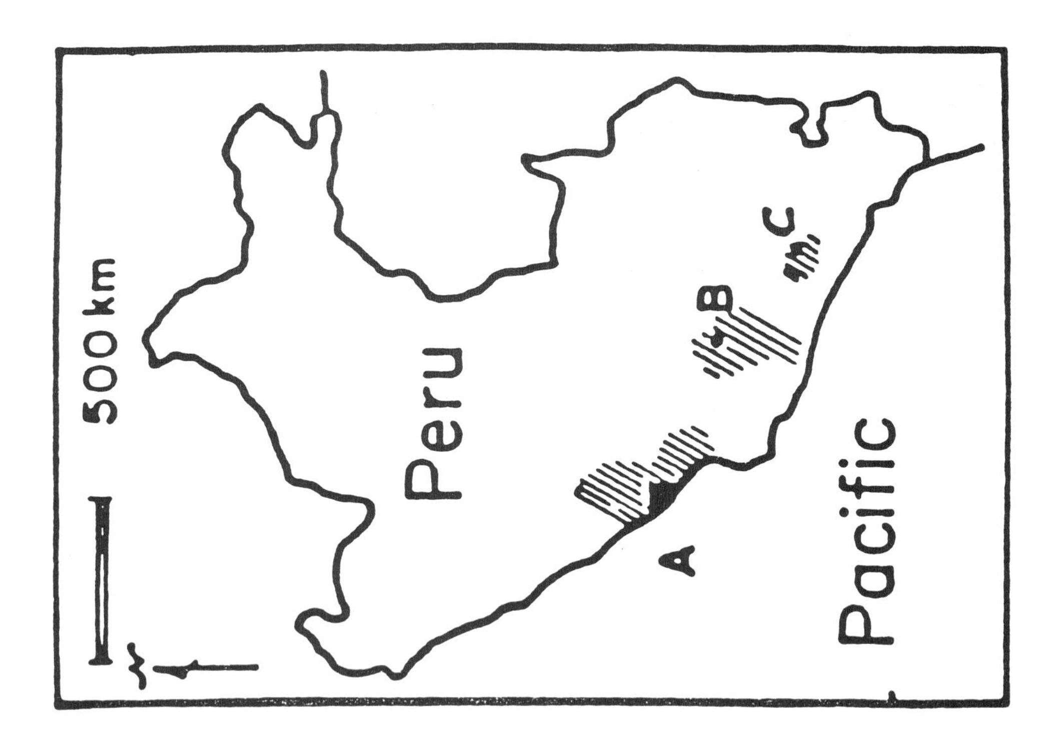

AREAS WHERE WATER FIESTAS OCCUR

A: SANTA EULALIA VALLEY, HUAROCHIRI

B: AYACUCHO

C: COLCA VALLEY, AREQUIPA

and is viewed in terms of a larger Andean cosmology; water management is ritualized in terms of these local beliefs; the distribution process is conditioned by local social structures. Moreover, as we have seen in the case of Cabanaconde, local models of irrigation management do not exist in a temporal vacuum. Although certain cultural frameworks remain in place, the history of these models is dynamic.

Just as the many local models found in the Andes share important characteristics, so, too, do the conflicts between local communities and the state. The power to impose a foreign cultural model through land and water use systems is at the center of these conflicts. As Bourdieu has put it in a different context, "the specifically symbolic power to impose the principles of the construction of reality–in particular social reality–is a major dimension of political power"(1977:165). Symbolic and political power are at stake in the conflict between the model that the Peruvian state elaborates for local irrigation systems (based on a bureaucratic and "rational" Western model) and the decidedly non-Western cultural forms governing irrigation in many Andean communities.[34] In this sense, the maintenance of local models of irrigation is a form of cultural resistance, one that questions the power and legitimacy of the Peruvian state.

Andean peasants are dominated peoples within a nation-state that neither shares nor respects highland cultural values. National power-holders, who determine the state's policy toward Andean communities, live an urban *criollo* life-style in coastal cities and emulate Western lifeways. They generally disdain "indigenous" culture and view Andean peoples as backward, dirty, lazy members of an almost subhuman caste. Unfortunately, these attitudes find institutional and increasingly violent expression.[35] Bureaucratic attacks against the beliefs and rituals surrounding irrigation water and mountain worship must be examined in this context. State officials who denigrate and ignore Andean ethnohydrological principles are actually carrying out the agenda of domination.[36]

Andean peasants resist new forms of distribution not only because of their potential for greater inequalities in water use but because they constitute a form of cultural hegemony. Perhaps the clearest symbol of resistance to the meanings and authority that the state attempts to impose is the continued use of staffs of authority, which, as Rasnake (1988) has shown, are related to a wide range of semantic and political domains. The staffs carried by traditional water authorities in San Pedro de Casta, Cabanaconde, and elsewhere in the Andes are powerful expressions of local meaning, ethnic identity, and cultural resistance.

Rehabilitation of Abandoned Canals and Terraced Fields

Peruvian cultural politics (expressed in the stereotypes surrounding the peoples and relative productivity of the coast and the mountain regions), together with the political economy of the Peruvian state, condition the direction of contemporary hydraulic development: huge development projects oriented to large-scale agriculture are carried out on the coast in contrast to the relatively small, and often times nonexistent, efforts to develop the hydraulic and agricultural potential of the highlands.

This bias is well illustrated by the construction of the Majes canal through the Colca Valley. The communities of this area were steamrollered by this billion dollar project oriented entirely to developing coastal agriculture. With relatively little expense (as exemplified by the minimal costs of diverting water to Cabanaconde), this venture could have included the Colca communities in its plans, helping to expand the cultivable area in the valley. If the communities of this area have benefitted from the project, it is because of their own courageous initiatives.

Although up to now the Peruvian state has helped very little in the capitalization and expansion of highland productive systems, this may be changing. The accelerating pace of scholarly research on the rehabilitation of ancient terraces, canals, and reservoirs has sparked national interest in expanding the terraced lands of Andean peasants. In the last decade considerable progress has been made in understanding ancient technologies and the reasons for their abandonment (de la Torre and Burga 1986; Denevan 1986, 1987, 1988). Improved water management is also recognized as necessary to create sustainable agricultural intensification.[37]

Until recently, however, "there has been an overevaluation of landform" in studies of terrace restoration that "conceptually severed terraces from their contemporary social and economic contexts" (Treacy 1989b:380). The central question today is how can terracing be successfully reintroduced within contemporary socioeconomic frameworks (Treacy 1989b:380).[38] The question of bringing new lands into production in Andean communities by acquiring new or greater sources of water is in part contingent upon the organizational power of the community and on management efficiency (Denevan 1987, 1988; Treacy 1989a; Waugh and Treacy 1986). As I have argued for Cabanaconde, we must also look more closely at the cultural underpinnings of different forms of water management, forms that provide varying degrees of efficiency, equity, and local control. We also must examine the regional and national powers that often constrain the organizational capacity of the community.

Another conclusion that emerges from the data presented here is that attempts to rehabilitate terraces must consider the ways in which competing families and interest groups use different community institutions as sources of power; institutional struggle can frustrate land recovery. In the Andes, the peasant community often represents the interests of the majority. Accordingly, in the recovery of abandoned terraces, development agencies should work through the peasant community, not parallel to it as has generally been the case in the past. In addition to representing the majority, the peasant community has a democratic and effective means of deciding who should receive land and organizational tools such as communal work parties to bring lost infrastructure back into use.[39]

To date, primarily because of the politics of development in Peru, local models of irrigation and community have not received sufficient attention. Attempts to expand hydraulic resources and cultivable land that dismiss the cultural and social forms of peasant communities are doomed to failure. As indicated in this essay, local models of irrigation and community create the rules for, and condition the extent of, state intervention in the highlands. They cannot be ignored.

NOTES

Acknowledgments. The material presented here is based on fieldwork carried out in the central Peruvian Andes between 1982 and 1984, and in the southern Peruvian Andes between 1987 and 1988. The Tinker Foundation funded pre-dissertation summer research in 1985. The Fulbright-Hays Commission, the Social Science Research Council, and the Interamerican Foundation provided financial support for the 1987-1988 field season. During the different periods of research I was affiliated with the Anthropology Department of the Pontificía Universidad Católica of Lima. I would like to express my thanks to these institutions, and to the Ciriacy-Wantrup Post-Doctoral Fellowship of the University of California at Berkeley. All of the translations from Spanish and Quechua are mine, except where noted. The maps were drawn by Nancy Lambert-Brown. This essay, a rough version of which was presented to the symposium on "Canal Irrigation at High Altitudes" at the Americanists Congress in 1988 (Gelles 1988), has benefitted from the helpful suggestions of many individuals. I wish to also thank David Guillet and William P. Mitchell for their extended and critical comments on several versions of this paper. Finally, this paper is indebted to the thoughtful and generous comments offered by the late John Treacy.

1. The models described in this paper are "ideal types" (Gudeman 1986; Leach 1954; Weber 1947), and are a "means of 'seeing' something, of knowing, interpreting, even doing something" (Gudeman 1986:37). These models are explicitly referred to by the Cabaneños as different "systems" and "forms" of irrigation management. I am both reifying their interpretations of these models and reinterpreting the models for my own purposes. For other theoretical perspectives that view irrigation as culturally constructed, see Geertz (1980), Lansing (1991), Pfaffenberger (1988), and Sherbondy (1982b). See Gelles (1990) for a more detailed theoretical discussion.

2. The embroidered hats, as well as other elements of the elaborate female dress, are important ethnic markers that differentiate Cabanaconde from other Colca Valley communities (see also Femenias 1987). Beyond this very visible marker, there are differences in many practices, such as funeral rituals (see Paerregaard 1987), sowing rituals, weaving, and, to a certain degree, language (Quechua pronunciation and vocabulary). Although ethnic differentiation between the upper and lower valley is an ancient phenomenon (see Ulloa Mogollón 1965 [1586]; see also note 8), the ethnic composition of Cabanaconde has changed considerably over time.

3. The population is in fact constantly shifting. There is a huge outflux when many of the young people leave to work for the summer. During the time of the harvest and the Patron Saint's *fiesta,* which occur between May and July, there is a large influx of people, mostly Cabaneños who have migrated to Lima and Arequipa. Many of them remain in the community for extended periods during these visits. Population growth in Cabanaconde is part of a general demographic trend that the Colca Valley has experienced in the last two centuries (see Cook 1982). The 1940 and 1981 censuses of Cabanaconde includes the population of Pinchullo (an annex of Cabanaconde); therefore, I have combined the population figures of Cabanaconde and Pinchullo in the 1876 census. Because of Cabanaconde and Pinchullo's strong nucleated settlement pattern (as opposed to that of Tapay, for example, where there is dispersed settlement), the absence of a "rural" head count in the 1876 census of Cabanaconde is not significant.

4. There has been a considerable amount of seasonal and permanent migration since at least the early part of this century, as evidenced by the regional associations (*clubes provinciales*) of Cabanaconde operating in both Lima and Arequipa by the 1940s. Today, these migrant colonies are estimated to be 1,000 and 3,000 respectively. The colony in Washington, D.C., is recent, but is rapidly growing. The Regional Associations formed by migrants in these cities participate in the life of the rural community, and they have intervened decisively in conflicts between the community and outside interests.

5. Ecologically, Cabanaconde differs from upper-valley Colca communities. Cabanaconde *directly* exploits a greater range of "ecological floors" and "natural regions" (see Murra 1975; Shimada 1985) than do the upper-valley communities. The main extension of cultivated fields has a much gentler slope than that of many of the upper-valley villages, allowing a much higher incidence of "broadfield" terraces (Denevan 1986). Cabanaconde has "by far the most land in cultivation and the most land in grains," as well as the largest population of the Colca Valley villages (Denevan 1987:21).

6. Cabanaconde's maize is known throughout a good part of the southern Andes for its taste and quality. During harvest time, truckloads of peasants, as well as herders on foot, accompanied by large llama and alpaca caravans, descend on Cabanaconde from the high provinces to trade meat, wool, *chuño*, and many other products for the highly valued *maíz Cabanita*. Part of the maize harvest is commercialized in the community and in the city of Arequipa. This specialization in one crop also distinguishes Cabanaconde from the other communities in the Colca Valley.

7. The Hualca-Hualca River could be considered a stream during most of the year. Even in the rainy season, when its volume swells, it resembles a large mountain stream rather than a river. Abril Benavides (1979:15), in his study of the hydrography of the Hualca-Hualca River basin, has shown that river volume can reach more than 1,500 liters per second during periods of intense precipitation. Since it is considered a river by the Cabaneños and maps of the region, I will retain the word "river." Virtually all of Cabanaconde's agriculture is irrigated. In years of abundant rain, a few farmers try to dry-farm additional fields (usually fodder crops). These dry-farmed fields (*terrenos de secano*), which constitute well under 1 percent of the total cultivated area, usually receive irrigation water at some point in their growing cycle. This essay does not examine the small irrigation systems in the lower reaches of the community lands, which are based on small springs. See Gelles (1990, n.d.) for more details on the hydrology and physical characteristics of irrigation in the community.

8. Together with differences in language and cranial deformation, myths about origins in different mountains distinguished the Cavanas from the nearby Collagua ethnic group during the sixteenth century. This supports Sherbondy's assertion that in the colonial period different ethnic groups claimed that their origins lay in mountains, rivers, springs, caves, and other elements of "sacred geography" (Sherbondy 1982b). According to her work, irrigation water has traditionally been conceived of as an extension of subterranean waters, which unite highland lakes, rivers, and mountains with the ocean. Together these form a hydraulic network through which the gods and ancestors travel, and from which humans and their world originated. As Sherbondy has put it, "the ancient Andean peoples did not only possess practical knowledge concerning subterranean hydrology, developing techniques for using these subterranean waters, but also elaborated a cosmology based on this knowledge which was useful for expressing concepts about ethnicity and political units" (1982b:24). For a complementary view that stresses mountains, rather than hydrological processes, as the primary religious symbol, see Reinhard (1985). Today there are many other political and religious ideologies in Cabanaconde. Marxism, Catholicism, and Peruvian nationalism are the most obvious of these.

9. Although the task of irrigating usually falls on the male head of household, women participate to a great extent, especially when they are heads of household or when their husbands are involved in some other activity. Women are often present in nocturnal irrigation, canal cleanings, and even in physical conflicts over water. For other descriptions of women in Andean irrigation, see Bourque and Warren (1981) and, especially, Lynch (1991).

10. The changing flow of the river during the year (and during the course of the day as more snow melts) is indicated by the number of canals in use at any one moment, from one to four. The volume of water is generally highest in March and lowest in July (Abril Benavides 1979:58).

11. Many Cabaneños believe that the political parties have corrupted the communal spirit. This is part of an ongoing discussion among the Cabaneños about the political changes the community has experienced in the last thirty years. One interpretation is the following: "Before, we were all equal, there was more order and respect, and everyone would go to the communal work projects (*faena*). Now only the most obedient attend." Other Cabaneños associate this "obedience" with subservience. This view asserts that the people in Cabanaconde "used to be obedient" but have since learned better. Several of the Cabaneños contrasted this self-perception with the people of Tapay and the upper-valley communities, who are viewed as still "obedient" and subject to abuse. The lack of "order" today, although associated with a more democratic political structure, is also recognized as providing a good deal of room for powerful peasants to maneuver and manipulate the system.

12. Irrigators councils (*juntas de regantes*) became irrigators commissions (*comisiónes de regantes*) in the 1970s. Before this change, there also existed a water sub-administrator (*sub-administrador de aguas*) with a secretary and two aldermen (*vocales*).

13. Instead of paying for the expensive *fiestas* that used to be performed on each of the saint's

days, the *mayordomos* are now obligated to pay for a much less expensive mass. They also pay a small nominal fee to the Peasant Community for the use of communal land. Nonetheless, the Peasant Community is still institutionally weak compared to such organizations in many other communities. This is due in large part to its brief existence and to the lack of a preceding *comunidad indígena*. Unlike many other officially recognized Peasant Communities, a significant portion of the resident population (approximately 20 percent) is not *comuneros*; that is, they do not belong to the Peasant Community, even though they possess irrigated land. At the same time, some *comuneros* do not even reside within the community. Despite these inconsistencies, the Peasant Community represents the interests of the great majority of peasants in Cabanaconde. The Peasant Community has had to defend communal territory and other interests threatened by aggresive mining and bus companies, by neighboring communities, and by the Majes project (BPC).

14. See Gelles (n.d.) for greater detail on the conflicts surrounding the Majes and Huataq canals.

15. Despite food aid, tools, and other support by a German development agency (COPASA), attendance at the work parties averaged about 50 percent. This was largely because the different lots varied in quality, providing different incentives, levels of enthusiasm, and work commitment. Another problem was that for the years 1987 and 1988 irrigation water for the lots had not been approved by the Irrigators Commission. Moreover, in 1987, when many of the plots were sown with barley during the rainy period, some of the envious lottery losers grazed their animals on the sprouting plants; the abrupt and early end of the rainy season ended all hope for the fields that had survived. In the years 1988-89, however, more than half of these fields yielded an excellent harvest of barley. This was due to a fairly prolonged rainy season and to the fact that these fields were irrigated during the period when maize no longer required water. See Smith (1989) for another example of land recovery.

16. This money was acquired largely through government money distributed throughout the highlands in 1987 ("Rimanakuy") and through communal funds generated from such things as grazing taxes.

17. The 600 l.p.s. is the minimum. During the rainy season this volume is much greater and can reach 1,120 l.p.s. (Estudio de Huataq 1985).

18. My ethnographic focus also differs from recent attempts to look at state and community relations (for example, Orlove et al. 1989; Roseberry 1989,) in that it isolates a particular domain: irrigation and community politics. It also differs in that it places emphasis on local social and cultural forms, on local perceptions of state models of organization, and on the ways that these forms and perceptions condition interaction between the state and the community.

19. They are also called *alcaldes de agua, alcaldes de campo, regidores de agua,* and *repartidores de agua*. The office is clearly a product of what Fuenzalida (1970) has called the "colonial matrix," in which indigenous and European social, cultural, and political forms came together to constitute a unique entity, the Andean Peasant Community.

20. I would speculate that the way that duality appears to have been institutionalized by the Inca–with the axis running vertically down the middle of the cultivated area, avoiding in this way a division of physically higher and lower fields (see also Mitchell 1981)–provided each moiety with relatively equal access to different ecological zones.

21. The ideological content of the local model is part of a much larger set of beliefs about the sacred properties of mountains, the Earth, and highland lakes. One's well-being depends on how well one maintains the required reciprocity with these larger forces (see also Allen 1988). Obviously, the beliefs and the larger semantic fields associated with "Cabaneño ethnohydrology" constitute only one of several ideologies that influence the religious, political, social, and cultural life of the Cabaneños. Nevertheless, the beliefs associated with Hualca-Hualca Mountain and the local model of irrigation are an important structuring force in community life. They permeate many activities other than irrigation, such as herding, health care, and gender, as well as communal and family ideologies (Gelles 1990).

22. As has been well-documented, Inca organization expanded upon those social and symbolic forms already present in pre-Incaic ethnic groups and polities. They reworked earlier forms of ecological complementarity, gender ideologies, reciprocal labor, kinship, ancestor worship, and the worship of water and mountains (see for example Alberti and Mayer 1974; Conrad and Demarest

1984; Murra 1975; Patterson 1991; Rowe 1946; Sherbondy 1982b; Silverblatt 1987; Spalding 1984; Urton 1990; Wachtel 1977; Zuidema 1964). A good deal of evidence (see for example, Duvoils 1973; Julien 1983; Netherly 1984; Ossio 1973) suggests that the Inca system of dual organization (*anan / urin*) incorporated and altered dualisms found in subject ethnic groups. Duality was firmly entrenched in the physical layout of the imperial city, Cuzco: there was a clear separation of social groups, as well as of land and water rights, along dual lines (Sherbondy 1982a, 1986, and *intra*). Most important, for our purposes, the "Incas implanted their own organizational principles in the conquered territories . . . in the likeness and image of what existed in Cuzco" (Rostworowski 1983:16). The most salient of these was the *anan/urin* distinction. The Inca presence was strong in Cabanaconde (de la Vera Cruz 1987), and we can be fairly certain that, together with a triadic *ayllu* division, an Incaic dual administrative system was in place there. This is supported by the way in which the Spaniards partitioned off the indians in Cabanaconde and throughout the Colca Valley (see note 23).

23. *Encomiendas* were grants of indian labor given to certain Spaniards in the 16th century (mostly the *conquistadores,* and influential Spaniards who arrived after the conquest) for their service to the Crown. However, what was put under their control "were not territories, nor even the Indians in a strict sense, *but rather the kurakas,*" or ethnic lords (Trelles 1983:158, as cited in Manrique 1985, emphasis mine). The population of the Cavana polity in 1572 was 5,846: *urinsaya* had 2,364 and *anansaya* had 3,482 (Cook 1982:17). Juan de la Torre and Hernando de la Cuba Maldonado each controlled one-half of Cavana: the former was *encomendero* of Cavana *urinsaya,* the latter, *encomendero* of Cavana *anansaya.* For the importance of the Cavbana *encomiendas,* see Barriga (1955).

24. During the early twentieth century a new system of "quarters" (*cuarteles*), spatially bisected by the old dividing line of *anan* and *urin,* was instituted. The dual divisions and quarters coexisted for some time. By the 1930s, the dual divisions were no longer used to organize *fiestas* and communal labor. This change was accompanied by other significant changes in the community's authority structure, some of which were introduced by the Peruvian state.

25. Although they are required to be fair in meting out water, the water mayors will often favor family and friends with a more thorough flooding of their fields. They will also occasionally divert water to fields of friends and family when irrigators miss their designated turn. They are watched closely during the distribution process, however, and are publicly denounced if an inordinate amount of favoritism is perceived.

26. One example is found in a 1956 entry in the Minutes of the Irrigators Commission: "for the moralization of administration of the agricultural fields, and to escape from the old and steadfast custom of Tiacc and Muycco . . . with the good will of the irrigators, this custom is now done away with" (BIC). The lack of any order and the frequent redirecting of water that continues today is wasteful and inefficient. As one Cabaneño put it, "people are shifting the water over here and over there, and nobody ends up irrigating." The relative anarchy of the informal system in Cabanaconde is in part responsible for the early entry of a police station in the community and for the eventual adoption of the state model during part of the rainy season. This informal model is similar to what has been described for the informal distribution system in Quinua, Ayacucho (Mitchell 1981).

27. A water users association has existed in Cabaconde since the 1940s, much earlier than has been reported for other highland communities (see Guillet *intra*; Lynch 1988). The relatively early adoption of this model was intimately linked to local politics and was generated internally. It has had ambivalent effects. Sometimes it has stemmed the abuses of powerful peasants, and at other times it has facilitated them. In many respects, the state model has become another local model. The "bureaucratic transition," like the systems it encounters in different Andean communities, is not uniform and varies significantly from one location to another.

28. Today, because this short period falls under the "Emergency Measures" of the Peruvian water law (Ley General de Aguas 1969), the *de canto* state model is employed, and an extension agent (*sectorista*) of the Ministry of Agriculture has the power to make decisions regarding distribution. The controllers are physically in charge of distribution and provide the functional equivalent of the water mayors during this relatively short period of the irrigation cycle.

29. Given the success of this system year after year, the rationality that it embodies is clearly

"instrumental" as well. The classic formulation of the question of rationality is, of course, Weber (1947). For overviews of this topic, see Hollis and Lukes (1984), Tambiah (1990), Ulin (1984), Wilson (1984). There appears to be a similar conflict of rationalities related to irrigation management in other world areas (see, for example, Lansing 1991).

30. The influence of the local model on the state model's sequence itself is clear: the controllers, like the water mayors, follow a dual sequence in that they trade shifts of four days and four nights. So, too, the rationale of the local model has symbolically incorporated the water recently obtained from the billion dollar Majes project. Tomanta, the place where the water departs the Majes canal for the Hualca-Hualca River, is often invoked during water-related rituals. Even a multimillion dollar irrigation project, the technical correlates of which the Cabaneños are fully cognizant, is incorporated into local belief.

31. A 1947 entry in the minutes of the Irrigation Council (BIC), for example, reveals that the *anan / urin* system was abandoned at this time in favor of the *de canto* system and that six full days of water were gained in the entire round (meaning it went six days quicker). The *anan / urin* system, however, was quickly reinstated. As mentioned above, the community opposed the new system in large part because it was initiated by local elites intent on using the added water to irrigate unauthorized fields. Different fields of contention–between members of a differentiated community and between state and local models of irrigation–are intimately tied together.

32. As I have detailed elsewhere (Gelles 1990), there is a historical disjunction between the conceptual realm of duality and the dual social forms found in the community, as well as a semantic conjunction between these exploitive social forms and local cultural categories. Today it is resources that are classified in terms of the dual divisions, not people.

33. Today the eldest (age generally coincides with the number of *cargos* passed) are given priority when the order of their plots roughly coincides with the most water-efficient route. When it is extremely inefficient to irrigate the plots of the elders first, the *de canto* method is used. The extreme population growth of the last century throughout the highlands has created internal pressure within highland communities to adopt more water-efficient distribution systems. However, as the case of Cabanaconde demonstrates, the pressure can in fact come from ambitious individuals wanting to expand their irrigated lands. The extent to which any community *as a whole* benefits from supposedly more efficient management systems is an empirical question.

34. Starn (1991) has recently called into question the usefulness of the Western/non-Western dichotomy when speaking of Andean society. Clearly, Andean "peasant communities" are the product of a mixture of indigenous and European social and cultural forms and are linked to national (and international) political and economic forces. In this sense I agree with Starn's observations that many Andean peasants have "plural identities" and that some anthropologists have "essentialized" Andean culture, creating an impression of this as being timeless and remote (what he calls "Andeanism"). However, the approach that Starn advocates falls into the opposite error of "occidentalizing" the highland community in particular, and Andean cultural identity in general, by neglecting the strength of Andean culture in both rural and urban areas. It also neglects the role of cultural resistance, and of Peruvian cultural politics and racist stereotypes, in the generation of the plural identities he alludes to. Although Andean peoples inhabit different urban and rural realities, they constantly negotiate their identities, partly by means of the beliefs and models for action that they regard as ancient (such as mountain and earth worship) and that they continue to use to give meaning to their lives. Part of this meaning has to do with cultural resistance and ethnic identity and derives from the ways that local belief and practice oppose the culture of a dominant minority. The dominant minority attempts to impose its worldview, its Western "rational" model, on Andean peoples through many means, including land use systems. There is a large and growing literature on cultural resistance in the Andes (for example, Allen 1988; Ansion 1987; Dillon and Abercrombie 1988; Flores Galindo 1987; Mannheim 1991; Murra 1982; Rasnake 1988; Salomon 1987; Wachtel 1977).

35. An extreme example of this is the "dirty war" being fought today in Peru, in which Andean peasants are caught in the crossfire and are sometimes the object of massacres by the military and by Shining Path because of their ethnicity. See also Bourque and Warren (1989) and Montoya (1987).

36. Other features of Andean ethnohydrology are a vast store of knowledge about water flows,

filtration, canal and terrace construction, and the changing chemical properties of water at different times of the year. Mountain and water worship, and the ways that Andean peoples understand hydrological processes, vary from one community to another and within communities as well (see Gelles 1990). In Cabanaconde, for example, one person has written a geology thesis about the hydrology of the Hualca-Hualca River basin (Abril Benavides 1979).

37. There is a growing literature on terrace recovery and the role of water management in this. See, for example, Benavides 1987;de la Torre and Burga 1986; Denevan 1987; Erickson 1988; Erickson and Candler 1989; Fonseca and Mayer 1979; Fonseca 1983; Gelles 1986; Guillet 1987; Lynch 1988; Malpass 1986; Masson 1982, 1987; Mayer 1985; Mitchell 1981; Portocarrero 1986; Treacy 1987, 1989a.) See Hunt and Hunt (1976) for related issues in the analysis of irrigation organization.

38. William Denevan, who has directed a multidisciplinary study on terraces and the causes of abandonment, points to this same problem when he states, "[even] if more water were available it would be very difficult for people in a community to agree on who would have access to it as well as to the new land which would be irrigated" (1987:35).

39. There are, of course, exceptions. The institution of peasant community is also subject to the manipulations of powerful families. Nevertheless, the regulations (*reglamento interno*) of the Peasant Community provide clear guidelines for democratic process. A fairly democratic decision-making process is found in the communal assembly, which is considered "the maximum authority."

ARCHIVAL SOURCES

BIC Books and documents of the Irrigators Commission, Cabanaconde.
BMC Books and documents of the Municipal Council, Cabanaconde.
BPA Books and documents of the Parish Archive, Cabanaconde.
BPC Books and documents of the Peasant Community, Cabanaconde.

REFERENCES CITED

Abril Benavides, Dionicio Nilo
1979 *Estudio hidrogeológico de la cuenca del río Hualca-Hualca*. B.A. thesis, Department of Geology, Universidad Nacional San Agustín de Arequipa, Arequipa.
Alberti, Giorgio, and Enrique Mayer
1974 *Reciprocidad e intercambio en los Andes peruanos*. Lima: Instituto de Estudios Peruanos.
Allen, Catherine
1988 *The Hold Life Has*. Washington, DC: Smithsonian Institution Press.
Andaluz, Antonio, and Walter Valdez
1987 *Derecho ecológico peruano: inventario normativo*. Lima: Editorial Gredes.
Ansión, Juan
1987 *Desde el rincón de los muertos: El pensamiento mítico en Ayacucho*. Lima: Editorial Gredes.
Arguedas, José María
1964 "Puquio, una cultura en proceso de cambio." In *Estudios sobre la cultura actual del Perú*. Jose María Arguedas, editor, pp. 221-272. Lima: Universidad Nacional Mayor de San Marcos.
Autodema (Autoridad Autónoma de Majes)
n.d. *Esto es Majes . . . Un Sueño Hecho Realidad*. Arequipa: Autodema.
Barriga, Victor M.
1955 *Documentos para la historia de Arequipa, Volume 3*. Arequipa: Editorial La Colmena.

Benavides, María
1983 *Two Traditional Andean Peasant Communities under the Stress of Market Penetration: Yanque and Madrigal in the Colca Valley, Peru.* M.A. thesis, Department of Anthropology, University of Texas, Austin.
1987 "Análisis del uso de tierras registrado en las visitas de los siglos XVI y XVII a los Yanque Collaguas, Arequipa, Perú." In *Pre-Hispanic Agricultural Fields in the Andean Region*. William Denevan, Kent Mathewson, and Gregory Knapp, editors, pp. 129-145. Oxford: B.A.R. International Series.
1988 "La división social y geográfica Hanansaya/Hurinsaya en el valle de Colca y la provincia de Caylloma." *Boletín de Lima* 60:49-53.

Bourdieu, Pierre
1977 *Outline of a Theory of Practice*. Cambridge: Cambridge University Press.

Bourque, Susan, and Kay Warren
1981 *Women of the Andes: Patriarchy and Social Change in Two Peruvian Towns*. Ann Arbor: University of Michigan Press.
1989 "Democracy Without Peace: The Cultural Politics of Terror in Peru." *Latin American Research Review* 24 (1):7-34.

Brush, Stephen
1977 *Mountain, Field, and Family*. Philadelphia: University of Pennsylvania Press.

Conrad, Jeffrey, and Arthur Demarest
1984 *Religion and Empire: The Dynamics of Aztec and Inca Expansionism*. Cambridge: Cambridge University Press.

Cook, David N.
1982 *The People of the Colca Valley: A Population Study*. Boulder, CO: Westview Press.

Costa y Cavero, Ramón
1939 *Legislación de aguas e irrigación*. Lima: Editorial Comercio del Peru.

De la Torre, Carlos, and Manuel Burga
1986 *Andenes y camellones en el Perú andino*. Lima: Concytec.

De la Vera Cruz, Pablo
1987 "Cambios en los patrones de asentamiento y el uso y abandono de los andenes en Cabanaconde, Valle del Colca, Perú." In *Pre-Hispanic Agricultural Fields in the Andean Region*. William Denevan, Kent Mathewson, and Gregory Knapp, editors, pp. 89-128. Oxford: B.A.R. International Series.

Denevan, William, editor
1986 *The Cultural Ecology, Arqueology, and History of Terracing and Terrace Abandonment in the Colca Valley of Southern Peru*. Technical Report, Volume 1, Madison: University of Wisconsin.
1988 *The Cultural Ecology, Archeology, and History of Terracing and Terrace Abandonment in the Colca Valley of Southern Peru*. Technical Report, Volume 2, Madison: University of Wisconsin.

Denevan, William
1987 "Terrace Abandonment in the Colca Valley, Peru." In *Pre-Hispanic Agricultural Fields in the Andean Region*. William Denevan, Kent Mathewson, and Gregory Knapp, editors, pp. 1-44. Oxford: B.A.R. International Series.

Dillon, Mary, and Thomas Abercrombie
1988 "The Destroying Christ: An Aymara Myth of Conquest." In *Rethinking Myth and History: Indigenous South American Perspectives on the Past*. Jonathan D. Hill, editor, pp. 50-77. Urbana: University of Illinois Press.

Duvoils, Pierre
1973 "Huari y Llacuaz: una relación prehispánica de oposición y complementaridad." *Revista del Museo Nacional* (Lima) 39:153-191.

Erikson, Clark
1988 *An Archaeological Investigation of Raised Field Agriculture in the Lake Titicaca Basin of Peru*. Ph.D. dissertation, Department of Anthropology, University of Illinois at Urbana-Champaign.

Erickson, Clark, and Kay Candler
1989 "Raised Fields and Sustainable Agriculture in the Lake Titicaca Basin of Peru." In *Fragile Lands of Latin America: Strategies for Sustainable Development.* John Browder, editor, pp.230-249. Boulder, CO: Westview Press.
Estudio de Huataq
1985 Cabanaconde: Files of Paul Gelles.
Femenias, Blenda
1987 "Regional Dress of the Colca Valley, Peru: A Dynamic Tradition." Paper read at the symposium "Costume as Communication," Brown University.
Flores Galindo, Alberto
1987 "In Search of An Inca." In *Resistance, Rebellion, and Consciousness in the Andean Peasant World, 18th to 20th Centuries.* Steve J. Stern, editor, pp. 193-210. Madison: University of Wisconsin Press.
Foley, Michael, and Karl Yambert
1989 "Anthropology and Theories of the State." In *State, Capital, and Rural Society: Anthropological Perspectives on Political Economy in Mexico and the Andes.* Benjamin Orlove, Michael Foley, and Thomas Love, editors, pp. 39-67. Boulder, CO: Westview Press.
Fonseca, César
1983 "El control comunal del agua en la cuenca del río Cañete." *Allpanchis* (Cuzco) 19 (22):61-74.
Fonseca, César, and Enrique Mayer
1979 *Sistemas agrarios en la cuenca del río Cañete, Departmento de Lima.* Lima: Impreso ONERN.
Fuenzalida, Fernando
1970 "La matriz colonial." *Revista del Museo Nacional* (Lima) 35:92-123.
Geertz, Clifford
1980 *Negara: The Theatre State in Nineteenth Century Bali.* Princeton, NJ: Princeton University Press.
Gelles, Paul H.
1984 *Agua, faenas, y organización comunal en los Andes: el caso de San Pedro de Casta.* M.A. thesis, Department of Anthropology, Pontificia Universidad Católica (Lima).
1986 "Sociedades hidraúlicas en los Andes: algunas perspectivas desde Huarochiri." *Allpanchis* (Cuzco) 27:99-147.
1988 "Irrigation, Community, and the Agrarian Frontier in Cabanconde (Caylloma, Arequipa), Peru: The Relevance of Socio-Cultural Research to the Rehabilitation of Indigenous Technologies in Highland Peru." Paper presented to the Congress of Americanists, Amsterdam.
1990 *Channels of Power, Fields of Contention: The Politics and Ideology of Irrigation in an Andean Peasant Community.* Ph.D. dissertation, Department of Anthropology, Harvard University.
n.d. "The Political Ecology of Irrigation in an Andean Peasant Community." In *Indigenous Irrigation.* Jonathan Mabry, editor. In review.
Glick, Thomas
1970 *Irrigation and Society in Medieval Valencia.* Cambridge, MA: Harvard University Press.
Gudeman, Stephen
1986 *Economics as Culture: Models and Metaphors as Livelihood.* London: Routledge & Kegan Paul.
Guillet, David
1985 "Irrigation Management Spheres, Systemic Linkages and Household Spheres in Southern Peru." Paper presented at the Annual Meeting of the American Anthropological Association, Washington DC.
1987 "Terracing and Irrigation in the Peruvian Highlands." *Current Anthropology* 28 (4):409-430.

Hollis, Martin and Steven Lukes, editors
1984 *Rationality and Relativism*. Cambridge, MA: MIT Press.
Hunt, Robert and Eva Hunt
1976 "Canal Irrigation and Local Social Organization." *Current Anthropology* 17:389-411.
Hurley, William
1978 *Highland Peasants and Rural Development in Southern Peru: The Colca Valley and the Majes Project*. Ph.D. dissertation, Department of Anthropology, Oxford University.
Isbell, Billy Jean
1974 "Kuyaq: Those Who Love Me. An Analysis of Andean Kinship and Reciprocity Within a Ritual." In *Reciprocidad e intercambio en los Andes peruanos*. Giorgio Alberti and Enrique Mayer, editors, pp. 110-152. Lima: Instituto de Estudios Peruanos.
Julien, Catherine J.
1983 "Inca Decimal Administration in the Lake Titicaca Region." In *The Inca and Aztec States, 1400-1800*. George Collier, Renato Rosaldo, and John Wirth, editors, pp. 121-151. New York: Academic Press.
Lansing, Stephen
1991 *Priest and Programmers: Technologies of Power in the Engineered Landscape of Bali*. Princeton, NJ: Princeton University Press.
Leach, Edmund R.
1954 *Political Systems of Highland Burma: A Study of Kachin Social Structure*. London: Athlone Press.
Lynch, Barbara
1988 *The Bureaucratic Transition: Peruvian Government Intervention in Sierra Small Scale Irrigation*. Ph.D. dissertation, Department of Rural Sociology, Cornell University.
1991 "Women and Irrigation in Highland Perú." *Society and Natural Resources* 4:37-52.
Malpass, Michael
1986 "Prehistoric Agricultural Terracing at Chijra, Coporaque." In *The Cultural Ecology, Arqueology, and History of Terracing and Terrace Abandonment in the Colca Valley of Southern Peru*. William Denevan, editor, pp. 150-166. Technical Report, Volume 1, Madison: University of Wisconsin.
Mannheim, Bruce
1991 *The Language of the Inka since the European Invasion*. Austin: University of Texas Press.
Manrique, Nelson
1985 *Colonialismo y pobreza campesina: Caylloma y el valle del Colca, Siglos XVI-XX*. Lima: DESCO.
Masson, Luis
1982 "La recuperación de los andenes como alternativa ecológica para la ampliación de la frontera agrícola." Lima: ONERN.
1987 "La ocupación de andenes en Perú." *Pensamiento Iberoamericano* (Madrid) 12:179-200.
Maybury-Lewis, David
1989 "The Quest for Harmony." In *The Attraction of Opposites: Thought and Society in the Dualistic Mode*. David Maybury-Lewis and Uri Almagor, editors, pp. 1-18. Ann Arbor: University of Michigan Press.
Mayer, Enrique
1985 "Production Zones." In *Andean Ecology and Civilization*. Shozo Masuda, Izumi Shimada and Craig Morris, editors, pp. 45-84. Tokyo: University of Tokyo Press.
Ministry of Agriculture, Perú
1987 *Diagnóstico de Cabanaconde*. Arequipa: Government Press.
Mitchell, William P.
1981 "La agricultura hidraúlica en los Andes: implicaciones evolucionarias." In *Tecnología del mundo andino, Volume 1*. Heather Lechtman and Ana Maria Soldi, editors, pp.145-167. Mexico: U.N.A.M.
Montoya, Rodrigo
1987 *La cultura quechua hoy*. Lima: Hueso Húmero Ediciones.

Murra, John
1975 *Formaciones económicas y políticas del mundo andino.* Lima: Instituto de Estudios Peruanos.
1982 "The Cultural Future of the Andean Majority." In *The Prospects for Plural Societies.* David Maybury-Lewis, editor, pp. 30-39. Washington: American Ethnological Society.
Netherly, Patricia
1984 "The Management of Late Andean Irrigation Systems on the North Coast of Peru." *American Antiquity* 49 (2):227-259.
ORDEA
1980 *Diagnóstico del distrito de riego no. 49: Colca.* Arequipa: Sub-dirección nacional de aguas y suelo.
Orlove, Benjamin, Michael Foley, and Thomas Love, editors
1989 *State, Capital and Rural Society: Anthropological Perspectives on Political Economy in Mexico and the Andes.* Boulder: Westview Press.
Ossio, Juan M.
1973 "Guamán Poma: nueva corónica o carta al Rey. Un intento de aproximación a las categorías del pensamiento del mundo andino." In *Ideología mesiánica del mundo andino.* Juan Ossio, editor, pp. 153-213. Lima: Edición Ignacio Prado Pastor.
1976 *El simbolismo del agua en la representación del tiempo y el espacio en la fiesta de la acequia en Andamarca.* Mimeograph, Pontificia Universidad Católica (Lima).
Paerregaard, Karsten
1987 "Death Rituals and Symbols in the Andes." *Folk* (Copenhagen) 29:23-42.
Patterson, Thomas
1991 *The Inca Empire: The Foundation and Disintegration of a Precapitalist State.* New York: Berg Press.
Peru
1969 *Ley General de Aguas.* Lima: Editorial Olimpico.
Pfaffenberger, Bryan
1988 "Fetishised Objects and Humanised Nature: Towards an Anthropology of Technology." *Man* 23 (2):236-252
Portocarrero, Javier, editor
1986 *Andenería, conservación de suelos y desarollo rural de los Andes peruanos.* Lima: Tarea.
Rasnake, Roger
1988 *Domination and Cultural Resistance: Authority and Power among an Andean People.* Durham, NC: Duke University Press.
Reinhard, Joseph
1985 "Chavin and Tiahuanaco: A New Look at Two Andean Ceremonial Centers." *National Geographic Research Reports* 1 (3):395-422.
Roseberry, William
1989 *Anthropologies and Histories.* New Brunswick, NJ: Rutgers University Press.
Rostworowski, María
1983 *Estructuras andinas de poder.* Lima: Instituto de Estudios Peruanos.
Rowe, John
1949 "Inca Culture at the Time of the Spanish Conquest." In *Handbook of South American Indians.* Bulletin 143, volume 2, pp. 183-330. Washington, DC: Bureau of American Ethnology.
Salomon, Frank
1987 "Ancestor Cults and Resistance to the State in Arequipa, ca. 1748-1754." In *Resistance, Rebellion, and Consciousness in the Andean Peasant World, 18th to 20th Centuries.* Steve J. Stern, editor, pp. 148-165. Madison: University of Wisconsin Press.
Seligmann, Linda
1987 *Land, Labor, and Power: Local Initiative and Land Reform in Huanoquite, Perú.* Ph.D. dissertation, Department of Anthropology, University of Illinois at Urbana-Champaign.

Sherbondy, Jeanette
1982a *The Canal Systems of Hanan Cuzco*. Ph.D. dissertation, Department of Anthropology, University of Illinois at Urbana-Champaign.
1982b "El regadío, los lagos y los mitos de origen." *Allpanchis* (Cuzco) 17 (20):3-32.
1986 "Los Ceques: Código de Canales en el Cusco Incaico." *Allpanchis* (Cuzco) 27:39-74.
Shimada, Izumi
1985 "Introduction." In *Andean Ecology and Civilization*. Shozo Masuda, Izumi Shimada, and Craig Morris, editors, pp. xi-xxxii. Tokyo: University of Tokyo Press.
Silverblatt, Irene
1987 *Moon, Sun, and Witches: Gender Ideologies and Class in Inca and Colonial Peru*. Princeton, NJ: Princeton University Press.
Smith, Gavin
1989 *Livelihood and Resistance: Peasants and the Politics of Land in Peru*. Berkeley: University of California Press.
Spalding, Karen
1984 *Huarochirí: An Andean Society under Inca and Spanish Rule*. Stanford: Stanford University Press.
Starn, Orin
1991 "Missing the Revolution: Anthropologists and the War in Perú." *Cultural Anthropology* 6 (1):63-91.
Sven, Herman
1986 *Tuteños, chacras, alpacas y Macones*. M.A. thesis, Department of Anthropology, University of the Netherlands.
Tambiah, Stanley J.
1990 *Magic, Science, Religion, and the Scope of Rationality*. Cambridge: Cambridge University Press.
Treacy, John M.
1987 "Canal Irrigation in Corporaque." Unpublished MS. University of Wisconsin.
1989a "Agricultural Terraces in Peru's Colca Valley: Promises and Problems of an Ancient Technology." In *Fragile Lands of Latin America: Strategies for Sustainable Development*. John Browder, editor, pp. 209-229. Boulder, CO: Westview Press.
1989b *The Fields of Coporaque: Agricultural Terracing and Water Management in the Colca Valley, Arequipa, Peru*. Ph.D. dissertation, Department of Geography, University of Wisconsin, Madison.
Trelles, Efraín
1983 *Lucas Martínez Vegazo: Funcionamiento de una encomienda peruana inicial*. Lima: Emprenta de la Pontificía Universidad Católica.
Ulin, Robert
1984 *Understanding Cultures: Perspectives in Anthropology and Social Theory*. Austin: University of Texas Press.
Ulloa Mogollón, Juan de
1965[1586] "Relación de la Provincia de los Collaguas para la discripción de las [1586] Indias que su majestad manda hacer." In *Relaciones Geográficas de Indias, Volume 1*. Marcos Jiménez de la Espada, editor, pp. 326-333. Madrid: Biblioteca de Autores Españoles.
Urton, Gary
1990 *The History of a Myth: Pacaritambo and the Origin of the Inkas*. Austin: University of Texas Press.
Valderrama, Ricardo, and Carmen Escalante
1988 *Del Tata Mallku a la MamaPacha: Riego, sociedad y ritos en los Andes peruanos*. Lima: DESCO.
Wachtel, Nathan
1977 *Vision of the Vanquished*. New York: Barnes & Noble.

Waugh, Richard, and John Treacy
1986 "Hydrology of the Coporaque Irrigation System." In *The Cultural Ecology, Archeology, and History of Terracing and Terrace Abandonment in the Colca Valley of Southern Peru*. Technical Report, Volume 2. William Denevan, editor, pp. 116-150. Madison: University of Wisconsin.

Weber, Max
1947 *The Theory of Social and Economic Organization*. New York: The Free Press.

Wilson, Bryan, editor
1984 *Rationality*. Worcester: Billing & Sons Limited.

Zuidema, Tom R.
1964 *The Ceque System of Cuzco*. Leiden: E.J. Brill.

CHAPTER ELEVEN

Dam the Water: The Ecology and Political Economy of Irrigation in the Ayacucho Valley, Peru

William P. Mitchell
Monmouth College

Nature determines how much water is available to an area, but humans decide whether or how it will be used. They make these decisions in an ecological, economic, and political context that selects for certain behaviors over others. People build and expand irrigation systems when it pays them to do so. Aridity is only one factor considered. Even under conditions of demographic growth, agriculturalists may decide against expanding an irrigation system–even though that may seem the "sensible" thing to do–if ecological conditions make the costs too high or if economic forces favor nonfarm work.

The Ayacucho Valley illustrates some of the forces militating against the expansion of farming and irrigation in rural Peru. Because this valley is lower and farther to the west (see Winterhalder *intra*) than many other inter-Andean valleys, such as Cuzco, it is drier and much poorer in agricultural resources (Díaz 1969: 23-82).[1] Ayacuchanos moderate this aridity by means of village and interhamlet irrigation systems, using them to cultivate maize and provide drinking water. These systems irrigate only a small portion of the potentially arable area, however, and large portions of the valley remain undercultivated.

Since at least 1966, when I first traveled to Ayacucho, the people of the district of Quinua (located 34 kilometers by road to the northeast of the city of Ayacucho) have spoken of enlarging their irrigation systems in order to increase agricultural production. Similar aspirations are common throughout the Andes (Dobyns 1964:31-33). Even in highland Ecuador, a region much moister than the Ayacucho Valley, many farmers report that water scarcity is their chief source of stress (Stadel 1989:44-45, 1991).

Thus far, however, Quinuenos and most highland peasants have been unsuccessful in most of their attempts to expand their water supply, in spite of the need for increased food production to provide for growing numbers. Quinuenos have been able to wrest control of water from semifeudal *haciendas* (which had monopolized much of the water until the mid-1960s), but they have been unable to mobilize labor to build bigger and more efficient irrigation systems. This inability to expand irrigation is largely a result of ecological and economic costs. The aridity of the areas to be irrigated, and the consequent labor costs of irrigation, are high. At the same time, demographic and national economic conditions have favored out-migration and nonfarm work.

Since the 1940s the population of Quinua has grown at an ever increasing rate, while the value of peasant farm production has declined in national markets in comparison to the price of manufactured goods. Traditional income, from cyclical labor migration to coastal cotton plantations, that was used to supplement farming also disappeared in the 1960s. To cope with these stresses, Quinuenos have entered nonfarm occupations (craft manufacture, petty trade, highway repair, et cetera), and farming has become increasingly supplemental to total household income.

Nonfarm work and migration, in turn, have created serious labor shortages, forcing Quinuenos to carefully allocate their energies. After comparing the relative value and costs of different productive strategies, they have opted to construct schools–the infrastructure for their new commercial roles–rather than irrigation–the infrastructure of farming.

THE NATURAL ENVIRONMENT AND THE DISTRIBUTION OF WATER

Two natural factors determine the availability of water in Quinua: seasons and altitude. I have discussed the ecology and climate of Quinua in detail elsewhere (Mitchell 1976a, 1976b, 1991a, and n.d.). Suffice it here to say that the Ayacucho Valley experiences three seasons based on precipitation (Rivera 1967, 1971:37-45; see Table 11.1). September through November (sometimes called "spring") is characterized by light rain that gradually increases in intensity. The rainy season proper usually begins in December, tapers off in March, and ends in April. In a normal year it rains torrentially for an hour or so most days at its peak in January and February. The rainy period is also the warmest season, because temperatures do not drop as rapidly at night, but it often seems colder because of the clouds and high humidity (Winterhalder and Thomas 1978:23-24). The dry season lasts from May to August. At its height in July and August there may be no precipitation; the

TABLE 11.1 VARIABILITY OF RAINFALL, CITY OF AYACUCHO 1961-1970

	Jan.	Feb.	Mar.	Apr.	May	June	July	Aug.	Sept.	Oct.	Nov.	Dec.	YEARLY TOTAL
Rainfall Amount (mm)													
Mean	113.8	107.1	104.2	33.4	15.4	4.5	6.3	7.0	26.9	40.6	49.2	82.4	**592.4**
Minimum	43.7	55.2	54.5	4.0	0.0	0.0	0.0	0.0	5.6	12.8	3.5	29.0	**458.5**
Maximum	287.5	241.9	194.3	58.9	41.2	24.6	21.8	21.8	55.0	109.0	131.7	141.4	**917.8**
Days With Rain													
Mean	21	20	20	10	4	2	4	6	10	13	12	19	**147**
Minimum	11	12	11	4	0	0	0	3	5	6	6	14	**117**
Maximum	31	28	29	15	9	3	10	12	15	21	16	23	**174**

Source: Rivera (1971:41-43). Data on days with rain include the years 1963-1970 and April through December of 1962. In Rivera's table providing the data on days with rain, I counted a dash as meaning "insufficient data" (January-March, 1962, July-September, 1969, and June 1970). Only those years marked with a "0" were counted as 0. The "yearly total" of days with rain is based only on the years 1963-1969.

terrain is exceedingly dry, and the soil is as hard as stone. This period of time is also characterized by lower, even freezing, temperatures at night.

These generalizations about climate do not preclude substantial differences from year to year. The amount of annual and monthly precipitation varies considerably, and droughts appear "without predictable regularity" (Winterhalder and Thomas 1978: 13; see Winterhalder *intra*). Between 1961 and 1970, it rained an average of 147 days in the city of Ayacucho, producing an annual mean of 592.4 millimeters of rain (Table 11.1).[2] Variability from year to year, however, was substantial. In 1963, 174 days of rain produced 917.8 millimeters, but in 1968, the 117 days of rain provided only 568.4 millimeters, about half as much. I do not have the figures on the harvest, but I am sure that 1969 was a poor year for farming.[3]

Even when total annual precipitation is high, monthly variation may adversely affect agriculture. Precipitation from September through December is crucial to the germination of the main crop (*hatun tarpuy*), but monthly differences from year to year are substantial. In the city of Ayacucho from 1961 to 1979, for example, an average 40.6 millimeters of rain fell in the month of October, but the deviation from this mean was high. In 1966, 109 millimeters of rain fell in October, but only 12.8 millimeters fell in 1961. Insufficient rain during germination harms the crop, a phenomenon that endangered the 1988-89 and 1989-90 harvests in Quinua (Mitchell n.d.a).[4]

Altitude also influences the distribution of water (Table 11.2; Winterhalder and Thomas 1978:9). The lands of Quinua extend from about 2500 meters above sea level to more than 4,100 meters, and the higher altitudes are colder, cloudier, and moister than those lower down (Arnold 1975; Holdridge 1947; Rivera 1971:29; Tosi 1960; Tosi and Voertman 1964). High elevations are moist largely because of abundant rain; annual precipitation is nearly double that found at low elevations (500-1000 compared to 250-500 millimeters per year). The low rate of evapo-transpiration also increases moisture at high altitudes, although these evapo-transpiration potentials are not as low as those proposed by Tosi (Frére et al. 1975:133; Knapp 1988:28-29 and personal communication September 14, 1989).

As in the rest of the tropics, Quinua has a diurnal temperature climate (Troll 1968) in which changes from day to night are greater than seasonal variation and the day's heat escapes quickly at night. Elevation also affects temperature; high altitudes are colder than low ones. Low altitudes are also sunnier, as clouds are fewer and usually dissipate earlier in the day than in high altitudes. Temperatures decrease 0.5 degree centigrade for every 100 meters of elevation in the Ayacucho region (Rivera 1967, 1971:36; Winterhalder and Thomas 1978:20).[5] The valley bottom, however, sometimes experiences nocturnal temperature inversion, since cold air is denser and sinks to the valley at night.

These ecological constraints restrict the zones suitable for agriculture. Most of the community is either too high and cold or too low and dry for intensive farming. In 1972, 57.6 percent of Quinua was natural pasture and 23.7 percent woods and mountains, whereas only 12.2 percent of the total land was permanent farmland (Mitchell 1991a:43, n.d.).[6] The situation on the western side of the Ayacucho Valley is even worse. Similar to Winterhalder's western escarpment (Winterhalder *intra*),

TABLE 11.2 QUINUA'S ECOLOGICAL ZONES: CLIMATOLOGICAL AND PRODUCTIVE CHARACTERISTICS

Zone	Altitude (Meters)	Annual Rainfall (mm)	Mean Annual Temp. (°C)	Productive Regime
Rain Tundra/ Wet Paramo	4100+	500-1000	0-6	Herding
Prairie	4000-4100	500-1000	6-12	Herding/farming
Moist Forest	3400-4000	500-1000	6-12	Herding/farming
Savannah	2850-3400	500-1000	12-24	Irrigated farming
Thorn Steppe	2500-2850	250-500	12-24	Unirrigated farming
Valley Bottom	2500	250-500	12-24	Irrigated farming

Source: Meteorological data are from Arnold (1975) and Tosi (1960)

it is in a rain shadow and significantly less productive than the eastern valley (Díaz 1969:61-74; Rivera 1971:39).

IRRIGATION FARMING AND PRODUCTIVE CAPACITY

The abundant moisture at high elevations in Quinua favors agriculture, but the cold temperatures and increased frost risk at these altitudes limit the kind of crops that can be grown. As cold temperatures retard plant growth, crops mature more slowly and farmers can grow only fast-maturing and/or frost-resistant ones. In the savannah (2,850-3,400 meters above sea level), the thorn steppe (2,500-2,850 meters), and the valley bottom (about 2,500 meters), temperature favors agriculture, but aridity seriously limits production.[7] Maize and many other crops require irrigation in these zones, but irrigation water is limited, except in the valley bottom.

Savannah farmers obtain irrigation water from small streams, lakes, and springs in the moist, high altitude grasslands above the village (see Mitchell 1976a and n.d. for a complete description of the irrigation system in Quinua). These sources also provide drinking water for people and animals. The volume of water available for agriculture is small, but that water is nonetheless crucial to farming (Mitchell 1976a; Murra 1960).

In the upper savannah (3,050-3,400 meters above sea level) the rainy period is too short for the cultivation of maize, so that irrigation is used to extend the growing time of the rainy-season crop (*hatun tarpuy*) and thereby the altitude of maize cultivation. Farmers only soak their fields once, but this water is generally sufficient to plow, plant, and germinate the seedlings (Mitchell 1976a). The sporadic rain at this time of year (see the months of September through December in Table 11.1) normally keeps the seedlings moist without further irrigation, although in years of low rainfall, this is not the case and crops fare poorly for lack of water. Irrigation is also sometimes used to grow an early crop (the *michka*), planted in August in the dry season, but little water is available to plant this water-demanding and labor-intensive crop cycle, and few people do so. Irrigation, however, is used for households and kitchen gardens throughout the dry season.

In the lower savannah (2,850-3,050 meters), the growth cycle of crops and the duration of the rainy season generally coincide. Irrigation is used here to supplement rainfall during lulls in the rainy season, but there is not enough water to do so on a sustained basis in any area of the savannah. Since little irrigation water is left by the time it reaches the lower savannah, the dry-season cycle is rarely planted here, although farmers do maintain households and kitchen gardens with the water.

The thorn steppe lies just below the savannah. It is a dry woodland with cactus, and in most areas there is not enough water for even drinking purposes. Farmers who cultivate here must bring drinking water to their fields in pots in the dry season and dig cisterns to collect rain water in the rainy season. Quinuenos rely entirely on unirrigated farming in the thorn steppe, not a very successful enterprise. Maize, for example, does not produce here in most years because of aridity. Potatoes do not grow here at all. Instead of these two crops (the most important in Quinua), farmers plant this zone with wheat (the third most important crop) and chick peas. This farming, however, is nonintensive, and farmers must often let their land lie fallow for long periods to store water for a later sowing.

The valley bottom is similar environmentally to the thorn steppe. It is too dry for most unirrigated farming, but water is available from the Rio Pongora, the Rio Chacco, and the Rio Occopa (Díaz 1969:39-52, 79; Rivera 1971:21-23). This abundant irrigation water supplements rainfall during the rainy season and permits a second crop during the dry season. As in the savannah, irrigation also supplies households, kitchen gardens, and drinking water for people and animals (Mitchell 1976a).

These ecological realities constrain production and thereby help organize the distribution of Quinua's population along the mountain slope. The two irrigated maize zones are the most densely populated areas (Table 11.3). In 1961 the savannah contained 83.8 percent of the inhabitants. They lived in 55 settlements with an average density of 81 people per settlement. The irrigated valley bottom had the next highest density with an average of 52 people in each of the six settlements. Since the valley is narrow (Rivera 1971:18), however, the total population is small, and in 1961 the valley bottom accommodated only 5.9 percent of the district's population.

The unirrigated thorn steppe and the high altitude zones are the least densely populated. In 1961, the thorn steppe, larger in area than the savannah, housed only 5.9 percent of the inhabitants living in 8 settlements with an average density of 40 people each. The density was even lower in the cold prairie and moist forest. Only 0.5 percent of the population lived in a single settlement in the prairie (density of 26 people), while only 2.1 percent of the population lived in three settlements in the

TABLE 11.3 PERCENTAGE OF DWELLINGS, SETTLEMENTS, INHABITANTS AND MEAN SETTLEMENT SIZE BY ECOLOGICAL ZONE DISTRICT OF QUINUA - 1961

	Prairie	Moist Forest	Savannah	Thorn Steppe	Valley Bottom	Location Unknown	TOTAL NUMBER
Dwellings	0.4%	1.9%	85.0%	6.1%	4.7%	1.8%	1,370
Inhabitants	0.5%	2.1%	83.8%	5.9%	5.9%	1.8%	5,348
Settlements	1.3%	3.8%	69.6%	10.1%	7.6%	7.6%	79
Mean Settlement Size (No. of People)	26	37	81	40	52	16	68

Note: The mean settlement size is the number of inhabitants divided by the number of settlements.

Source: Peru 1966:345-347 and informant interviews.

moist forest (average settlement size was 37 people).[8]

These patterns of population and land use are characteristic of those found throughout the semiarid eastern Ayacucho Valley. The thorn steppe is one of the largest zones found in the eastern valley, but because of limited water, it is only sparsely cultivated and inhabited.[9] More irrigation water here would result in a significant increase in agricultural production that would help sustain the continuously increasing population of the valley.

In sum, irrigation is necessary for the intensive cultivation of maize and most other crops in the savannah, thorn steppe, and valley bottom, but in the savannah such water is scarce, and in the thorn steppe it is absent. In the savannah irrigation provisions households, livestock, and kitchen gardens. In farming, the limited water is used primarily at the beginning of the main planting (*hatun tarpuy*) to extend the growing season. There is not enough water to irrigate during prolonged droughts or for most people to plant a dry-season crop (*michka*). To do so, Quinuenos would have to enlarge their water sources and improve the canals to prevent filtration.[10]

The unirrigated thorn steppe, which lies just below the savannah, is too dry to be farmed intensively. Quinuenos would have to extend their irrigation system into this zone to grow maize and most other crops here. The labor and material costs of this irrigation, however, would be higher than those needed to irrigate the savannah. The aridity of this zone increases the need for water and thereby the number of irrigation turns, while the sources of water are farther away, requiring a greater investment in canals.

The valley bottom is the only area with adequate water for both households and farming. This water is near at hand, so that, although the irrigation turns are many, the canals are short and the energy investment less than that in the thorn steppe.

Quinuenos often speak of increasing their water supply and improving their canals to stabilize agriculture in the savannah, to increase the dry-season planting and double cropping in the savannah, and to bring irrigation water to fields in the thorn steppe. They fantasize about the crops they could grow and the population they could sustain if they did so. The archeological site of Huari lies in the thorn steppe, and people sometimes point to the ancient stone-lined canals that once fed the large population of that city but that are no longer used. Every now and then officials of the community have attempted to organize to tap into high altitude lakes to increase their water supply and thereby realize their dreams, but so far they have been unsuccessful in most of these efforts. They have been unsuccessful, not because the people of Quinua are unwilling to work in a sustained fashion (as some officials maintain), but because the people have decided that the energy costs of that irrigation are too high and the return on the resulting production too low. That low return is a consequence of government policy that has undermined the value of peasant production in comparison with the return on nonfarm wage work.

THE SOCIAL ALLOCATION OF WATER

Until 1973, when Quinua was reorganized by the agrarian reform of President Velasco, the community consisted of both independent peasants and semifeudal *haciendas*. The independent community was located in the upper and lower savannah around the central town and received water from the Hanan Sayoc-Lurin Sayoc irrigation network. *Haciendas* were found in parts of the lower savannah and in most of the thorn steppe and valley bottom.

Quinua never appears to have contained as many *haciendas* as did other areas of Ayacucho. Even in 1876, the first period for which we have data, only 15.5 percent of the population was classified as living on a *hacienda* (Table 11.4). By 1961, only 3.7 percent of the population was doing so. At the time of the Agrarian Reform in 1973, there were eight large *haciendas* and six small ones in Quinua. Most of their owners lived in the central town of Quinua or Ayacucho. In 1973 the Agrarian Reform created cooperatives out of all but two of those *haciendas* that still remained.[11] Over time, the number of cooperatives was reduced as the former peons divided the lands into private parcels. In 1981 two cooperatives remained, with a total of 177 inhabitants (Peru 1983).

In spite of their small numbers, *haciendas* dominated Quinua production through control of land and–most especially–water. Until 1966 not only was irrigation water scarce in Quinua, but *haciendas* had preferential rights to the water, which they used to obtain a peasant work force. In the Hanan Sayoc irrigation network, one *hacienda* took the entire flow every Saturday. In Lurin Sayoc, *haciendas* controlled the water on Mondays and Tuesdays, while in the Susu network they had the water every day but Sunday. In the valley bottom, water is

TABLE 11.4 NUMBER AND POPULATION OF HACIENDAS DISTRICT OF QUINUA

	1876	1940	1961	1972
Number of Haciendas	9	16	11	14
Number of Families	ND	78	51	ND
Resident Population	539	330	199	ND
HACIENDA RESIDENTS AS PERCENTAGE OF DISTRICT POPULATION	**15.5**	**5.8**	**3.7**	**ND**

Note: ND = No data

Source: Peru 1878, 1948, 1966, 1983 and informant data for 1972.

abundant and therefore open to all, but *haciendas* controlled most of the land as they did in the thorn steppe.

Although independent peasants owned most of the land in the savannah, *haciendas* nonetheless captured their labor through this monopoly control of water. The independent peasants of this densely populated area had to purchase the water they needed for farms and households with work. The amount of water given depended on the type of chore and amount of energy required. A man might have to plow an *hacienda* field for an entire day (supplying his labor, plow, oxen, and animal food) in return for one day's water. *Haciendas* especially needed peasant workers during the plowing and crop cultivation. If the *hacendado* (the owner) discovered someone taking water without paying for it, he would seize some valuable item (a tool, a piece of clothing, an animal, et cetera), known as a *prenda*, that he held until the person worked for him (see also Skar 1982: 289). In the thorn steppe and valley bottom, *haciendas* used control of land to get laborers, giving peasants cultivation or grazing rights in return. (See Díaz 1969 for brief descriptions of such arrangements throughout the department of Ayacucho.)

On days not monopolized by *haciendas*, water in the savannah was supposed to be allocated among peasants equally, but since the amount of water was insufficient, it was (and still is) rationed (Mitchell 1976a). In spite of the egalitarian ethos surrounding water and other resources ("we are all equal," "we are all poor"), however, the resulting distribution of water was (and is) far from equal (Mitchell 1991b).[12] Because water is distributed to plots of land, not people, the rich (who own more land) get more water than the poor. Since work on the irrigation corvée is assigned to households (rather than apportioned according to land ownership), however, the poor must contribute more work for every unit of water.

Aside from Sundays, which are always reserved for the first-come, first-served distribution of household water throughout the year, Quinuenos use two systems of distribution depending on the crop cycle. During the sowing of the rainy-season crops (September-December), demand is high and the irrigation schedule tight. Peasants respond to the tight schedule by implementing the patacha, a formal system of irrigation turns administered by water officials. In this procedure, the officials assign water first to the higher fields and descend field by field until the onset of the full rains, usually in December, when everyone is able to plant. Apportioning water in this manner is an adaptation to the ecological realities of high altitude farming: the higher the altitude the longer the maturation period and the earlier crops need to be planted (Mitchell 1976a).

During the rest of the year, peasants distribute water in the savannah on a first-come, first-served basis. At this time, water is needed for household use, livestock, kitchen gardens, the dry-season planting (the *michka*), and double cropping. It is also sometimes used to irrigate rainy-season crops during drought. However, there is rarely enough water for more than household use, livestock, and kitchen gardens. Only a few people, mostly powerful farmers who cash crop potatoes and who are able to mobilize sufficient moral and physical force to obtain water, are able to plant the dry-season crop. Little water is available to irrigate fields during droughts.

Quinuenos often fought (and still fight) over water during the first-come, first-served distribution system, sometimes coming to blows. During conflicts, they consider a person's communal service in assigning moral rights to the water. Previously, people who sponsored religious fiestas had preferential rights, that even allowed them to take water out of turn during the *patacha*. Since the decline of the fiesta system, Quinuenos legitimize water rights primarily through work on irrigation corvées and other communal obligations.

But moral rights do not necessarily give a person water. Powerful townspeople and peasants often lay claims to a myriad of vague communal service, obtaining water ahead of others even when they have not participated in the irrigation corvées. In some cases, people may decide that so-and-so has the right to water, but a competitor may mobilize more force and the farmer not receive it. In the patacha, however, the intervention of political authorities and the clear altitude rule of distribution alleviate much of this strife.

AUGMENTING THE WATER SUPPLY: ATTACKING *HACIENDA* PREFERENCES

In the 1960s Quinuenos began to increase the water available to them by attacking the preference given to *haciendas*, using existing legal prohibitions on the private ownership of water to do so (see also Guillet *intra*). They initiated legal action to form irrigation boards (*juntas de regantes*) to administer the water in the savannah systems (see Bolin *intra*) and to introduce the altitude distribution system (the *patacha*) administered by an irrigation judge (*juez de agua*). *Hacienda* owners came to be treated like everyone else and lost any rights to special days.

Instead, field altitude determined water allocation during times of great demand. During the rest of the year the first-come, first-served system continued, but *haciendas* now had to compete with everybody else.

The Susu network switched to an irrigation board in July 1966. Before the change, five *haciendas* had received water six days of the week. The change to the standard peasant system was initiated by the mayor of Quinua. He was well connected to Popular Action (*Acción Popular*), the political party of President Belaunde. Belaunde, an elected president, had campaigned in Ayacucho and had received the support of the mayor and the people of Susu. The mayor used his connections to the party and to a distant relative in the national senate to petition the Ministry of Agriculture in Cuzco, which granted Susu the right to elect an irrigation board.

The affected *hacendados* countered with legal proceedings in Cuzco against the action, but they lost in this and in all subsequent litigation. A peasant invasion of their lands also forced them to temper their actions in town. Undertaken in retaliation for *hacienda* legal maneuvers against the irrigation board, the invasion increased *hacendado* fears that they might lose their properties in some agrarian reform. By 1973, when the Agrarian Reform formally arrived in Quinua, peasants in Susu were still distributing water by means of the irrigation board and irrigation judge.

AUGMENTING THE WATER SUPPLY: THE DAM AT YANAQUCHA

At the same time that Quinuenos were moving against the control of water by the *haciendas*, the same mayor initiated a project to expand the irrigation system of Hanan Sayoc. He and other authorities petitioned the government to construct a new dam at Lake Yanaqucha Chica to increase the water in the system and to build new canals lined with cement to limit filtration. Quinua was already using this lake for water, and no other community had claims to it or the surrounding watershed.[13] Quinuenos wanted to increase the capacity of the lake to stabilize the water supply for the hydroelectric plant and potable water system that had been built earlier in 1966 for the central town. They also hoped to increase the amount of double cropping in the savannah and to expand the irrigation system into the thorn steppe to permit more intensive cultivation of that large zone. The government provided $10,000 US (250,000 *soles*) for the venture, and the community began work on it in 1967.

Although a new dam was eventually constructed, budgetary constraints limited its size. It held only 90,000 cubic meters of water (compared to an expected 280,000 cubic meters), too little to significantly increase irrigation capacity. The supply was sufficient to ensure a steady source of tap water and hydroelectric power but not to expand the amount of farmland under irrigation or to increase double cropping. This project had been funded by Popular Cooperation (*Cooperación Popular*), a political arm of President Belaunde. Possibly because of its political mission, Popular Cooperation spent its limited budget on funding as many "showy" initiatives as possible, spreading its largesse among several communities rather than seriously increasing the productive potential of a single one or even a few. In Quinua's case, electricity and tap water were emphasized over farming.

According to its mandate, Popular Cooperation was to provide the materials and engineering expertise for the dam, while the people of Quinua were supposed to provide the labor to build it. That mandate was more easily stated than realized, however. The limited utility of the dam created a major furor about the labor to be provided, and the resulting dispute ultimately led to a general decline in the community's acceptance of corvée labor.

At first, every family in the central district (defined by the Hanan Sayoc-Lurin Sayoc irrigation networks) was assigned to work two days on dam construction, to be followed by a second round after all families had completed a turn. The dam, however, was designed only to provide water for the central town and the barrio of Hanan Sayoc. The people from Lurin Sayoc barrio, which has a separate irrigation system, were to derive no direct benefit from the dam.[14] They were adamantly opposed to the work, for it violated the peasant sense of justice that people work on only those parts of the irrigation system that they use. The authorities responded to their complaints by arguing that everyone should unite to construct the Hanan Sayoc dam, after which they would cooperate to get money to build a new dam for Lurin Sayoc.

This argument proved ineffective, and many peasants demanded to be paid for their labor. Peasants were also upset at the fines set for missing the work. To secure enough workers, the municipality had set a very high fine of $1.00 U.S. (25 *soles*) for a missed day or a total of $2.00 U.S. (50 *soles*) for the two days. Replacement peons were charging $0.80 U.S. (20 *soles*) per day. At the various meetings to discuss the project, women without husbands complained loudly that they were unable to pay such high fines and fees. They also protested the decision to exclude children under 15 as workers, even though families often sent their children for such corvée labor.

Peasants refused to show up at the construction site. The authorities of the central town split on the issue of forcing them to work. The population of Quinua often divides into two loose political factions that fight over control of community resources and, consequently, money and jobs (see also Gelles *intra*).[15] In the case of the Yanaqucha dam undertaking, the factions coalesced around national party affiliations. The mayor of the town,[16] a *populista,* that is, a member of Belaunde's political party, had been appointed guard of the construction site, earning a fabulous $2.40 U.S. (60 *soles*) a day. His job was the only compensated position on the project other than those filled by outside experts.

The governor (the town official in charge of police and public order), a member of the *Aprista* party and opposite town faction, declined to help the *populista* mayor. He was embittered by the earlier reforms in which he had lost preferential rights to water. He and others of his group were also angered at being cut out of any paid work on the dam. Consequently, he fomented the opposition to the dam and used his post as governor to stymie the construction. He would neither fine nor jail those who refused to work on the corvée (as he was supposed to do), and he sometimes refused to call public assemblies to discuss the problems on the project (as he was also supposed to do).

Frustrated by the lack of cooperation and inability to get workers, the German engineer in charge threatened to cancel the undertaking. After several months of debate, the mayor and town council bent to popular will and the engineer's threat. Henceforth only the people using the Hanan Sayoc irrigation system were required to work on it. The people of Hanan Sayoc did so reluctantly. It was widely believed that only the residents of the central town would truly benefit from the dam. To create an incentive for the work, therefore, the engineer and the mayor began to distribute "Food for Peace" parcels[17] to the workers–even though this payment was contrary to the ideology of Popular Cooperation, which held that work on public projects was to be donated freely, that such constructions were "made by the people" ("*el pueblo lo hizo*").

CHANGES IN CORVÉE LABOR

Not only did this dam fail to bring water to the thorn steppe but it marked a serious shift in the acceptance and allocation of corvée labor. Local authorities had previously organized annual corvées to repair the general infrastructure: irrigation canals, reservoirs, roads, bridges, public buildings, schools, the plaza, chapels, the church, and the cemetery. In addition to these annual tasks, they periodically employed the corvée for new construction: schools, the hydroelectric plant, the communal hall, and the dam at Yanaqucha (Mitchell 1991b).

Corvée labor is the tax obligation of the conjugal family, and each family has to supply a male worker or pay a fine. The family might send a member of the conjugal unit to the work, often a younger son, or hire a day laborer (peon) as a replacement (Mitchell 1991b). Townspeople rarely work on the corvée: they send peons in their stead, pay fines, or donate alcohol and coca leaves to the peasant workers.

Although peasants still work on the corvée, they do so less frequently. In 1966 families contributed seven to nine days of free labor each year just to maintain the infrastructure. In 1988, however, they were contributing only four to six days, as they no longer repaired roads and bridges.[18] After their experience with the dam, moreover, Quinuenos have come to reject most claims for work on new projects. If they do work on such ventures, they generally want to be paid. Schools are an important exception.

SCHOOLS: THE INFRASTRUCTURE OF COMMERCE

Quinuenos still use the corvée to construct new schools, and their efforts to extend irrigation appear truly halfhearted when compared to their efforts to expand education (Table 11.5). In the 1930s, few Quinuenos went to school (Mitchell 1991a:119-123). There were only four schools in the entire district, and none taught beyond the second grade. In the 1940s, school attendance began to accelerate. In 1953, to meet this increased demand for instruction, Quinuenos built a central school (*nucleo escolar*) containing all primary grades. Since then they have

TABLE 11.5 NUMBERS OF STUDENTS AND TEACHERS DISTRICT OF QUINUA–1986

	Nursery School	Primary School	High School	Technical School	Ceramic School	Other	TOTAL
Central Town							
Students	57	623	192	39	22	31	**964**
Teachers	2	17	11	6	2	2	**40**
Hamlets							
Students	63	925	0	0	0	219	**1,207**
Teachers	2	28	0	0	0	7	**37**
District Total							
Students	120	1,548	192	39	22	250	**2,171**
Teachers	4	45	11	6	2	9	**77**

Note: The ceramic school is run by the Ministry of Industry and Tourism, rather than by the Ministry of Education.

Source: Peru, Ministerio de Educación. 1986. Resumen Estadístico de Supervivencia, al junio de 1986. Ministerio de Educación, Oficina de Presupuesto y Planificacción Educativa, Dirección de Estadística.

constructed fifteen more schools with corvée labor: nine primary schools in the hamlets and a secondary school, three nursery schools, and two artisan schools in the central town. In 1986 the 77 teachers employed by these 16 schools were educating 2,171 students.

The government has contributed the materials for school construction, but the people of Quinua have provided free labor for both construction and maintenance. This corvée work on the schools is popular and sustained. The people themselves have urged the construction of the schools, and most households have willingly (more or less) donated one to two days of free work each year toward maintenance. Even when peasants have built schools that few of their children would use (as in the construction of a nursery school in the central town in 1973), they have tackled the tasks without the fierce opposition to the work found in the construction of the Yanaqucha dam.[19] Quinuenos do this work and encourage their children to go to school, even though education is costly. Not only do parents lose the help of their children on the farm and in the household but they must buy them clothing and school supplies (Mitchell 1991a:119-123).

While Quinuenos have willingly worked on school construction, they have also frequently attacked teachers for poor performance and drinking on the job. This conflict further supports my argument: Quinuenos are often unhappy because they believe the teachers are not adequately instructing Quinua children.

In contrast to their struggle for education, the efforts of Quinuenos in expanding irrigation appear irresolute. They would like to have more water, but not at the expense of diverting significant energy from schools and other more important paid work.

The failure of the dam to expand agriculture, of course, lies not only with the political decision of Popular Cooperation to underfund the original proposal but also with the truly limited funds available for such projects in Ayacucho (McClintock 1984, 1988). The lack of funds, however, does not explain why Quinuenos had first pushed for electricity and potable water over irrigation in 1966 or why they failed to unite against the decision of Popular Cooperation to build a smaller dam. Nor does it explain why they refused to cooperate in providing the labor on this undertaking or why they have been unable to agitate collectively for funds to improve irrigation since then. In 1987, for example, a number of persons (mostly townspeople) met in an attempt to launch the construction of a new and even larger dam at Yanaqucha. After an initial flurry of activity spurred by the efforts of one migrant who often dreamt of public projects to improve his pueblo and who frequently cash-cropped potatoes on his Quinua fields, interest subsided and the scheme was abandoned.

POPULATION PRESSURE AND THE POLITICAL ECONOMY OF PRODUCTION

The changes in the corvée and the inability of Quinuenos to expand irrigation stem from the combined pressures of population growth and adverse rural-urban terms of trade. These pressures have undermined agricultural production (and its irrigated infrastructure) and have encouraged peasants to enter nonfarm work. These same pressures, however, had also undermined the *hacienda* system, facilitating the peasant victory in wresting the water from them.

It is a serious error to think of peasants as self-sufficient (Mitchell 1987). They rarely have enough land to provision themselves, and they are tied into the larger economy in very complex ways. Quinua households generally cultivate only between one-quarter to one-half hectare of land, an amount insufficient to support most of them. Single adults need about one-eighth hectare of irrigated land during good agricultural years just to provide food, which leaves them no income to provide for fiestas, coca leaves, clothing, medicines, and supplies (Mitchell 1991a:81-86). In poor years, they need even more land. Since the size of the household averaged 4.4 persons in 1981, many households were unable to grow enough food for their members. Like rural people elsewhere in Peru (Collins 1988; Figueroa 1984), Quinua farmers must use a variety of strategies to buy additional food. Some grow early potatoes as a cash crop (the *michka,* see Zimmerer *intra*), a lucrative but risky and energy-demanding activity. Others produce crafts for sale. Still others work on the highway or in some other occupation. Many have migrated to obtain work.

Until the 1940s nonfarm work and labor migration were generally temporary.

TABLE 11.6 RATIO OF BIRTHS TO DEATHS DISTRICT OF QUINUA–1955 TO 1985

	1955	1960	1965	1970	1975	1980	1985
Births	145	163	218	231	227	249	260
Deaths	83	97	70	125	80	86	78
RATIO BIRTHS/DEATHS	**1.75**	**1.68**	**3.11**	**1.85**	**2.84**	**2.90**	**3.33**

Source: Municipal Records, District of Quinua

Peasants worked on craft production or travelled to the coast to work on the cotton harvest after their own crops had been planted. They used their earnings to buy seed and food, to obtain money for ceremonial expenses, and to purchase additional land in Quinua. In the 1940s several processes began to undermine this rough equilibrium.

The first and perhaps most important catalyst was the rapid growth in population resulting from reduced infant mortality (Mitchell 1991a:29-33). The birth rate jumped from 12.6 births per 1,000 in 1940 to 42.0 per 1,000 in 1981. If we consider a death the opening of a "social space" and a birth as a claim on that "space," the ratio of births to deaths clearly illustrates the pressures of population growth. In 1955 there were 1.75 births for every death, a figure that more than doubled to 3.33 in 1985 (Table 11.6). In 1981 the 2.8 percent annual growth rate would cause the population to double in only 25 years.

This rapid increase in population has created profound pressures on production. There is not enough arable land to feed all the people. The increased numbers have also exacerbated the water problem. Since the irrigation system provides both agricultural and drinking water, the near doubling of inhabitants of the central town between 1961 and 1986 (Table 11.7) has seriously diminished the water available for farming. The rainy-season sowing consequently had to be delayed by two weeks in Hanan Sayoc in 1987. This increased consumption of drinking water may also have made the thorn steppe less amenable to cultivation in recent years.

The stresses of population growth have been exacerbated by national economic pressures on farm prices. Since the 1950s Peruvian farm products have declined in value vis-à-vis manufactured ones because of the measures taken by successive governments to subsidize the food consumption of the export sector and urban masses (Appleby 1982; Collins 1988:20-21; de Janvry 1981; Franklin et al 1985; Long and Roberts 1984:60-63; Thorp and Bertram 1978; see Meillassoux 1981 for a description of similar processes elsewhere in the world). Food prices are often controlled directly. Foreign exchange rates are also manipulated to favor cheap food

TABLE 11.7 POPULATION GROWTH OF QUINUA 1876-1986

	1876	1940	1961	1972	1981	1986	Percent Increase 1961-1986
Town	200	824	394	465	745	778	97.5
Hamlets	3278	4825	4954	5057	5184	5361	7.6
TOTAL	**3478**	**5649**	**5348**	**5522**	**5929**	**6139**	**14.8**

Note: I discount the 1940 figures because they are too high, even after one takes into account the fact that the census combined the data for the hamlets of Lurin Sayoc and Hanan Sayoc with those for the central town. Informants report a consistent increase in the size of the central town, not a decline.

Source: Peru 1878, 1948, 1966, 1974, 1983. The 1986 figures are from the listado de establecimientos de salud, Ministerio de Salud.

imports. At the same time, major agricultural exporters such as the United States have subsidized Peruvian and other Third World purchases of their grains (Caballero 1981:182-183; Contreras 1984; de Janvry 1981:148-149; Hall 1985; Lazo and Morgan 1985; Sanchez 1984:9; Valderrama and Moscardi 1977), further undermining peasant prices and therefore production. Ayacucho had been a major producer of wheat since the colonial period (Urrutia 1985: 112), but that production declined significantly after the United States began subsidizing Peruvian purchases of United States wheat in 1954 through the cheap credit provided by PL 480 (the so-called Food for Peace law).[20]

Low urban food prices translate to reduced farm income. Using 1961 as a baseline, the value of the principal sierra products (maize, potatoes, barley, wheat, beef, mutton, and milk) dropped 15.2 percent in 1972 compared with the price of the principal goods consumed (rice, cooking oil, fats, noodles, sugar, beer, cane alcohol, soda, textiles, school supplies, detergents, soaps, candles, kerosene, plastics, and salt) (Alvarez 1979, as cited in Caballero 1981:212). In consequence, peasants today need more sacks of potatoes to buy kerosene (and even coca leaves) than they did in 1950.[21]

Farmers in Ayacucho are also relatively isolated from Lima, the nation's largest market. Other areas of Peru (such as the Mantaro Valley) have lower transport costs for their produce, increasing the relative economic pressures on Ayacucho farmers. (See Zimmerer *intra,* for another example of the importance of transportation costs in cash cropping.)

In the 1960s, moreover, traditional sources of income from cyclical migration to

TABLE 11.8 PERCENTAGE OF UNOCCUPIED HOMES DISTRICT OF QUINUA 1961-1981

	1961	1972	1981
Town Center	5.4	2.7	23.0
Hamlets	0.3	7.2	16.4
District Total			
Percent	0.9	6.8	17.2
N (Unoccupied Homes)	12.0	116.0	279.0

Source: Peru 1966, 1974, 1983

the coastal cotton plantations also disappeared, the result of reduced cotton demand on world markets and disruptions in production caused by a poorly executed agrarian reform (Mitchell 1991a:98-101, 1993). Quinuenos and other peasants had used this migratory income to supplement their own farm production, buying food, seeds, and other goods with their earnings.

These combined economic pressures have made farming in Ayacucho increasingly marginal. Even the experimental agricultural station of the University of Huamanga did not show a profit in the early 1960s, in spite of its favorable ecological location in the valley bottom and its access to the technological innovations and resources of the university (Díaz 1969:48-50).

Quinuenos have experienced these demographic and economic pressures in personal terms. They wonder how they are to feed themselves and their children. Most have rejected the idea of putting more energy into farming because of its poor returns compared to other occupations. Many have left the community. Only 34.6 percent of the people born in one hamlet had left to live elsewhere in 1967, but 55.4 percent of them had done so in another hamlet in 1987 (Mitchell n.d.).[22]

Migrants maintain myriad relationships with Quinua. Many male migrants, especially highway workers and those living in Ayacucho, have wives and children in Quinua to whom they return frequently. Still other migrants have taken their entire families with them to the coast, leaving their homes unoccupied but retaining their lands, which they actively cultivate by using sharecroppers or hired hands. Data on home occupancy (Table 11.8) show that the number of unoccupied houses increased considerably between 1961 and 1981. Some migrants, however, have abandoned their homes and fields entirely. In 1972, 16.3 percent of irrigated lands had lain fallow for more than one year (Mitchell 1991a:72, Table 17). Since farmers cultivate their irrigated fields whenever they can, some of these fallow lands had

TABLE 11.9 MALE EMPLOYMENT–DISTRICT OF QUINUA PERCENTAGE OF FARM AND NONFARM WORKERS 1955-1985

	1955	1960	1965	1970	1975	1980	1985	TOTAL
Farm Work	83.6%	85.0%	87.5%	80.5%	81.8%	69.9%	63.5%	**77.7%**
Nonfarm Work	16.4%	15.0%	12.5%	19.5%	18.2%	30.1%	36.5%	**22.3%**
N=	146	160	216	231	203	249	260	**1465**

Source: Data on the occupation of fathers, Municipal birth records, District of Quinua.

probably been abandoned by their owners.

Instead of migrating, some Quinua farmers have responded to the squeeze on their income by entering the nonfarm economy to a greater and greater extent (Mitchell 1991a:103-106). They have become artisans, truckers, and petty entrepreneurs of all sorts. In 1966 fewer than 50 households produced the famous Ayacucho churches and other ceramics for sale. Today more than 500 do so, using the income to purchase food, manufactured goods, and other items. Data from birth records demonstrate that the percentage of fathers considered nonfarm workers leapt from 16.4 percent in 1955 to 36.5 percent in 1985 (Table 11.9).[23] Farming has become increasingly an ancillary occupation that subsidizes the low wages and profits of the nonfarm sector (de Janvry 1981; Meillassoux 1981).

Some Quinuenos, however, have responded to these demographic and economic pressures by more intensive cash cropping. Many have substituted cash crops for those they had previously bartered. A few devote all their fields to the commercial production of early potatoes, buying their own potatoes and other food in local markets (in Quinua and surrounding villages) at a cheaper price after the main harvest is in. Only a smattering, however, have sufficient capital, water, and labor to be cash-crop specialists.

Increased cash cropping raises the question of why peasants have failed to expand irrigation to stabilized potato production (low rainfall is a major risk of early potato production) and to permit more intensive cultivation of the thorn steppe. Anecdotal evidence suggests that those most involved in commercial agriculture favored the work on the dam and other attempts to expand irrigation. But these farmers are few in number. The majority of Quinuenos have decided against agricultural expansion. People earn more money with less risk producing ceramics for market than they do cash cropping potatoes.

LABOR SCARCITY AND PEASANT PRODUCTION

Labor scarcity also severely constrains Quinua agriculture. Most farmers are unable to field enough hands to cash crop, to expand irrigated farming, or even to cultivate the farms they already have. It is ironic that Quinua suffers from labor scarcity at the same time its population is growing. Nonetheless, Quinua's population structure is characterized by few adult men and high dependency ratios.

Men migrate (both seasonally and permanently) more often than do women, creating persistent male labor shortages (see also Brush 1977:32-39, 57; Caballero 1981:125; Collins 1988; Degregori 1986:45, 119; Díaz 1969:62, 65). In 1981, Quinua's population of working adults (ages 20 to 39) had 41 percent more women than men (Mitchell n.d.). A similar ratio of women to men is found throughout the department of Ayacucho (Díaz 1969; Rivera 1971:87-88).[24] Those men who remain, moreover, are often too busy with nonfarm occupations to work as much as they once did in agriculture. Women, therefore, are obliged to take on additional farmwork, adding to an already heavy labor schedule (Mitchell 1991b:197-200).

High birth and migration rates have also created high dependency ratios. In 1981, 47 percent of the population was under fourteen years of age, a significant burden on those adults who remain. These children, moreover, generally attend school today, and are therefore unavailable to care for livestock and help in the home and fields as they once did.

Census data demonstrate that many households are without an effective work force. Many Quinuenos live alone or with an unrelated companion (see also Brush 1977:32-39, 57). In 1972, 15.7 percent of the households in the district (total N = 1,382) were single-person households, and in 1981, 19.3 percent of the households (total N = 1,347) contained 1.5 people. The majority of people who live by themselves are women. Since the sexual division of labor is considered essential to the effective functioning of the household, women live alone primarily because of adverse life circumstances: they are widows, their husbands and children have migrated, or they and their children were abandoned by migrating men.[25]

The inadequate size of the adult male work force is further demonstrated by a census of 25 households in the savannah in 1987. In this rural hamlet, 12 percent of the households consisted of single persons, another 12 percent were grandparent-grandchild households, and 8 percent were two generation mother-daughter households. Although the average size of the households was 4.4 persons, 44 percent of them consisted of two people or less. Since the productive unit is usually the household, most households are too small for much of the agricultural work.

Labor stress has been intensified by the Shining Path guerrilla war that began in 1980 (Mitchell 1991a:10-13). Young men are those most likely to be killed by the military and paramilitary, but young women are also assaulted, raped, and killed. Parents worry, moreover, that Shining Path partisans might murder or injure their children, or that their children might join the revolutionaries or be compromised by them in some way. Many young adults, therefore, have been sent by their parents to live with kin in the city of Ayacucho or in Lima, further depriving the community of workers.

It is this labor scarcity, together with the poor rural-urban terms of trade, that explains why Quinuenos have been unable to organize effectively to expand irrigation. Farming and irrigation require work. Irrigated cultivation not only takes an initial large investment of energy to build the system but also requires heavy labor throughout the year (about 38 additional person days a year per hectare) (Mitchell 1991a:76, 207-209, 216-219).[26]

Quinuenos are forced to allocate scarce labor. They must evaluate the advantage of farm work in comparison to the value of alternative occupations, and many have opted for nonfarm work. They are therefore willing to put time and effort into building the infrastructure for commerce (education) but not into expanding that for farming (irrigation).

Similar processes are common elsewhere in the world. In Yemen, for example, labor scarcity has encouraged peasants to put less effort into the maintenance of agricultural terraces and sometimes to abandon them completely (Vogel 1988). In Ecuador, peasants identify labor scarcity as an important source of stress (Stadel 1989:47, 1991).

THE DECLINE OF THE *HACIENDAS*

Quinuenos were successful in organizing against the *haciendas* in the 1960s because the *hacienda* system had already been weakened by the same economic forces affecting the peasantry and because peasants were also acting to free their labor.

The decline of Quinua's *haciendas* was part of a national trend that began in the 1930s and accelerated in the 1950s and 1960s (Caballero 1981:83, 239-240, 313-332; Maltby 1980; van den Berghe 1980). *Haciendas* are systems of labor recruitment in which land and water are exchanged for work. As population grew, *haciendas* replaced these semifeudal mechanisms of labor recruitment with cash labor (de Janvry 1981:82; Skar 1982:256-261). Peasants also began to demand cash remuneration for their work at the same time that *haciendas* were replacing human energy with fossil fuel energy. Low commodity prices affected *haciendas* and peasants alike, decreasing the value of *hacienda* production along with peasant production.

The everpresent risk of peasant invasion and of agrarian reform that dominated Peruvian life during the 1960s (Skar 1982:53) further increased the anxieties of ownership, prompting *hacendados* to sell or abandon their property. Wealthy Quinuenos were very nervous during this period. In 1964, for example, a rumor that guerrillas were marching on Quinua caused many of the rich to hide their possessions and flee the town temporarily.

The children of *hacendados,* often reluctant to become tied to the land, preferred to migrate to department capitals and Lima. Some *haciendas* sold marginal lands to tenants to capitalize their better parcels with the proceeds. Others sold out altogether, abandoning agriculture for the better returns of nonfarm work. When the University of Huamanga offered to buy a high-altitude *hacienda* in 1964, they received more than 20 offers (Díaz 1969:67)! Some *haciendas*–even those with irrigated lands–were simply abandoned after tenants agitated for a more just system (Díaz 1969:77).

As a result of these processes, since 1876 Quinua *haciendas* have declined in number as has the size of their resident population (Table 11.4). Some *hacendados* divided and sold their property, often to peasants who had earned the capital by migrating to work on coastal cotton plantations. In at least one case in Quinua in the early 1960s, the heirs to a *hacienda* had abandoned the area of the estate occupied by tenants and sold only the portion that the *hacienda* had cultivated directly. They were rightfully concerned that the peasants would have rebelled if they had tried to sell the lands occupied by the former peons (see also Díaz 1969:43-44).

Informants report that Quinua peasants had always resented the labor service on *haciendas* and (not unexpectedly) took whatever opportunity they could to minimize that service. (See Mitchell 1991a:179-184 for a life history of attempts by peasants to escape *hacienda* service.) By the mid-1960s, the underlying economic forces allowed them to act on this resentment to free their own labor.

CONCLUSIONS

Land, labor, and water are inextricably linked in Quinua's agricultural production and carrying capacity. Intensive farming in the Ayacucho Valley requires irrigation. Farming produces calories, but that production itself takes work. Irrigated agriculture requires more energy than rainfall farming, not only for the construction of the infrastructure but for its maintenance and for the actual cultivation of fields. Irrigation in the thorn steppe would require even more effort than that in the savannah. Quinuenos cannot expend labor thy do not have. They make decisions about that labor depending on its relative value.

Quinuenos have decided to devote more and more resources to nonfarm production. A trucker is able to buy food in the market for his children. Once farmers consume all their harvest, they would starve if they depended only on farm earnings. A ceramicist can buy seeds, supplemental food, school supplies, a radio, or a phonograph with his earnings, but a farmer cannot. Others have been unsuccessful in the commercial economy, an economy that has nonetheless become vital to their survival. Men or women who simply sell their labor are only able to eke out a living, but they rely on that work to survive nonetheless.[27]

Quinuenos would like to extend the irrigation system into the thorn steppe, but they are unwilling to divert energy from more productive enterprises to irrigate this ecologically difficult zone. They are willing to build and maintain schools, however, because they consider education the key to participation in the commercial economy. Irrigation–despite its importance to farming–now takes a secondary place to nonfarm activities.

Similar forces have discouraged heavy investment in irrigation throughout the Ayacucho Valley (see also Díaz 1969:29). The aridity of Ayacucho makes the expansion of irrigation more difficult than it would be in moister areas of the Andes. The city of Ayacucho itself has had a shortage of drinking water since at least the early 1970s. This aridity increases the costs of irrigation, but the problem is not always one of water. Rather, as in Quinua, many areas have water available; people only need to construct the dams and canals to get it to their fields.

Ayacuchanos have developed new irrigation systems and in 1968 fifty such projects had been planned or were in construction throughout the whole department[28] (Rivera 1971:153-171). These activities, however, only touch the surface of what could be done to extend agriculture if peasants found value in doing so. Instead, Ayacuchanos have put considerable energy into education. In 1961, only 28.8 percent of the total population and 47.3 percent of children aged 15 to 19 were able to read (Degregori 1986:133). By 1981, only twenty-years later, these figures had changed. In that year, 55.0 percent of the total population and 82.8 percent of children 15 to 19 years of age were literate (Degregori 1986:139).

Responding to the same ecological and economic constraints as exist in Quinua, the people of Ayacucho generally have been preparing themselves and their children for migration and commerce, de-emphasizing farm work and irrigation to do so.

NOTES

Acknowledgments. I very much appreciate the thoughtful comments of Dean Arnold, Jane Freed, David Guillet, Barbara Jaye, Gregory Knapp, and Barbara Price on earlier versions of this paper. The funding for my research has been provided by the Freed Foundation, the National Science Foundation, the Fulbright-Hayes Foundation, and the Monmouth College Grants and Sabbaticals Committee.

1. Ayacucho is also a very narrow valley (Rivera 1971:30-31), further limiting its agricultural potential.

2. Data on days of rain only cover April-December 1963 and the years 1964-1970. See explanatory note in Table 11.1.

3. Rainfall data are from the city of Ayacucho. The city is located on the valley floor and consequently much more arid than most of Quinua. The meteorological experiences of the two areas, however, are similar, and Rivera (1971:39-45) classifies them together on the basis of rainfall in his "Andean Type" ("*tipo andino*").

4. Very little rain fell in November and December 1989. I have no data on what has taken place since December of that year. Excessive rain from September through December is an additional risk to the harvest, for it damages maize seedlings, Quinua's most important crop.

5. Elsewhere in the Andes a drop of 0.8 degrees centigrade has also been recorded (Winterhalder and Thomas 1978:20).

6. The remaining 6.6 percent was devoted to other uses.

7. My terminology for the ecological zones of Quinua is based on Arnold's (1975) analysis of Tosi's (1960) classification for Peru. The English terms for the zones are derived from Holdridge (1947), a source for Tosi's terms. The terms are artificial ones (Gregory Knapp, personal communication, 1990), and my classification is etic rather than emic. The people of Quinua distinguish terminologically only between the high altitudes called *sallqa* or *urqu* (the moist forest and above) and the low altitudes called *quechua* (the savannah and lower zones), a classification that corresponds to Pulgar Vidal's (n.d.). See Mitchell (1976b, 1991a:39-46) and Mitchell (n.d.) for other discussions of Quinua's

ecological zones.

8. The rain tundra/wet paramo areas have always been (at least in recent times) uninhabited. Since the Shining Path revolutionary movement began in 1980, many people have abandoned all the high altitude zones, fleeing for safety to the savannah and the central town or leaving the community entirely.

9. See Rivera (1971) and MacNeish (n.d.) for a description of the ecology of the Ayacucho Valley. The members of the Ayacucho Archeological-Botanical project organized by MacNeish use a different terminology to describe the ecological zones of the valley, but their "thorn scrub" is roughly equivalent to my "thorn steppe." Table 5 in Mitchell (n.d.) depicts the differences in terminologies.

10. Considerable water is lost from the canals through seepage and evaporation. Evaporation is difficult for Quinuenos to solve, but cement canals would limit seepage.

11. Two *haciendas* around the archeological site of Huari had not been turned into cooperatives, at least as late as 1980, but were considered part of the national patrimony.

12. It is unwise to accept informant accounts of equality in water distribution at face value as some scholars have done (cf. Lees 1973, but see my critique of this position in Mitchell 1977). When water is tied to land and land is owned unequally, water is *ipso facto* distributed unequally. Both the rich and poor (if they pass the dress code) are equally free to enter banks in the United States, but they are restricted in what they withdraw. Quinuenos are similarly free to go to the water distribution, but the rich and powerful withdraw the most water. See Mitchell (1991b) for an extended discussion of inequality in Quinua in spite of the egalitarian ethos.

13. At least no other community was exercising any claims to the lake or watershed.

14. Hanan Sayoc and Lurin Sayoc irrigation systems are connected, and some water is shared between the systems but only in a minor way (Mitchell 1976a; n.d.).

15. These factions are led by townspeople but include rural peasants. They are informal alliances that often use kin and *compadrazgo* (fictive kin) ties (Mitchell 1972, 1979).

16. The mayor subsequent to the one who had initiated the project.

17. This food caused considerable confusion. The labels were mostly in English and without any instructions on how to use such unknown foods as powdered eggs. Some Quinuenos suspected the food was poisoned or contained chemicals to sterilize the users.

18. With the advent of heavy truck traffic and bus service since the late 1960s, Quinuenos are no longer using foot and animal paths extensively, allowing them to neglect their repair.

19. A proposal to build a high school (colegio) in Quinua in 1967 did engender enormous anger. The anger, however, was not at the school construction but at the deception on the part of the townsman who had obtained signatures on a petition from illiterate peasants without telling them they were asking to be taxed to pay for the school.

20. The United States subsidy lasted until 1973. These cheap prices for imported foods are probably an important factor in the dietary shift from potatoes and maize to imported wheat (in the form of bread and noodles) throughout the Andes (Crissman and Uquillas 1989: 10; see also Contreras 1984 and Hall 1985). The imported wheat was also better suited to milling than was Peruvian wheat, which created additional pressures for foreign rather than domestic purchases (Valderrama and Moscardi 1977).

21. Peasants have begun to organize against these unfavorable terms of trade. Indeed, peasant protest against low agricultural prices in Ayacucho was one of the factors that led to the arrest of the president of the Agrarian Federation of Ayacucho (*Federación Agraria de Ayacucho*) in 1989 on charges of "terrorism" (Amnesty International 1989: 27-28). I have no data on the price of coca leaves since the development of the cocaine trade in the 1980s, but I presume that the price of the leaves to the peasant also must have increased relative to other prices.

22. Unfortunately, I do not have comparable data from the same hamlet. The increased percentage of permanent migrants suggested by these data from different hamlets, however, is supported by informant reports of such increases.

23. The percentages of nonfarm workers are probably higher than indicated, as I counted the men labeled *obreros* in the birth records as farm laborers, even though many of them undoubtedly work in manual labor rather than in agriculture.

24. In the areas of the coast to which Quinuenos migrate in large numbers, there is a

corresponding increase of males over females: the rural areas surrounding the city of Lima had a ratio of 125 men to 100 women in 1961, a figure that declined to 111 to 100 in 1981 (Maletta et al. n.d.: I, 48, 60).

25. Women, of course, are independent actors who sometimes live alone out of choice or who do so because they have left abusive husbands. Nonetheless, most people try to live with somebody else not only for emotional and sexual companionship but to share the workload.

26. My data on the increased labor intensity of irrigated farming are at variance with data collected by Gregory Knapp in Ecuador, where irrigated farming is less labor demanding than rainfall farming (personal communication, 1990). The difference may rest on the inclusion of the time needed to get and guard water in my data. The actual time spent irrigating a field is small.

27. These economic disparities are not inevitable. Change, however, requires effective political action to provide resources for the peasant infrastructure and higher returns on peasant production. These steps require modifying the political and economic environment favoring the coast over the sierra. In addition, action must be taken to control population. In my research, I have found that Quinuenos are interested in limiting family size. A number of men and women have spoken to me of wanting smaller families, but they have neither the knowledge nor resources to effect their desires. Women, nonetheless, are already bearing fewer children than did their mothers (Mitchell 1991a:33).

28. The department is significantly larger in area than is the Ayacucho Valley.

REFERENCES CITED

Amnesty International
1989 *Peru: Human Rights in a State of Emergency*. New York: Amnesty International.

Alvarez, Elena
1979 "Política agraria y estacamiento de la agricultura, 1969-1977." Ponencia presentada al Primer Semenario sobre Agricultura y Alimentación en el Peru. (Cited in Caballero 1981:212).

Appleby, Gordon
1982 "Price Policy and Peasant Production in Peru: Regional Disintegration During Inflation." *Culture and Agriculture* 15:1-6.

Arnold, Dean E.
1975 "Ceramic Ecology in the Ayacucho Basin, Peru: Implications for Prehistory." *Current Anthropology* 16:183-203.

Brush, Stephen B.
1977 *Mountain, Field and Family; the Economy and Human Ecology of an Andean Valley*. Philadelphia: University of Pennsylvania Press.

Caballero, José María
1981 *Economía agraria de la sierra peruana; antes de la reforma agraria de 1969*. Lima: Instituto de Estudios Peruanos.

Collins, Jane
1988 *Unseasonal Migrations: The Effects of Rural Labor Scarcity in Peru*. Princeton, NJ: Princeton University Press.

Contreras, Willy
1984 "Comercialización de granos básicos." In *Comercialización de productos básicos*. Fernando Sanchez Albavera, editor, pp. 107-123. Madrid: Ediciones de Cooperación Iberoamericana, Centros de Estudios y Promoción del Desarrollo, DESCO.

Crissman, Charles C., and Jorge E. Uquillas
1989 *Seed Potato Systems in Ecuador: A Case Study*. Lima: International Potato Center.

Degregori, Carlos I.
1986 *Ayacucho, raíces de una crisis*. Ayacucho: Instituto José María Arguedas.

de Janvry, Alain
1981 *The Agrarian Question and Reformism in Latin America*. Baltimore, MD: Johns Hopkins University Press.

Díaz Martinez, Antonio
1969 *Ayacucho: hambre y esperanza*. Ayacucho (Peru): Ediciones Waman Puma.

Dobyns, Henry
1964 *The Social Matrix of Peruvian Indigenous Communities*. Cornell Peru Project Monograph. Ithaca: Department of Anthropology, Cornell University.

Figueroa, Adolfo
1984 *Capitalist Development and the Peasant Economy in Peru. Cambridge Latin American Studies*, vol 47. Cambridge: Cambridge University Press.

Franklin, David, Jerry B. Leonard, and Alberto Valdes
1985 "Consumption Effects of Agricultural Polices: Peru; Trade Policy, Agricultural Prices and Food Consumption: An Economy Wide Perspective." Report Prepared for USAID/PERU. Raleigh, NC: Sigma One Corporation.

Frére, M., et al.
1975 Estudio *agroclimatológico de la zona andina (informe tecnico)*. Rome: Food and Agriculture Organization (Proyecto Interinstitucional FAO/UNESCO/OMM en Agroclimatologia).

Hall, Lana L.
1985 "United States Food Aid and the Agricultural Development of Brazil and Colombia, 1954-73." In *Food, Politics and Society in Latin America*. John Super and Thomas Wright, editors, pp. 133-149. Lincoln: University of Nebraska Press.

Holdridge, L. R.
1947 "Determination of World Plant Formations from Simple Climatic Data." *Science* 105:367-368.

Knapp, Gregory W.
1988 "Ecología cultural prehispanico del Ecuador." *Bibliografía de Geografía Ecuatoriana* (Quito: Banco Central del Ecuador) 3:28-29.

Lazo L., Manuel, and Mariluz Morgan T.
1985 *Reforma agroalimentaria: cambios en la producción, en el consumo y en los precios relativos*. Lima: Publicaciones CIPESA, 62, Pontificía Universidad Catolica del Peru, Departamento de Economía.

Lees, Susan
1973 *Sociopolitical Aspects of Canal Irrigation in the Valley of Oaxaca. Prehistory and Human Ecology of the Valley of Oaxaca, 2. Memoirs of the Museum of Anthropology, University of Michigan, 6.* Ann Arbor: Museum of Anthropology.

Long, Norman, and Bryan R. Roberts
1984 *Miners, Peasants, and Entrepreneurs; Regional Development in the Central Highlands of Peru*. Cambridge: Cambridge University Press.

MacNeish, Richard S.
n.d. *The Ayacucho Archeological-Botanical Project*, volume 1. Ann Arbor: University of Michigan Press. In press.

Maletta, Héctor, Alejandro Bardales, and Katia Makhlouf
n.d. *Perú: Las provincias en cifras, 1876-1981*, vols. 1-3. Lima: Universidad del Pacifico, Ediciones AMIDEP.

Maltby, Laura
1980 "Colonos on Hacienda Picotani." In *Land and Power in Latin America; Agrarian Economies and Social Processes in the Andes*. Benjamin S. Orlove and Glynn Custred, editors, pp. 99-112. New York and London: Holmes and Meier Publishers, Inc.

McClintock, Cynthia
1984 "Why Peasants Rebel: The Case of Peru's Sendero Luminoso." *World Politics* 37:48-84.
1988 "Peru's Sendero Luminoso Rebellion: Origins and Trajectory." In *Power and Popular Protest: Latin American Social Movements*. Susan Eckstein, editor, pp. 61-101. Berkeley: University of California Press.

Meillassoux, Claude
1981 *Maidens, Meal and Money; Capitalism and the Domestic Community*. New York: Cambridge University Press.

Mitchell, William P.
1972 *The System of Power in Quinua: A Community of the Central Peruvian Highlands*. Ph.D. dissertation, University of Pittsburgh.
1976a "Irrigation and Community in the Central Peruvian Highlands." *American Anthropologist* 78:25-44.
1976b "Social Adaptation to the Mountain Environment of an Andean Village." In *Hill Lands: Proceedings of an International Symposium*. John Luchok, John D. Cawthon, and Michael J. Breslin, editors, pp. 187-198. Morgantown: West Virginia University Press.
1977 Review of Susan Lees: *Sociopolitical Aspects of Canal Irrigation in the Valley of Oaxaca. Prehistory and Human Ecology of the Valley of Oaxaca, 2. Memoirs of the Museum of Anthropology, University of Michigan, 6. American Anthropologist 79:731-733.*
1979 "Inconsistencia de status social y dimensiones de rango en los andes centrales del Peru." *Estudios Andinos* 15:21-31.
1987 "Comment on 'Terracing and Irrigation in the Peruvian Highlands' by David Guillet." *Current Anthropology* 28:422.
1991a *Peasants on the Edge: Crop, Cult, and Crisis in the Andes*. Austin: University of Texas Press.
1991b "Some are More Equal than Others: Labor Supply, Reciprocity, and Redistribution in the Andes." *Research in Economic Anthropology* 13:191-219.
1993 "Pressures on Peasant Production and the Formation of Regional Culture." In Migrants and Regional Cultures in Latin *American Cities*. Teofilo Altamirano and Lane Hirabayashi, editors. Publication Series of the Society for Latin American Anthropology, vol. 20. Washington, DC: American Anthropological Association. In press.
n.d. "Multizone Agriculture in an Andean Village: The Ecological Basis of Peasant Farming in the Ayacucho Valley." In *The Prehistory of the Ayacucho Basin, Peru*, Volume 1. Richard S. MacNeish, editor. Ann Arbor: University of Michigan Press. In press.

Murra, John V.
1960 "Rite and Crop in the Inca State." In Culture in History. Stanley Diamond, editor, pp. 393-407. New York: Columbia University Press.

Peru
1878 *Resumen del censo general de habitantes del Peru, hecho en 1876*. Lima: Imprenta del Estado.
1948 *Censo nacional de población de 1940*, vol. 6. Lima: Ministerio de Hacienda y Comercio, Dirección Nacional de Estadística.
1966 *Sexto censo nacional de población; primer censo nacional de vivienda, 2 de julio de 1961, tomo I, volumen de centros poblados*. Lima: Dirección Nacional de Estadística y Censos.
1974 *Censos nacionales, VII de población, II de vivienda, 4 de junio de 1972, departamento de Ayacucho*. Lima: Oficina Nacional de Estadística y Censos.
1983 *Censos nacionales, 1981; VIII de población, III de vivienda, resultados definitivos, vol. A, 3 tomos*. Lima: Instituto Nacional de Estadística.

Pulgar Vidal, Javier
n.d. *Las ocho regiones naturales del Perú*. Lima: Editorial Universo S.A.

Rivera, Jaime
1967 "El clima de Ayacucho." *Universidad, Organo de Extensión Cultural de la Universidad Nacional de San Cristobcl de Huamanga (Ayacucho)*, Año 3, No. 9.
1971 *Geografía general de Ayacucho*. Ayacucho: Universidad Nacional de San Cristobal de Huamanga, Dirección Universitaria de Investigación.

Sanchez Albavera, Fernando
1984 "Presentación." In *Comercialización de Productos Básicos*. Fernando Sanchez Albavera, editor, pp. 7-10. Madrid: Ediciones de Cooperación Iberoamericana, Centro de Estudios y Promoción del Desarrollo, DESCO.

Skar, Harald O.
1982 *The Warm Valley People: Duality and Land Reform Among the Quechua Indians of Highland Peru*. Oslo: Universitetsforlaget.

Stadel, Christoph
1989 "The Perception of Stress by *Campesinos*: A Profile from the Ecuadorian Andes." *Mountain Research and Development* 9:35-49.
1991 "Environmental Stress and Sustainable Development in the Tropical Andes." *Mountain Research and Development* 11:213-223.
Thorp, Rosemary, and Geoffrey Bertram
1978 *Peru: 1890-1977; Growth and Policy in an Open Economy*. New York: Columbia University Press.
Tosi, Joseph A.
1960 *Zonas de vida natural en el Perú. Instituto Interamericano de Ciencias Agrícolas de la OEA, Zona Andina, Boletín Técnico, No. 5*. Lima: Organizaci6n de Estados Americanos.
Tosi, Joseph A., and Robert F. Voertman
1964 "Some Environmental Factors in the Economic Development of the Tropics." *Economic Geography* 40:189-205.
Troll, Carl
1968 "The Cordilleras of the Tropical Americas: Aspects of Climatic, Phytogeographical and Agrarian Ecology." In *Geo-Ecology of the Mountainous Regions of the Tropical Americas*. Carl Troll, editor, pp. 15-56. Bonn: Ferd Dummlers Verlag.
Urrutia, Jaime
1985 *Huamanga: region e historia, 1536-1770*. Ayacucho (Peru): Universidad Nacional de San Cristobal de Huamanga.
Valderrama, Mario, and Edgardo Moscardi
1977 "Current Policies Affecting Food Production: The Case of Wheat in the Andean Region." In *Proceedings of the World Food Conference 1976*. Ames: Iowa State University Press.
van den Berghe, Pierre L.
1980 "Ccapana: The Demise of an Andean Hacienda." In Land *and Power in Latin America; Agrarian Economies and Social Processes in the Andes*. Benjamin S. Orlove and Glynn Custred, editors, pp. 165-178. New York, NY: Holmes & Meier Publishers, Inc.
Vogel, Horst
1988 "Deterioration of a Mountainous Agro-Ecosystem in the Third World Due to Emigration of Rural Labor." Mountain *Research and Development* 8:321-329.
Winterhalder, Bruce, and R. Brooke Thomas
1978 *Geoecology of Southern Highland Peru: A Human Adaptation Perspective*. Occasional Paper No. 27, Institute of Arctic and Alpine Research. Boulder: University of Colorado.

NOTES ON CONTRIBUTORS

Inge Bolin teaches anthropology at Malaspina University College in British Columbia. She combines academic research with applied work on locally initiated small-scale development projects. As Visiting Research Fellow at the Institute of Southeast Asian studies in Singapore, she undertook cross-cultural comparisons of Peru and the highlands of West Sumatra. Her recent publications include "Upsetting the Power Balance–Cooperation, Competition, and Conflict along an Andean Irrigation System" (*Human Organization,* 1990) and "The Hidden Power of Women–Highland Peru and West Sumatra Compared" (*Development and Cooperation,* 1990).

Stephen G. Bunker is Professor of Sociology at the University of Wisconsin, Madison. He has carried out studies of peasantry, rural development, and agriculture in Uganda, Brazil, Guatemala, and Peru. His most recent book is *Underdeveloping the Amazon: Extraction, Unequal Exchange, and the Failure of the Modern State* (University of Illinois Press, 1985). At present he is doing research on the ecological, demographic, and economic effects of large-scale mineral extraction in Brazil.

Paul H. Gelles is Assistant Professor of Anthropology at the University of California, Riverside. He was formerly a Postdoctoral Fellow in the College of Natural Resources, University of California, Berkeley. He has done fieldwork in the central and southern Peruvian highlands, studying, among other things, the way that ethnic identity and cultural politics condition water management. In addition to several published articles that have appeared in *Allpanchis* and other journals, he is currently rewriting his doctoral dissertation (*Channels of Power, Fields of Contention: The Politics and Ideology of Irrigation in an Andean Peasant Community*) into book form.

David Guillet is Professor of Anthropology at Catholic University, Washington D.C. He has maintained teaching and research interests on the economic and ecological aspects of the Andes for many years. Among his recent publications is *Covering Ground: Communal Water Management and the State in Highland Peru* (University of Michigan Press, 1992). At present, he is conducting an ethnographic and historical study of the irrigation systems of the Río Orbigo in the province of León, Spain.

William P. Mitchell is Freed Professor in the Social Sciences at Monmouth College in New Jersey. He has worked in the Andes since 1966, studying the ecology, economy, and social systems of both highland and coastal people. He is the author of "Irrigation and Community in the Central Peruvian Highlands" (*American Anthropologist,* 1976), among other papers. His recent book, *Peasants on the Edge: Crop, Cult, and Crisis in the Andes* (University of Texas Press, 1991), analyzes the ecological and economic changes in Peru over the last 40 years and the impact of these changes on social and religious organization.

Karsten Paerregaard is a postdoctoral fellow at the Institute of Anthropology of the University of Copenhagen. He has conducted research in Peru over a period of five years on ecology, irrigation, ritual, migration, and social change. He is the author of *Nuevas organizaciones en comunidades campesinas* (Universidad Catolica del Perú, 1987) and his publications include articles in *Ethnology, Folk,* and *Antropológica*.

Linda J. Seligmann is Assistant Professor of Anthropology at James Madison University. She was previously Associate Director of the Latin American and Iberian Studies Program, Adjunct Assistant Professor at the University of Wisconsin, Madison, and Visiting Fellow in the Program in Agrarian Studies and the Department of Anthropology at Yale University. In addition to her work on irrigation, she has studied Peruvian agrarian reform, market women, and issues of class and ethnicity. She is author of "To Be in Between: The *Cholas* as Market Women in Peru" (*Comparative Studies in Society and History,* 1989) and "The Burden of Visions amidst Reform: Peasant Relations to Law in the Peruvian Andes" (*American Ethnologist,* 1993).

Jeanette Sherbondy teaches anthropology at Washington College in Chestertown, Maryland. A former fellow in pre-Columbian studies at Dumbarton Oakes in Washington, D.C., she has also taught anthropology at the University of Iowa and Indiana University. She has studied the ethnohistory of the Andes for many years, focusing on irrigation and the organization of Cuzco and the Inca empire. Among other works, she is the author of *Cuzco: Aguas y poder* (Cuzco, Centro de Estudios Rurales Andinos Bartolomé de Las Casas, 1979) and "Water Ideology in Inca Ethnogenesis" (*Andean Cosmologies through Time,* Indiana University Press, 1992).

John M. Treacy died suddenly and unexpectedly during the preparation of this volume. At the time of his death, he was Assistant Professor of Geography at George Washington University. Aside from research on irrigation and terracing in the Colca Valley village of Coporaque, on which his dissertation and the contribution to this volume are based, Treacy also had carried out studies of land use in the Amazon. A number of his articles have been published or are in the process of being published. His dissertation is being published in Lima by the Instituto de Estudios Peruanos.

Bruce Winterhalder is Professor of Anthropology and chair at the University of North Carolina, Chapel Hill. His research has focused on the foraging strategies of hunter-gatherers and on the ecological adaptations of Andean agriculturalists. He is author, with Brooke Thomas, of *The Geoecology of Southern Highland Peru: A Human Adaptation Perspective* (Institute of Arctic and Alpine Research, University of Colorado, Boulder, 1978), as well as articles in *Human Ecology, American Naturalist, Journal of Anthropological Archaeology,* and other journals. With E. A. Smith he has edited and contributed papers to two volumes on human evolutionary ecology ("Hunter-Gatherer Foraging Strategies," 1981, and "Evolutionary Ecology and Human Behavior," 1992).

Karl S. Zimmerer is Associate Professor of Geography at the University of Wisconsin, Madison. His research focuses on Andean crop and soil resources. He recently completed work on the history of soil erosion in Cochabamba, Bolivia, which is published in *World Development* (1993) and *Economic Geography* (1993). Among other works, he is the author of "Wetland Production and Smallholder Resistance: Agricultural Change in a Highland Peruvian Region" (*Annals of the Association of American Geographers,* 1991) and "Labor Shortages and Crop Diversity in the Southern Peruvian Sierra" (*Geographical Review,* 1991).